FUNDAMENTALS OF HORTICULTURE

ABOUT THE AUTHORS

Bhaskar Chandra Das (b.1964) graduated from the College of Agriculture, Odisha University of Agriculture and Technology, Bhubaneswar in 1986 and joined OUAT service in 1987 as Research Assistant. He completed M.Sc. (Ag) in Horticulture with specialization in vegetable science in 1993 from OUAT and continued serving under OUAT. He has been associated with different research projects like NARP (presently Regional Research and Technology Transfer Sub-Stations), All India Co-ordinated Potato Improvement Project, Seed Technology Research, NSP (Crops) and PG Department of Microbiology during his 22 years of service career. He qualified National Eligibility Test conducted by ASRB during 1997 and 1998 in the subject Horticulture and Vegetable Science respectively. He has more than three decades of teaching and research experience in the field of Agriculture and Horticulture. He has published more than twenty-five research papers, seven abstracts, two technical bulletins, five practical manuals and three popular articles. He joined as Assistant Professor in College of Horticulture, OUAT, Chiplima, Sambalpur, Odisha during 2009 and engaged in teaching courses like Fundamentals of Horticulture, Post-Harvest Management of Horticultural Crops, Post-Harvest Management and Value addition of Fruits and Vegetables, Processing of Horticultural Crops, Medicinal and Aromatic crops and Spices & Condiments. He is a life member of Odisha Horticultural Society, Indian Society of Vegetable Science and Trends in Biosciences.

Prof. Bhimasen Naik (b.1961) hails from an agri-horticultural family of Western Odisha (Village: Jamtalia, Tehsil: Sundargarh, District: Sundargarh). He obtained B.Sc. (Ag. & A.H.) and M.Sc. (Ag.) in Genetics and Plant Breeding from Chandra Shekhar Azad University of Agriculture and Technology, Kanpur in 1981 and 1983, respectively, and Ph.D. from Utkal University, Bhubaneswar in1999. He has been awarded ICAR Junior Research Fellowship and CSIR Senior Research Fellowship during his studies. He is a Fellow of the Indian Society of Genetics & Plant Breeding, New Delhi; and life member of the Crop Improvement Society of India, Ludhiana; the Indian Society of Plant Breeders, Coimbatore; the Society for Advancement of Rice Research, Hyderabad; the Indian Society of Oilseeds Research, Hyderabad and the Society for Plant Biochemistry & Biotechnology, New Delhi. He has worked on breeding and genetics of ginger, turmeric, mungbean and linseed, and presently working on rice. He has 36 years of research and teaching experience, and is presently working as Professor of Plant Breeding and Genetics at Regional Research and Technology Transfer Station of the Odisha University of Agriculture and Technology at Chiplima, Sambalpur,

Odisha. He has 52 research papers published in peer-reviewed national and international journals, 23 abstracts of symposia, one book chapter (Springer-Nature) and 20 popular articles to his credit. He has guided one M.Sc. (Ag.) and one Ph.D. student. He is associated in the development of five cultivars in ginger (Suruchi), turmeric (Surama, Ranga and Rashmi) and linseed (Arpita); collaborated in development of three cultivars in rice (Ashutosh, Gobinda and Hasanta) and registered one genetic stock of mungbean (BSN 1, INGR 00011).

Dr. Ranjan Kumar Tarai (b.1977) obtained his B.Sc. (Ag.) degree in 1999 from College of Agriculture, Chiplima under Odisha University of Agriculture and Technology; M.Sc. (Hort.) in Fruits and Orchard Management and Ph.D. (Hort.) in Fruits and Orchard Management in 2002 and 2006 respectively from Bidhan Chandra Krishi Viswa Vidyalaya, Mohanpur, Nadia, West Bengal. He successfully qualified NET in Fruit Science during 2004 conducted by ASRB, New Delhi. He started his career in 2006 as Subject Matter Specialist (Horticulture) at Krishi Vigyan Kendra (Gajapati) and later on joined as Junior Scientist (Horticulture) at Regional Research and Technology Transfer Station, Bhawanipatna and as Programme Coordinator at KVK, Kalahandi under Odisha University of Agriculture and Technology, Bhubaneswar. Dr. Tarai is basically a Pomologist inclined to fruit research and teaching. He has an experience of over 14 years of teaching, research and extension. He has published more than 53 research papers in national and international journals, 03 Books (Plant growth regulators in Tropical & SubTropical Fruit Crops Part-I & Part-II and Plant growth regulators in Temperate & Underutlized Fruit Crops), 26 popular articles, 18 book chapters, 20 extension booklets & Newsletters to his credit. He has guided two M.Sc. Ag. (Hort.) students. He is a life member of several societies in the field of Horticulture. He has devoted his works in development and transfer of technologies in fruits (especially minor fruit crops), vegetables and horticulture-based farming system suitable for resource poor area. Currently he is working as Associate Professor (Fruit Science) at College of Horticulture, Chiplima under Odisha University of Agriculture and Technology, Bhubaneswar.

Bijaya Kumar Sethy (b.1975) born in village: Bamara, Tehsil: Garadpur, District: Kendrapara. He obtained B.Sc. (Ag.) and M.Sc. (Ag.) in Horticulture (Fruit Science) from Odisha University of Agriculture and Technology, Bhubaneswar in 1996 and 1999, respectively and successfully qualified National Eligibility Test (NET) in 2001 conducted by Agricultural Scientist Recruitment Board (ICAR), New Delhi. He started his career in 2006 as Subject Matter Specialist (Horticulture) at Krishi Vigyan Kendra (Nabarangpur) and later on joined as Assistant Professor (Horticulture) in 2010 at College of Agriculture, Chiplima under Odisha University of Agriculture and

Technology, Bhubaneswar. He is basically a Horticulturist inclined to various horticultural research, teaching and extension. He has 15 years of research, teaching and extension experience and is presently working as Assistant Professor (Horticulture) at College of Agriculture, Chiplima under the Odisha University of Agriculture and Technology, Bhubaneswar, Odisha. He has published more than 12 research papers in various journals of national and international repute, 15 abstracts and 10 popular articles (Odia and English), 2 book chapters, 1 book and 3 booklets (Odia) and Newsletter to his credit. He is also life member of several societies in the field of horticulture and reviewer of different journals.

FUNDAMENTALS OF HORTICULTURE

Bhaskar Chandra Das
Assistant Professor of Post-Harvest Management
Odisha University of Agriculture and Technology
Chiplima Campus, Sambalpur-768 025, Odisha

Bhimasen Naik
Professor of Plant Breeding and Genetics
Odisha University of Agriculture and Technology
Chiplima Campus, Sambalpur-768 025, Odisha

Ranjan Kumar Tarai
Associate Professor of Fruit Science
Odisha University of Agriculture and Technology
Chiplima Campus, Sambalpur-768 025, Odisha

Bijaya Kumar Sethy
Assistant Professor of Horticulture
Odisha University of Agriculture and Technology
Chiplima Campus, Sambalpur-768 025, Odisha

NEW INDIA PUBLISHING AGENCY
New Delhi-110 034

NEW INDIA PUBLISHING AGENCY
101, Vikas Surya Plaza, CU Block, LSC Market
Pitam Pura, New Delhi – 110 034, India
Email: info@nipabooks.com
Web: www.nipabooks.com

For customer assistance, please contact
Phone: + 91-11-27 34 17 17 Fax: + 91-11- 27 34 16 16
E-Mail: feedbacks@nipabooks.com

ISBN 978-9390175-17-8

Composed and Designed by NIPA.

Digitally printed all across the globe.

"Errors, like straws, upon the surface flow;
He who would search for pearls, must dive below"

—John Dryden (1631–1700): "All for Love" Prologue

To
Abhipsa
Dibyajyoti
Abhilash
Biswajit

Preface

The Sixth Deans' Committee of Indian Council of Agricultural Research has recently revised the syllabus of B.Sc. (Hons) Horticulture which is uniform throughout the country. The course 'Fundamentals of Horticulture' is taught in the first semester. The present textbook covers the entire syllabus sequentially in 19 chapters. Simple and lucid language has been followed for easy understanding of the beginners. The information contained in the book has been gathered from various published sources and internet websites which are mentioned at the end of each chapter under references. Attempts have been made to provide latest information; still some valuable information might have been missed. Questions are set at the end of each chapter to assess the understanding of the students.

We have tried our best to remove the errors, typographical or otherwise, from the text; still there might be some. We would highly appreciate if it is brought to our notice for rectification in next edition. We cherish the encouragement and cooperation received from our family members during preparation of the manuscript. We congratulate M/S New India Publishing Agency, New Delhi for their support and publishing in a short time.

Though the book is primarily written for B.Sc. (Hons) Horticulture students, the counterparts of B.Sc. (Hons) Agriculture also may be benefited. It may serve as a help book for post-graduate students.

We seek suggestions from learned teachers and students for further improvement of the book.

Chiplima **Authors**

Contents

Chapter 1

Scope and Importance of Horticulture

1.1 INTRODUCTION

India is the seventh largest country in the world in total geographical area and has second largest population next to China. Around 55 to 60 per cent of its population depend on agriculture and allied activities. Horticultural crops constitute a significant component of total agricultural production in the country. They require very intensive care in planting, carrying out different cultural operations, manipulating growth, harvesting, packing, transportation, marketing, storage and processing.

1.2 AGRICULTURE

The term agriculture broadly refers to the technology of raising plants and animals. Agriculture is the science, art or practice of cultivating soil, producing crops and raising livestock in varying degree of proportion followed by marketing. In other words, it is the activity of man to produce three basic needs of human being, i.e., food, fuel and fibre (3F) by using terrestrial resources optimally. The term agriculture is derived from two Latin words *ager* meaning field or land and *cultura* meaning cultivation or growing.

1.3 HORTICULTURE

It is the branch of plant agriculture which is concerned with cultivation of garden plants. According to Jules Janick, 'Horticulture can be defined as the branch of agriculture which is concerned with the intensively cultured plants, directly used for food, fuel, fibre, medicinal purpose or for aesthetic gratification by the people'. Horticulture is not only production of different garden plants but also utilization, improvement and value addition of those crops.

The term "horticulture" first appeared in written language during 17th century. It is found mentioned in Peter Leuremburg's treatise written in 1631. In English Horticulture was first mentioned in the book "The New World of English Words" written by E. Philips in 1678 published in London. M.H. Marigowda is popularly known as the Father of Horticulture and Dr. K. L. Chadha as the Father of Modern Horticulture in India. Liberty Hyde Bailey (1858–1954), an American, is the Father of American Horticulture and Dean of Horticulture throughout the world.

The term 'horticulture' is derived from two Latin words; *hortus* and *colere*. The meaning of *hortus* is garden and that of *colere* is to cultivate. As horticulture is a branch of agriculture, it is concerned with cultivation of garden plants because the meaning of garden is cultivation of plants in an enclosed area. It implies that horticultural crops are grown in an enclosed area in contrast to the field crops. In ancient times gardens were protected by enclosing with high walls or similar structures all-around the houses. It implies that horticulture refers to cultivation of garden plants in protected enclosures.

In present days, horticulture may be defined as the science and technique of production, processing and merchandizing of fruits, vegetables, flowers, spices & condiments, plantation crops, medicinal& aromatic crops and mushroom. As a science, it involves the application of physics, chemistry and other fundamental sciences, and plant sciences *viz.* biochemistry, plant physiology, botany, genetics, plant breeding etc. As an art it involves the artistic application of technical knowledge gained in raising flowering plants in different containers, budding and grafting of plants for various colour of flowers and fruits, different methods of training and pruning for giving different shapes, designing of gardens, growing of flowers for colour combinations and according to season etc. After all, horticulture is the science, art, technology and business-oriented study.

Difference Between Horticultural Crops and Field/Agricultural Crops

Horticultural Crops	Agricultural/Field Crops
Horticultural crops are grown in intensive manner.	Field crops are grown in extensive manner.
Horticultural crops are highly perishable in nature.	Field crops are less or non-perishable in nature and have a high dry matter content.
Horticultural crops are used in fresh/living form.	Produces are often utilized in dry state.
Different cultural operations need high skill.	Requires less skill for cultivation.
Horticultural crops are richest source of vitamins and minerals.	Field crops are rich in carbohydrate or protein.
Aesthetic sense of gratification is an exclusive phenomenon in horticultural crops.	No such phenomenon.

1.4 SCOPE

- Due to diverse climatic and edaphic conditions in India a good number of horticultural crops can be grown.
- India is bestowed with tropical, sub-tropical and temperate climates for which humid, semi-arid and arid crops are also grown successfully.
- Similarly, soils of different types like loamy, alluvial, laterite, medium black, rocky shallow, heavy black, sandy, etc. are also available. For this, large crop areas can be grown with very high level of adaptability.
- To meet the per capita daily requirements of vitamins and minerals, minimum of 85 g of fruits and 200 g of vegetables with population of above 1350million people, fruits and vegetables are to be grown on large scale.

- For providing raw material to small scale industries like silkworm, lac, honey, match, paper, canning, dehydration etc. horticulture has wide scope.
- In India, larger areas of land are wastelands, problematic soil, desert land which can be utilized for hardy fruits and medicinal plants.
- The fast development of communication and transport system creates wide scope for horticulture development particularly in transporting the perishable commodities and products.
- Now-a-days, there is increasing demand among the consumers for ready-to-serve and ready-to-eat processed products.
- There is scope for large scale production of plantation crops which serve as raw materials for industries.
- Large scale production of fruits and vegetables make the processing industries viable throughout the year.
- Due to surplus production of different horticultural crops, a sizable quantity with quality assurance is exported to foreign countries to earn foreign exchange.
- Different horticultural crops, particularly, fruits and vegetables are nutritionally superior to field crops and used to combat malnutrition among infants and children.
- Higher productivity level of horticultural crops improves the economic condition of the grower.
- Viable processing industries create employment opportunity for the rural youth and labourers.
- Organic production of horticultural crops not only protects the environment but also the human health.
- Due to viable processing industries, there is scope for import of raw material from other countries.
- The protected cultivation of high value crops can be taken up due to diverse climatic conditions and availability of appropriate seed materials.

1.5 IMPORTANCE

1.5.1 Source of Income Generation

- Horticultural crops like spices and condiments, medicinal and aromatic plants, and ornamental plants are high value crop.
- Crops like vegetables and fruits are prolific yielders.
- Most of the plantation crops are dollar-earning crops.
- All the horticultural crops fetch a higher price in the market than agricultural crops.

1.5.2 Contribution to National Economy

- It is understood in terms of area occupied and its contribution to national economy.
- Horticultural crops occupy only 13% of total cropped area but add 28% to national economy.
- The percentage of share of horticultural crops to agriculture is 30%.

- Different horticultural crops are exported in a sizable quantity to achieve annual foreign exchange earnings.
- It is reported that agricultural sector accounts for about 25% and horticultural sector over 56% of our export basket.
- It reveals that production of horticultural crops is much higher than the production of agricultural crops per unit area.
- As horticultural crops fetch more price or gold to the farmers, the development in horticultural sector is called "Golden Revolution".
- Different horticultural produces and products thereof are export-oriented including products from spices and condiments, plantation crops, ornamental flowers, cut flowers, and planting materials.

1.5.3 Employment Generation

- Horticultural sector provides a good opportunity for employment generation.
- Horticultural crops are tender and delicate. They require utmost care starting from the raising of crops till marketing and storage. Hence, they are considered as labour-intensive crops and raise the man-days for various operations.
- Horticulture is mother of several industries like canning, essential oil, dehydration, refrigeration, wine, cashewnut, transport etc. which provide work for many people.
- Farmers and labourers can keep themselves engaged throughout year. For example, cereal crops require143 man-days for a hectare of land per annum whereas 860 man-days are required in case of fruit production from same land area.
- Cashew processing plants in India provide employment to more than 5.5 lakhs workers annually.
- Similarly, labour requirement for grape, banana, vegetables, tea, coffee and cardamom are 2500, 1000, 200, 875,350and 190 man- days respectively.

1.5.4 Industrial Development

- Processing industries are the backbone of Indian economy. Hence, different horticultural crops make the industries to live long for year-round production.
- Different horticultural plants and produces serve as raw materials for the industries.
- Different plantation crops, indoor and outdoor ornamental plants, medicinal and aromatic plants come in the front line for survival of processing, perfumery and therapeutical industries.

1.5.5 Fruits and vegetables are protective foods

- Fruits and vegetables are food of the future.
- They provide adequate quantity of minerals like calcium, iron, phosphorus and vitamins like A, B-complex and C along with a variety of polyphenols. Polyphenols are regarded as protective substances against chronic diseases.
- The Nutrition Expert Group prescribes a daily requirement of:
  - Energy-2400-3900 calories

- Protein: 55 grams
- Calcium: 0.4 to 0.5 grams
- Beta carotene (precursor of vitamin-A): 3000 mg
- Thiamine:1.2 to 2.0 mg
- Riboflavin: 1.3 to 2.2 mg
- Nicotinic acid: 16 to 26 mg
- Ascorbic acid: 50 mg
- Folic acid: 100 mg
- Vitamin B_{12} : 1 mg

- To obtain all these above vitamins and minerals they recommend 90g fruits and 300g vegetables in the form of 125g green leafy vegetable, 100g root and tuber crops and 75g other vegetables per day.
- But in our country the actual per capita availability is 30g fruits and 92g vegetables which are very low. So, one should take enough fruits and vegetables to reduce the risk of diabetics, cardiovascular diseases, obesity and cancer.
- Indian Council of Medical Research recommends use of 120g fruits and 280g vegetables per capita per day. Consumption of balanced quantity of fruits and vegetables helps in maintaining good health and vigour.
- Fruits and vegetables have a key role in neutralizing the acid produced during digestion of protein rich and fatty foods.
- They provide valuable roughages which promote digestion and help in preventing constipation.

Calcium-rich sources:

Produces	Amount of Calcium
Litchi	0.21%
Karonda (dry)	0.16%
Wood apple	0.13%
Agasthi	1130mg/100 g
Coriander leaves	184mg/100g
Curry leaves	813mg/100g
Amaranthus	395mg/100g
Fenugreek leaves	395mg/100g
Radish leaves	265mg/100g

Phosphorus-rich sources:

Produces	Amount of Phosphorus
Cashewnut	0.45%
Walnut	0.38%
Litchi	0.30%
Wood apple	0.11%

Iron-rich sources:

Produces	Amount of Iron
Karonda (dry)	39.1%
Date	10.6%
Cashewnut	5.0%
Walnut	4.80%
Agasthi	83.90mg/100g
Amaranthus (tender)	25.5 mg/100g
Coriander leaves	1.5 mg/100g

1.5.6 Horticultural crops are highly remunerative and profitable

- Horticultural crops generate more income per unit land area and the cost of the produce is more as compared to the field crops. Hence, the profit is also more.
- Cost benefit ratio is also high.

1.5.7 High Food Value

- Fruit crops yield more calorific value.
- To achieve the required calorific value or energy, a smaller number of horticultural crops are required as compared to cereals and millets.
- Hence, it reduces the load on food grains like rice and wheat.
- Fruits like bananas and vegetables like potato and sweet potato are rich in carbohydrate.
- Almond, cashewnut and walnut are rich sources of fat and protein.
- About two dozen of banana would provide about 2400 calories of energy per day required for a sedentary man.
- They not only adorn the table but also enrich health from the most nutritive menu and tone up the energy and vigour of man.
- From energy point of view the fruit crops give very high number of calories per acre e.g., wheat 1034880 calories/acre and banana 15252800 calories/acre.

Carbohydrate-rich sources:

Produces	Amount of Carbohydrate (%)
Apricot(dry)	72.81
Date	67.30
Karonda(dry)	67.10
Banana	36.40
Bael	36.60
Custard apple	23.90
Cashewnut	22.37
Jamun	19.70
Jackfruit	18.90

Produces	Amount of Carbohydrate (%)
Tapioca	38.10
Sweet potato	28.20
Potato	22.60
Curry leaves	18.70

Protein-rich sources:

Produces	Amount of Protein (%)
Pod and tender leaves of Agasthi	8.40
Cluster beans	3.20
Peas	7.20
Curry leaves	6.10
Cashewnut	21.20
Almond	20.80
Walnut	15.60
Cowpea	4.30

Fat-rich sources:

Produces	Amount of fat (%)
Walnut	64.50
Almond	58.90
Cashewnut	46.90
Avocado	22.80
Coconut	41.60

Fibre-rich sources:

Produces	Amount of fibre (%)
Guava	6.90
Wood apple	5.20
Pomegranate	5.10
Walnut	2.60
Aonla	3.40
Grape	3.00
Amaranthus	1.00
Mustard leaves	0.80
Beet root leaves	0.70
Spinach leaves	0.60

Vitamin A-rich sources:

Produces	Amount of Vitamin A (per 100g)
Mango	4800 IU
Papaya	2020 IU
Beet root leaves	9770 IU
Mustard leaves	4370 IU
Spinach leaves	9300 IU
Fenugreek	6450 IU
Carrot	28129 IU
Colocasia leaves	10278 µg
Coriander leaves	6918 µg
Drumstick leaves	6730 µg

Vitamin B_1 (Thiamine)-rich Sources:

Produces	Amount of VitaminB_1 (per 100g)
Cashewnut	630 mg
Walnut	450 mg
Apricot (dry)	217 mg
Chilli	0.55 mg
Colocasia leaves	0.22 mg
Tomato (red)	0.12 mg

Vitamin B_2 (Riboflavin)-rich Sources:

Produces	Amount of VitaminB_2 (per 100g)
Bael	1.19 mg
Papaya	0.02 mg
Litchi	0.06 mg
Fenugreek leaves	0.31 mg
Amaranthus	0.30 mg

Vitamin C (Ascorbic Acid)-rich Sources:

Produces	Amount of Vitamin C (per 100g)
Barbadus cherry	1000-4000mg
Aonla	600 mg
Guava	299 mg
Lime	63 mg
Sweet orange	50 mg
Drumstick leaves	220 mg
Coriander leaves	135 mg

Produces	Amount of Vitamin C (per 100g)
Chilli	111 mg
Tomato	31 mg

1.5.8 Horticultural crops yield more produces per unit land area as compared to field Crops

Crops	Yield (tones/ha)
Fruits	8.4
Banana	40-60
Pineapple	45
Vegetables (in general)	12.2
Potato	22-25
Tapioca	30-40
Tomato	10-14
Spices	10-12
Paddy	5-6
Wheat	4-5
Maize	7.98

1.5.9 Horticultural crops provide raw materials for many industries

- Horticultural plants and their products provide raw materials for many industries directly or indirectly.
- Most of the plantation crops like tea, coffee, rubber, cardamom, coconut and oil palm are used throughout the year to run the industries.
- Medicinal plants are largely exploited in therapeutical industries for their active ingredients in medicare system.
- Aromatic plants bestowed with aroma principle and essential oil, and oleoresins are used largely in perfumery industries.
- Different horticultural plants also yield dye.
- Different plant-based products like soaps, shampoos, creams, lotions, ointments etc. are also manufactured.
- Coconut provides raw materials for coir, timber and cottage industries.
- Rubber is used in rubber industries, and tea and coffee for beverages.

1.5.10 Religious Importance and Sacred Values

- Many plants and plant parts like leaves, twigs, flowers, fruits etc. possess religious importance and used in many ceremonies.
- Flowers are used for worship of God and Goddess and adorning braid of women.
- Flowers are also used for decoration.

- Leaves of *tulsi* are offered to Lord Krishna and Jagannath and leaves of *bael* to Lord Shiva.
- Hanging of mango leaves and planting of banana plants at the entrance gate is worth mentioning.
- Coconut and arecanut are also used in religious and social functions.
- The leaves of *ber* and *doob* are used in social functions.
- Use of turmeric is worth mentioning in all social and religious functions of Hindu families.
- Leaves of banana are used in all religious and social functions and also used as plate for eating prasadam.
- Some trees are also worshipped.
- The tree of kadamba is associated with Lord Krishna.
- Different types of fruits and flowers are offered to God and Goddesses during worshipping.
- Cut flowers in vase is common now-a-days.

1.5.11 Horticultural Therapy

- People suffering from mental imbalance are given horticultural therapy to divert their attention towards flower, its colour, aroma and arrangement of petals, decoration and ornamental gardening. It is common in USA.
- Essential oils of flowers, leaves, roots and heartwoods of some plants act as antibacterial, antifungal and antiviral agents.

1.5.12 Aesthetic Value

- Fruit trees are considered symbol of life and youth. A young growing plant gives an impression of youth.
- The Moghuls were fond of trees and many Moghul rulers and emperors planted avenues with trees of mango, litchi, mahua, sapota, khirni etc. in their garden to enhance the beauty and aesthetic look of the trees.
- Now-a-days for aesthetic looks many trees are being planted in urban plantations, landscape, park etc.

1.5.13 Export Values

- Indian export basket is rich with different horticultural crops like mango, grape, potato, rose, black pepper, cardamom, ginger, turmeric, chilli, cashewnut, tea, coffee, coconut, arecanut etc.
- European and Gulf Countries are the major importers of Indian horticultural produces.

1.5.14 Orchard Tourism

- People in urban areas remain engaged in different horticultural activities like planting, training and pruning, grafting, budding, harvesting etc. and exercising different experience out of it.

- Visiting of orchards for recreation has become an important tourist activity for urban citizens.
- Making different activities in the gardens is a means of exercise.
- The flowers, ornamental plants and gardens play a very important role in refreshing the minds of people and reducing air pollution.

1.5.15 Attachment of Interest by Governments

Government of India and state governments are attaching more interest on horticultural development for overall development in horticulture sector.

1.5.16 Effective utilization of wasteland through hardy fruit crops and medicinal plants

Fruit crops like bael, ber, cashewnut and custard apple and medicinal plants like khus, sweet flag, citronella etc. can be grown successfully in wastelands.

1.5.17 Miscellaneous

- It helps in maintaining eco-sphere.
- It helps in transferring micro-climate.
- It provides shelter to birds, reptiles, and microorganisms and adds to the geo-zoological diversity on the land.
- It provides scope for the writers, poets, thinkers, analysists to write, compose, to think and analyze different beauty of the eco-sphere.
- It adds to the survival of the living entity.

REFERENCES

Kumar N 1990. *Introduction to Horticulture*. Rajyalaxmi Publications, Nagarcoil, Tamil Nadu.

Singh J 2002. *Basic Horticulture*. Kalyani publishers, Ludhiana.

Singh J 2018. *Basic Horticulture*. Kalyani publishers, Ludhiana.

e-resources

ecoursesonline.iasri.res.in

e-Krishi Shiksha

hortsci.ashspublications.org>content

https://shodhgangotri.inflibnet.ac.in

https://www.merriam-webster.com

https://en.m.wikipedia.org.>wiki

OUTCOMES ASSESSMENT

PART A

Answer the following questions (True or False).

1. Horticultural crops are grown in extensive manner. (True/False)
2. Cashewnut is rich in protein. (True/False)
3 Agriculture broadly refers to the technology of raising plants only. (True/False)
4. People suffering from mental imbalance are given horticultural therapy. (True/False)
5. Aonla is poor in vitamin C. (True/False)

PART B

Answer the following questions.

1 Horticultural crops are— source of vitamins and minerals.
2. The term 'horticulture' is derived from two Latin words; — and —.
3. Whether bael is rich in riboflavin?
4. What is the recommendation of ICMR for use of fruits and vegetables?
5. Name one iron-rich fruit.

PART C

Write a brief note on each of the following.

1. Scope of horticulture.
2. Contribution of horticulture to national economy.
3. Religious importance and sacred values of horticulture.

Chapter 2

Classification of Horticultural Crops

2.1 INTRODUCTION

Owing to a varied agro-ecological and eco- edaphic condition prevalent in one or other part of the country, quite a large number of crops are grown in different parts of India. India is rich in bio-diversity with four hotspots out of twenty-five in the world. About 356 domesticated plant species having economic and botanical importance and 326 species with their wild relatives are found in Indian sub-continent. Besides the above, 9500 other plant species are also available. More than 50 different types of fruits, 50 different types of vegetables, 95 types of spices and condiments, medicinal and aromatic plants, plantation crops and ornamental crop plants have been recorded. To avoid confusion and identity, the horticultural plants are grouped into different categories based on similarity or dissimilarities. Grouping of the crop plants with similar characteristics is referred as classification.

2.2 DIVISIONS OF HORTICULTURE

1. Pomology
2. Olericulture
3. Floriculture
4. Landscape Gardening
5. Plantation Crops
6. Medicinal and Aromatic Plants
7. Spices and Condiments
8. Silviculture
9. Sericulture
10. Apiculture
11. Mushroom
12. Post-Harvest Management and Preservation

2.3 POMOLOGY

The term Pomology is derived from the Latin word '*pomum*', meaning fruits and Greek word '*logy*' meaning science. Hence, Pomology is the study of science of production of fruit crops. Fruit plants are classified according to their nature of growth, climatic requirement, types of fruits, part of fruit used etc.

2.3.1 Classification of Fruit Plants

Based on Nature of Growth

a. Herbaceous: e.g., Pineapple, Banana
b. Shrubaceous: e.g., Pomegranate, Phalsa, Karonda
c. Woody trees: e.g., Mango, Ber, Sapota, Apple, Guava, Pear and many other fruit plants

Based on Continuation of Growth

a. Evergreen: e.g., Mango, Citrus, Litchi, Sapota, Olive, Loquat, Lemon, Pomegranate etc.
b. Deciduous: e.g., Grape, Apple, Pear, Peach, Plum, Apricot, Cherry, Nectarines, Avocado, Kiwi.

Based on Climatic Requirement

1. Tropical
   a. Major tropical: e.g., Mango, Banana, Papaya, Pineapple etc.
   b. Minor tropical: e.g., Sapota, Guava, Custard apple, Loquat, Soursop, Mangosteen etc.
2. Sub-tropical
   a. Citrus fruits: e.g., Lime, Lemon, Orange, Mandarins, Grape fruit etc.
   b. Non-citrus fruits: e.g., Litchi, Kiwi, Grape, Pomegranate, Fig etc.
   c. Nuts: e.g., Cashewnut, Pecan nut, Almond, Walnut.
3. Temperate
   i. Based on plant structure
      a. **Temperate tree fruits:** These fruits are borne on trees, e.g., Apple, Pear, Stone Fruits etc.
      b. **Temperate small fruits:** Fruits are generally borne on vines or herbaceous plants, e.g., Straw berry, Blue berry, Cranberry, Black berry etc.
      c. **Temperate nuts:** Nuts are characterized by hard shell outside, separating the kernel and husk of the fruits, e.g., Pecan nut, Hazelnut, Walnut etc.

   ii. Based on fruit morphology

   Depending upon the number of ovaries involved in fruit formation, fruits are of three groups. They are:

   A. *Simple fruits*: These fruits are derived from a single ovary of one flower. They are again classified as fleshy fruits and dry fruits.

   1. *Fleshy fruits*: These are the types of fruits in which the pericarp (ovary wall) becomes fleshy or succulent at maturity. Again, they may be either pome fruits or drupes.

   I. *Pomes*: These are inferior, two or more celled fleshy syncarpous fruits surrounded by the thalamus and referred as false fruits because the edible fleshy part is not derived from the ovarian tissues but from external ovarian tissues, thalamus, e.g., Apple, Pear, Quince. Here carpel is leathery and the edible portion is receptacle.

II. *Drupes*: These are also known as stone fruits. Fruits are developed from a single carpel whereas olive is an exception. In olive the flower has two carpel and four ovules but one carpel develops. Two ovules are borne in most of the drupes but one seed develops. In this type of fruit, the pericarp is differentiated into three distinct layers; thin exocarp or peel of fruits, the mesocarp which is fleshy and a hard-stony endocarp, enclosing seed. Examples of temperate drupe fruits are Cherry, Peach, Plum and Apricot. In Almond at maturity the exocarp and mesocarp get separated as leathery involucres and are removed before marketing and the only endocarp containing the edible seed is used, hence it is called nut.

2. *Dry fruits*: These types of fruits are classified on the basis of pericarp (ovary wall) at maturity. The entire pericarp becomes dry and often brittle or hard at maturity. They are dehiscent (in which the seeds are dispersed from fruits at maturity) and indehiscent (not split open when ripe). Nuts are typical example of indehiscent dry fruits.

I. *Nut*: A fruit in which carpel wall is hard or bony in texture. Fruit is derived from a hypogynous flower or an epigynous one and is enclosed in dry involucres (husk). It is only one seeded, but in most cases, it is derived from two carpels. Examples are Walnut, Almond, Chestnut, Pecan nut and Hazelnut. Dry fruits are not juicy or succulent when mature and ripe. When dry, they may split open and discharge their seeds (called dehiscent fruits) or retain their seed (called indehiscent fruits).

II. *Achene*: A one seeded fruit in which the seed is attached to ovary wall at one point, e.g., Straw berry.

B. *Aggregate fruits*: These fruits develop from numerous ovaries of the same flower. Individual ovary may be drupe or berry. Raspberry is included in this category.

C. *Multiple or composite fruits*: Multiple or composite fruits are produced from the ripened ovaries of several flowers crowded on the same inflorescence. The example of this type is Mulberry.

iii. Based on fruit growth pattern

Temperate fruit trees are classified into two groups basing on fruit growth pattern. They are:

1. S*igmoid pattern*: The combined growth of fruits results from cell division, cell enlargement and air space formation results in sigmoidal growth (S- shapes curve) when fruit weight is plotted as function of time, e.g., Apple, Walnut, Pecan nut, Strawberry and Pear.
2. *Double* sigmoid: The Fruit's slow growth period coincides with the period of pit hardening, during which lignification of the endocarp (stone) proceeds rapidly, while mesocarp and seed growth is suppressed. Near the end of pit hardening,

flesh cells enlarge rapidly until fruit is ripe, after which growth slows down and ceases, e.g., Peach, Plum, Cherry and Kiwi fruit.

Based on Type of Fruit

1. Pome: e.g., Apple, Pear, Quince.
2. Stone (drupe): e.g., Mango, Coconut, Apricot, Almond, Ber, Peach, Plum.
3. Berry: e.g., Grape, Banana, Date palm, Sapota, Guava, Papaya.
4. Hesperidium: e.g., All Citrus fruits.
5. Psorosis: e.g., Jackfruit, Mulberry, Pineapple.
6. Syconus: e.g., Fig.
7. Nuts: e.g., Litchi, Rambutan.
8. Capsule: e.g., Aonla, Carambola.
9. Aggregate fruit: e.g., Black berry, Raspberry, Straw berry, Custard apple.
10. Multiple fruit: e.g., Pineapple, Mulberry.

Based on Plant Parts Used

1. Aril: e.g., Litchi, Pomegranate.
2. Endosperm: e.g., Coconut.
3. Fleshy pericarp of individual berry: e.g., Custard apple.
4. Fleshy receptacle: e.g., Fig.
5. Fleshy receptacle and thalamus: e.g., Fig.
6. Fleshy layer of pericarp: e.g., Bael.
7. Fleshy bract, perianth, and seed: e.g., Pineapple.
8. Fleshy thalamus: e.g., Apple.
9. Juicy placenta: e.g., Citrus.
10. Mesocarp: e.g., Mango.
11. Mesocarp and endocarp: e.g., Banana.
12. Pericarp and placenta: e.g., Grapes.
13. Pericarp and thalamus: e.g., Jamun.
14. Seed or kernel: e.g., Almond, Walnut.
15. Fruit stalk and thalamus: e.g., Pear.
16. Thalamus and pericarp: e.g., Guava.

Based on Botanical Relationship

a) ***Monocots***

1. Musaceae: e.g., Banana.
2. Bromeliaceace: e.g., Pineapple.
3. Palmae: e.g., Coconut

b) ***Dicots***

1. Anacardiaceae: e.g., Mango, Ambada, Cashewnut.

2. Arecaceae: e.g., Date palm.
3. Apocynaceae: e.g., Karonda.
4. Annonaceae: e.g., Custard apple.
5. Caesalpinaceae: e.g.,Tamarind.
6. Caricaceae: e.g., Papaya.
7. Euphorbiaceae: e.g., Aonla.
8. Juglandaceae: e.g., Walnut.
9. Moraceae: e.g., Mulberry.
10. Myrtaceae: e.g., Guava.
11. Punicaceae: e.g., Pomegranate.
12. Rutaceae: e.g., Citrus fruits, Wood apple, Bael.
13. Ramnaceae: e.g., Ber.
14. Sapotaceae: e.g., Sapota, Khirni.
15. Sapindaceae: e.g.,Litchi.
16. Rosaceae: e.g., Apple, Pear, Peach, Plum.
17. Tiliaceae: e.g., Phalsa.
18. Vitaceae: e.g., Grapes.

Based on Bearing Habit

1. Terminal
   a. Current season or new growth: e.g., Jackfruit, Bael, Pecan nut.
   b. Old growth: e.g., Mango, Litchi.
2. Axillary
   a. New growth: e.g., Guava, Ber, Aonla, Sapota, Fig, Phalsa, Karonda.
   b. Old growth: e.g., Apple, Pear, Plum, Tamarind, Peach.
3. Mixed/Adventitious: e.g., Pomegranate, Citrus, Jack fruit.

Based on Branching Behaviour

1. **Heliotropic:** Plant with branches arising high up on the stem and growing upward, e.g., Hog Plum
2. **Geotropic:** Plant with branches arising near proximity to the ground and growing parallel to the ground surfaces, e.g., Banyan

Based on Rate of Ethylene Evolution

Group	Rate of ethylene evolution at 20°C (µ l/kg/hour)	Name of the fruit plant
Very low	< 0.1	Citrus, Grape, Pomegranate
Low	0.1 – 1.0	Pineapple
Moderate	1.0-10.0	Banana, Fig, Guava, Mango
High	10.0-100.0	Avocado, Papaya
Very high	> 100.0	Passion fruit, Sapota

Based on Ripening Behaviour

- *Climacteric*: Fruit where there is sudden rise in rate of respiration and ethylene evolution at the time of ripening, e.g., mango, Guava, Papaya, Jackfruit, Fig, Sapota, Passion fruit.
- *Non-climacteric*: These fruits produce very little quantity of ethylene and don't response to ethylene treatment. These fruits if harvested before ripening on the plant, don't ripen further in the room condition and don't responds to ethylene treatment. There is no characteristic increase in rate of respiration or production of carbon dioxide. There is no dramatic change in rate of respiration and ethylene production remains at a very low level, e.g., Citrus fruits, Grapes, Pine apple, Custard apple, Litchi, Pomegranate, Cherry, Raspberry, Black berry, Straw berry, Carambola, Cashew nut.

Based on Salinity Tolerance

Tolerance Limit	Concentration (milli mohs)	Examples
Tolerant	8	Date, Ber, Pomegranate, Aonla, Custard apple, Guava, Phalsa
Moderately tolerant	6–3	Fig, Orange, Lemon, Grapefruit, Grape, Mango etc.
Sensitive	3–1.5	Peach, Plum, Apricot, Avocado, Almond

Based on Growing Environment

Growing Condition	Characteristics	Examples
Lithophyte	Plants grow on stones, rocks	Sphagnum moss and Fern
Obligate lithophyte	Plants grow only on rocks or gravels	Liverwort and many Algae.
Facultative lithophyte	Plants grow on rocky substrate and else-where	Paphiopedilum orchid, Liverworts
Terrestrial	Plants grow on earth surface	Most of the fruit plants
Epiphyte	Plant grow on another plant (parasitic plant)	Sandalwood and Orchid
Psammophyte	Plants grow in soil having sand and gravels	Lavenders and Juniper
Petrophyte	Plants able to grow on rocks	Rock felt fern, Bird's nest fern, Pitcher plants
Chasmophytes	Plants grow on fissures in rocks where soil and organic matters are accumulated	Fern

Based on Water Requirement

Condition	Characteristics	Examples
Hydrophyte	Plants growing fully or partially submerged in water	Banana, Makhana, Water chestnut

Condition	Characteristics	Examples
Mesophyte	Plant growing on conditions of neither abundant nor scarcity of water	Mango, Papaya, Guava, Citrus, Aonla, Apple, Pear, Plum, Peach, Custard apple
Xerophyte	Plants growing on extreme scarcity of water	Ber, Datepalm

Based on Light Requirement

Condition	Characteristics	Examples
Heliophytes	Plant grow on open sunny condition	Coconut
Facultative heliophytes	Plant can grow in open sunny condition and can also grow in shade	Mango, Guava, Jackfruit
Obligate heliophytes	Plants always grow in sunny condition	White clover
Sciophytes	Shade loving fruit plants	Arecanut, Pineapple
Obligate sciophytes	Plants grow in complete shade	Acid Cherry
Facultative sciophytes	Plants can grow in shade but not so optimally	Goose berry

Based on Photoperiodic Requirement

Condition	Characteristics	Examples
Short day	Requiring light period of 12 hours or less	Strawberry, Pineapple
Long day	Requiring light period of 12 hours or more	Plantain
Day neutral	No effect of light	Most of the fruit plants

2.4 OLERICULTURE

Olericulture is the study of vegetable science. The term olericulture is derived from Latin word "*Oleris*" meaning pot herbs and the English word "*culture*" meaning raising of plants. Etymologically, the word 'Vegetable' has been developed from Latin word "*Vegetabilis*" meaning animated or enliven and from *"vegetare"* meaning enliven which denotes lively nature of vegetables. Vegetables are the herbaceous edible plants or part thereof which is commonly used for culinary purposes or as salad.

2.4.1 Classification of Vegetable Crops

Vegetables are classified in the following manner:

Classification Based on Botanical Relationship

A. Monocot

Family	Common Name	Scientific Name
Alliaceae	Onion	*Allium cepa*
	Garlic	*Allium sativum*
	Multiplier onion	*Allium cepa* var. *aggregatum*
	Top onion	*Allium cepa* var. *viviparum*
	Leek	*Allium porrum*
	Welsh onion	*Allium fistulosum*
	Shallot	*Allium ascalonicum*
	Chive	*Allium schoenoprasum*
Liliaceae	Asparagus	*Asparagus officinalis*
Araceae	Elephant foot Yam	*Amorphophallus campanulatus*
	Colocasia	*Colocasia esculenta*
Dioscoreaceae	Greater yam	*Dioscorea alata*
	Lesser yam	*Dioscorea esculenta*
	White yam	*Dioscorea rotundata*

B. Dicot

Family	Common Name	Scientific Name
Aizoaceae	New Zealand spinach	*Tetragonia expansa*
Chenopodiaceae	Palak	*Beta vulgaris var. bengalensis*
	Beet root	*Beta vulgaris*
	Spinach (Bilayati palak)	*Spinacia oleracea*
	Swiss chard	*Beta vulgaris var. cicla*
Compositae (Asteraceae)	Lettuce	*Lactuca sativa*
	Artichoke	*Cyanara scolymus*
	Endive	*Cichorium endivia*
	Chicory	*Cichorium intybus*
Convulvulaceae	Sweet potato	*Ipomoea batatas*
Cruciferae (Brassicaceae) Cole Crops	Cauliflower	*Brassica oleracea var. botrytis*
	Cabbage	*Brassica oleracea var. capitata*
	Knol-khol (Kohlrabi)	*Brassica oleracea* var. *gongylodes/coulorapa*
	Broccoli	*Brassica oleracea* var. *italica*
	Brussels sprout	*Brassica oleracea var. gemmifera*
	Rutabaga	*Brassica napus* var. *napobrassica*
	Turnip	*Brassica campestris var. rapa*
	Chinese cabbage	*Brassica pekinensis*
	Radish	*Raphanus sativus*
	Horse radish	*Armoracia rusticana*

Family	Common Name	Scientific Name
Cucurbitaceae	Pumpkin	*Cucurbita moschata*
	Summer squash	*Cucurbita pepo*
	Winter squash	*Cucurbita maxima*
	Water melon	*Citrullus lunatus/ Citrullus vulgaris*
	Musk melon	*Cucumis melo*
	Cucumber	*Cucumis sativus*
	Tinda	*Citrullus vulgaris var. fistulosus*
	Long melon	*Cucumis utilissimus*
	Ridge gourd	*Luffa acutangula*
	Sponge gourd	*Luffa cylindrica*
	Bottle gourd	*Lagenaria siceraria*
	Bitter gourd	*Momordica charantia*
	Sweet gourd	*Momordica cochin-chinensis*
	Spine gourd	*Momordica dioica*
	Pointed gourd	*Trichosanthes dioica*
	Snake gourd	*Trichosanthes anguina*
	Ash gourd	*Benincasa hispida*
	Ivy gourd/little gourd/small gourd	*Coccinia cordifolia/ Coccinia indica*
	Chow-chow/Chayote	*Sechium edule*
Euphorbiaceae	Tapioca	*Manihot esculenta*
Leguminosae	Garden pea	*Pisum sativum var. hortense*
	French bean	*Phaseolus vulgaris*
	Lima bean	*Phaseolus lunatus*
	Broad bean	*Vicia faba*
	Winged bean	*Psophocarpus tetragonolobus*
	Country bean/Dolichos bean	*Dolichos lablab*
	Cowpea	*Vigna unguiculata* subsp. *unguiculata* *Vigna unguiculata var. cylindrica*
	Asparagus bean	*Vigna unguiculata* subsp. *sesquipedalis*
	Soyabean	*Glycine max*
Malvaceae	Bhindi/Okra	*Abelmoschus esculentus*
	Rossel	*Hibiscus sabdariffa*
Polygonaceae	Rhubarb	*Rheum rhaponticum*
	Sorrel	*Rumex vesicarius*
Solanaceae	Potato	*Solanum tuberosum*
	Brinjal	*Solanum melongena*
	Tomato	*Solanum lycopersicon*
	Chilli	*Capsicum annuum*

Family	Common Name	Scientific Name
	Capsicum/Sweet pepper	*Capsicum annuum*
	Bird eye chilli	*Capsicum frutescense*
Umbelliferae (Apiaceae)	Carrot	*Daucus carota*
	Parsley	*Petroselinum crispum*
	Celery	*Apium graveolens*
	Parsnip	*Pastinaca sativa*

Based on Hardiness/Temperature Tolerance

Type	Condition	Examples
Warm season vegetables	These vegetables prefer higher temperature and whose fruits are consumed.	Brinjal, Cucurbits, French bean, Cluster beans, Cowpea, Okra etc. (exception- Beans, Cluster beans)
Cool season vegetables	These crops require low temperature for growth and development. These are vegetables of which roots, stems, leaves, buds, immature flowers and flower parts other than fruits are eaten.	Cabbage, Cauliflower, Knol-khol, Broccoli, Brussels sprout and Leafy vegetables etc.

Based on Plant Parts Used

Plant Parts	Examples
Root and tuber crops	Elephant foot yam, Sweet potato, Colocasia, Yam, Tapioca, Carrot, Radish and Turnip.
Bulbs	Onion, Garlic
Leafy vegetables	Amaranthus, Palak, Spinach, Portulaca, Basella, Bathua, Lettuce, Mustard, Radish leaves, Colocasia leaves etc.
Stem vegetables	Potato, Asparagus
Flower vegetables	Cauliflower, Broccoli, Globe artichoke
Fruit vegetables	Tomato, Brinjal, Chilli, Okra, all Cucurbits

Based on Method of Raising

Methods	Examples
Direct sown	Okra, Carrot, Radish, Peas, Beans, Cucurbits
Transplanted	Tomato, Brinjal, Chilli, Cabbage, Cauliflower

Based on Forcing

Conditions	Examples
Cool forcing	Asparagus, Beet, Carrot, Cauliflower, Cabbage, Radish, Pea, lettuce, Celery etc.
Warm forcing	Brinjal, Chilli, Cucumber, Beans etc.

Based on Rate of Respiration

Respiration Rate	Examples
Very high	Asparagus, Broccoli, Peas, Spinach
High	Bean, Lettuce, Lima bean
Moderate	Beet, Carrot, Celery, Cucumber
Low	Cabbage, Sweet potato, Turnip
Very low	Potato, Onion, Garlic

Based on Pigmentation

Colour	Pigments	Remarks
Green	Chlorophyll in leaves and fruits	Green colour becomes olive green in acidic condition and under alkali condition it becomes bright green in colour
Yellow	Xanthophylls and Carotenoid	
Red or blue	Anthocyanin	It looks purple in neutral medium, red in acidic and blue in alkali

Based on Soil Reaction (pH)

Basophyll	Vegetables which grow satisfactorily in alkaline condition
Acidophyll	Vegetables which grow satisfactorily in acidic condition
Neutrophyll	Vegetables which grow satisfactorily in neutral soil
Psamophytes	Vegetables which grow satisfactorily on sand and gravel
Petrophytes	Vegetables which grow satisfactorily on rocks

Based on Soil Acidity Tolerance Depending Upon Tolerance Limit

Condition	pH	Examples
Slightly tolerant	6.8 to 6.0	Asparagus, Broccoli, Beet, Cabbage, Cauliflower, Lettuce, Okra, Onion, Muskmelon, Spinach, Palak
Moderately tolerant	6.0 to 5.5	Brinjal, Cucumber, Knol-khol, Garlic, Pea, Pumpkin, Radish, Squash, Turnip, Tomato
Very tolerant	5.5 to 5.0	Potato, Sweet potato, Watermelon

Based on Salt Tolerance

Condition	Concentration of salt (milli mohs)	Examples
Sensitive	0.25	Tomato, Snake gourd
Medium	0.5	Cabbage, Chilli, Okra, Sweet potato
	0.75	Amaranthus, Bottle gourd, Cauliflower, Onion, Radish
Highly tolerant	1.0	Ash gourd, French bean
	1.25	Bitter gourd

Based on Photoperiodic Requirements

Condition	Examples
Short day	Sweet potato, Vegetable Soya bean, Bean
Long day	Radish, Cucumber, Spinach, Lettuce, Onion
Day neutral	Tomato

2.4.2 Difference between Fruits and Vegetables

S. No.	Fruits	Vegetables
1.	Fruit is a mature ovary developed from the flower or inflorescence.	Vegetables are fleshy edible products developed from herbaceous plant.
2.	Fruit plants are mostly perennial in nature. Except Banana, Papaya, Strawberry.	Mostly annual, some are biennial and few are perennial, e.g., Biennial- Pointed gourd, Spine gourd, Chayote. Perennial- Drumstick, Curry Leaf.
3.	Fruits are acidic in nature.	Vegetables are alkaline in nature.
4.	Study of fruit is Pomology.	Study of Vegetable is Olericulture.
5.	Fruit plants are mostly woody in nature.	Vegetable plants are mostly herbaceous, soft and succulent.
6.	Fruits developed from the ovary are the edible part.	Plant parts like fruits, flowers, flower buds, leaves, stems, roots etc. are edible parts.
7.	Fruit plants are deep rooted.	Vegetable crops are shallow rooted.
8.	Mostly propagated through vegetative means, exception Papaya, Phalsa etc.	Mostly propagated through seeds except perennial cucurbits.
9.	Fruit plants need special training and pruning.	No special training and pruning is required. Except- Tomato and Garden Pea.
10.	Fruits contain high amount of sugar.	Low in sugar.
11.	Fruit plants require wider spacing for planting.	Vegetable plants are narrow spaced as compared to fruit plants.
12.	Yield per unit land area is less as compared to vegetables.	More yield per unit land area.
13.	Some fruit plants can grow well in wastelands, e.g., Bael, Ber, Custard Apple, Cashewnut.	Vegetable crops need good and fertile soil.
14.	Fruit plants can reduce air pollution in massive plantations.	No such activities.
15.	Fruit plants contain less dietary fibre as compared to vegetables.	Vegetables add more dietary fibre to the food.

2.5 FLORICULTURE

The word floriculture is derived from two words *viz.* Latin word *flos, flor* meaning flower and English word *culture* meaning cultivation. It is the study of growing and marketing of flowers, foliage plants for commercial purposes. It is a discipline of horticulture concerned with cultivation of flowering and ornamental plants for gardens and floristry, comprising the floral industry. Floriculture can be defined as the art and knowledge of growing flowers for perfection.

2.5.1 Classification of Floriculture

Based on Season of Growing

Season	Plants
Summer annual	Zinnia, Portulaca, Tithonia, Gomphrena, Sunflower etc.
Rainy annual	Amaranthus, Balsam, Cock's comb, Gillardia etc.
Winter annual	Aster, Phlox, Cornflower, Verbena, Candytuft, Petunia etc.

Based on Nature of Growth

Nature of growth	Plants
Annual	Nasturtium, Hollyhock, Sweet pea, Chrysanthemum, Carnation, Cornflower, Sweet allysum, Dahlia, Marigold, Verbena, Phlox
Perennial	Rose, Jasmine, Chrysanthemum

Based on Growth Behaviour

Types	Plants
Herbs	Linum, Anchusa, Browallia, Verbena, Viola
Shrubs	Rose, Jasmine, Bougainvillea, Tecama, Nyctanthes
Trees	Champak, Anthocephalus, Gulmohar, Pride of India, Erythrina

Based on Growing Ability

Types	Characteristics	Plants
Climber	These plants are capable of growing over some supports through their climbing structures like tendrils, rootlets, thorns	Antigonon, Wild rose, *Ficus repens*
Twiner	These plants have no climbing structures but can manage to grow over some sort of supports	Asparagus, Madhulata
Rambler	These plants are unable to climb but try to manage to support on stems or branches over themselves	*Quisqualis indica*
Creeper	These plants are not capable of climbing vertically but can grow horizontally over the ground	Potato vines (*Solanum jasminoides*)

Based on Flower Colour

Colour of flower	Plants
White	Alyssum, Dianthus, China aster, Zinnia, Nigella, *Tabernamontana coronarea*
Purple, Lavender and Blue	Ageratum, Anchusa, Browallia, Clitoria, Delphinium, Petunia,Viola,Verbena
Yellow and Orange	Marigold, Calendula, Zinnia

Based on Purpose of Growing

Purpose	Plants
Rock garden/rockery	Ageratum, Alyssum, Phlox, Portulaca, Euphorbia.
Hanging basket	Sansevieria, Dwarf ageratum, Petunia, Portulaca, Verbena, Begonia.

Purpose	Plants
Hedging	Duranta, Tecoma, Bougainvillea, Thevetia, Hibiscus, Murraya, Lawsonia
Edging/path	Dianthus, Nigella, Portulaca, Alyssum, Dwarf ageratum.
Bedding	Dahlia, Marigold, Phlox, Verbena, Carnation, Petunia, Candytuft, Portulaca
Pots	Carnation, Antirrhinum, Petunia, Aglaonema, Alocassia, Anthurium, Aralia, Begonia, Chlorophytum, Dracaena
Fragrant flower	Rose, Jasmine, Tuberose, Sweet alyssum, Sweet sultan, Sweet pea, Stock, Phlox, Carnation
Aromatic flowers	Rose, Jasmine, Tuberose
Dry flower	Helichrysum, Lady's lace, Acrolinum, Nizella, Statice
Loose flower	Marigold, Chrysanthemum, Aster, Sunflower, Zinnia
Loose flower with stick	Rose, Gerbera

Based on Mode of Propagation

Mode of propagation	Plants
Seed method	Petunia, Marigold
Cuttings	Chrysanthemum, Rose, Duranta, Marigold
Rhizome	Canna, Iris
Corms and cormels	Gladiolus, Crocus
Bulbs	Lily, Tulips
Tuber	Dahlia

Based on Photoperiodic Requirement

Type	Plants
Short day	Salvia, Poinsettia
Long day	Aster, Calendula, Delphinium, Stock
Day neutral	Hibiscus, Carnation

2.6 LANDSCAPING

The term landscaping is originated from the Middle Dutch Landscap region; *land* meaning land area and *scap* meaning equivalent of ship. It is concerned with layout of lawns, gardens, planting of trees etc. for beautification.

Landscaping is done with a view to create a natural scene by planting of lawn, trees and shrubs. It is the imitation of nature in the garden and improves the total living environment of the people. We live in a vast planet which is characterized by different types of landscaping depending upon prevailing geographical and agro climatic conditions. There are mountains, hills, glens, valleys, seas, rivers, forests, plains, deserts, lakes, swamps, streams etc. which comprise major part of natural landscape. Man has copied the natural elements for improving landscape around him and converted certain areas in the form of garden for his pleasure.

2.6.1 Basic Principles

Actually, before drawing the master-plan, the following points should be kept in mind in home landscaping.

Background

The background in a garden, whether a wall, tall trees or a hedge should be neutral in nature; that is to say this should not become a distracting feature over the main features of the garden.

Contrast

The design should be such that it should break monotony. To achieve this, a variation in form, texture, or colour has to be brought in.

Balance or Proportion

A balance has to be maintained between different components (masses, forms, colour, etc.) of a garden. In a formal garden, this is achieved by balancing the quantity or by objects, whereas in an informal garden this can be achieved by planting a small mass of colour in front of a large neutral mass. Overcrowding of plants or other garden features should be avoided.

Open Centre

The central area of the garden should be left out of any items of major interest. The best way to achieve this is to have a lawn, which also gives an effect of largeness to the property. A specimen shrub in the centre of the lawn is unsuitable as this goes against the principle of spaciousness, but a tree branching at higher levels from the ground could be planted.

Repetition

The repetition or duplicating some features of a garden helps achieve rhythm, balance, and unity. In a formal garden, generally the same feature is repeated. But for an informal design this need not be so. Here one may repeat the colour tone without disturbing the texture, form, or quantity. If there is a circular path this can be repeated by having two or three consecutive circular shaped beds of annual flowers, hedges, and shrubbery border. Thus, though the shape is repeated, the variation in texture, colour, and form ensure that the design does not look monotonous.

Rhythm

A landscape designer should have an artistic sense to understand how to bring in rhythm in the design. Arranging different elements haphazardly, without harmony, does not enhance beauty. Harmonious lines, often artistically curved, bring in rhythm to the landscape. A group of shrubs in front of a rockery breaks the rhythm. Repetition of certain elements, such as form, enhances the rhythm.

Variety

To break the monotony in a garden, variety is essential. This is achieved by contrast of colour, form, and texture. Planting of different seasonal flowers all in red colour, does not necessarily mean variety.

Besides these, a design should be simple, easy to maintain, and provide comfort for inmates.

2.7 PLANTATION CROPS

Plantation is the large-scale farm or estate in a tropical or semi- tropical country meant for farming that specializes in cash crops. The crops taken in large scale as plantation are termed as plantation crops. Most of the plantation crops are dollar earning crops.

2.7.1 Classification of Plantation Crops

Based on Botanical Relationship

1. Monocot

Family: Arecaceae e.g., Coconut, Arecanut

2. Dicot

Family	Examples
Theaceae	Tea
Rubiaceae	Coffee
Moraceae	Rubber
Sterculiaceae	Cocoa
Piperaceae	Black pepper
Lauraceae	Cinnamon
Anacardiaceae	Cashewnut

Based on Growth Pattern

Type	Examples
Vine	Betel vine, Black pepper
Shrub	Coffee, Tea
Tree	Rubber, Oil palm, Arecanut

Based on Extent of Growing

Grown as	Examples
Homestead plantation	Coconut, Arecanut, Black pepper
Estate plantation	Tea, Coffee, Rubber

Based on Intensity of Cultivation

Type	Examples
Single storeyed	Clove, Nutmeg
Multi-storeyed	Cashewnut, Arecanut,Turmeric, Ginger

2.8 MEDICINAL AND AROMATIC PLANTS

Medicinal plants are also called medicinal herbs. They are being used in traditional medicines since pre-historic period. These plants synthesize hundreds of chemical compounds. Medicinal plants yield alkaloids which are used for natural source of drugs and medicines. Medicinal plants contain steroid principles both with preventive and curative properties. The secondary metabolites available with them are used for preparation of drugs. Medicinal plants may be defined as those plants that are commonly used in treating and preventing specific ailments and diseases and that are generally considered not to be harmful to human.

More than 1500 plant species contain aromatic principles. The aromatic principles are volatile in nature. Out of these only about 50 species serve as source of essential oils, oleoresins and aroma chemicals on commercial scale but when considered on large scale cultivation of aromatic plants, its number hardly exceeds two dozen. The aromatic plants are valued for their essential oils and oleoresins. These oils and oleoresins find their use in medicines, cosmetic industry, perfumery, soap-making and toiletry preparation.

2.8.1 Classification of Medicinal Plants

According to growth habit

(a) *Herbaceous or herbs*: Bramhi, Amhaldi, Turmeric, Datura, Nizella, Long pepper, Babchi

(b) *Shrubs*: Davana, Safed musli, Belladonna, Lavender, Sarpagandha, Chitrak

(c) *Trees*: Babul, Bael, Neem, Palas, Gugal, Olive, Arjun, Hirda, Nagakesar, Marking nut

According to duration

(a) *Annual*: Jangalimuli, Cock's comb, Red poppy, Bhuin amla

(b) *Biennial*: Bankulthi, Caper spurge, Catch fly

(c) *Perennial*: Chocholate vine, Malkunki, Hajodi, Khira, Gudmari, Kali mirchi

According to plant parts used

(a) *Tubers*: Satabari, Safed musli, Sakarkand

(b) *Pods*: Babul

(c) *Whole plant*: Bramhi, Jangalimuli, Bhuin amla, Catch fly, Gudmari, Hazodi

(d) *Leaves*: Babul, Bael, Neem, Palas, Olive, Davana, Belladonna, Chitrak, Datura, Bankulthi, Malkunki, Gudmari

(e) *Inflorescence*: Cock's comb

(f) *Bark*: Babul, Neem, Palas, Arjun, Bahala, Malkunki

(g) *Rhizome*: Amhaldi, Turmeric

(h) *Gum/latex*: Babul, Palas, Guggal, Poppy, Caper

(i) *Roots*: Bael, tuberous root of Safed musli, Belladonna, Sarpagandha, Chitrak, Pipali, Puskamal

(j) *Stem*: Chocolate vine

(k) *Fruits*: Bael, Bahara, Harida, Marking nut, Pipali, Chocolate vine, Black pepper

(l) *Flower*: Neem, Palas, Davana, Lavender, Datura, Red poppy

(m) *Seed*: Neem, Palas, Nizella, Poppy, Babchi, Bankulthi, Malkunki,

(n) *Oil*: Neem, Olive, Nagakesar

2.8.2 Classification of Aromatic Plants

According to Their Use

(a) Medicinal herb e.g., Basil, Eucalyptus

(b) Culinary herb e.g., Cinnamon, Thyme, Dill

(c) Aromatic herb e.g., Vetiver, Marigold, Lemon grass

(d) Ornamental herb e.g., Marigold

According to Active Constituent

(a) Aromatic herb

(b) Astringent herb (Tanin)

(c) Bitter herb (Phenols, Glucosides, Alkaloids, Saponins)

   (i) *Laxative*: Blood purifier, Bowl movement e.g., Aloe and Cassia

   (ii) *Diuretic*: Loss of fluid from body through urinary system

   (iii)*Saponin*: Produce frothing and foaming agent

(d) *Mucilagenous*: Polysaccharides give slippery feel to tongue

(e) *Nutritive*: Protein, Carbohydrate, Fat

According to Period of Life (Completion of Life Cycle)

(a) *Annual*: e.g., Basil, dill, fennel, Saffron, Chamomile, Ajowan, Davana etc.

(b) *Biennial*: e.g., Caraway, Vetiver, Celery

(c) *Perennial*: e.g., Allspice, Fennel, Lavender, Mint, Thyme, Lemon grass, Geranium, Palmarosa, Rose, Cinnamon, Tea tree

(d) *Aromatic foliage*: Onion, Garlic, Mugwort, Wormwort, Sea wormwort, *Lavender*, Salvia, Marigold etc.

According to Their Importance

(a) *Major aromatic plants*: These plants are exclusively grown for extraction of aromatic principles to be used in perfumery and cosmetic industries. They are Chamomile, Vetiver, Lemon grass, Patchouli, Tea tree, Eucalyptus.

(a) *Minor aromatic plants*: These are grown for their volatile oil and aroma principles which are the byproducts or secondary products. They are Cinnamon, Marigold, Dill, Ambrette, Celery.

According to Their Economic Parts

(a) *Herbs*: Patchouli, Citronella, Sweet basil, Geranium, Rosemary

(b) *Roots*: Vetiver, Sandalwood, Camphor

(c) *Wood/Heart wood*: Sandalwood, Camphor, Linaloe

(d) *Bark*: Cinnamon, Cassia

(e) *Leaves*: Eucalyptus, Tea tree, Lemongrass, Mint, Camphor

(f) *Flower*: Jasmine, Rose, Marigold, Chamomile, Champak, Tuberose

(g) *Flowering tops*: Davana, Palmarosa, Thyme

(h) *Fruit:* Dill, Linaloe
(i) *Seeds:* Ambrette, Ajowan, Celery

According to Growth Habit

(a) *Grasses:* Palmarosa, Rosa grass, Lemon grass, Vetiver, Citronella
(b) *Herbs:* Sweet basil, Tuberose, Thyme, Rosemary, Chamomile, Ajowan, Davana, Marigold, Mint
(c) *Shrubs:* Patchouli, Rose, Geranium, Jasmine
(d) *Trees:* Eucalyptus, Tea tree, Camphor, Champak, Cinnamon, Linaloe

According to Habitat

(a) *Temperate:* Chamomile, Ajowan, Fennel, Pepper mint, Spear mint, Bergamot mint
(b) *Sub-tropical:* Vetiver, Mint, Eucalyptus, Ajowan, Thyme, Rosemary, Citronella, Davana, Fennel, Japanese mint
(c) *Tropical:* Lemon grass, Ocimum, Cinnamon, Linaloe, Sandalwood, Eucalyptus, Citronella, Palmarosa, Patchouli

According to Method of Propagation

(a) *Sexually propagated:* Clarysage, Cumin, Davana, Camphor, Eucalyptus, Sandalwood, Palmarosa
(b) *Asexually propagated:* Citronella, Geranium, Jasmine, Patchouli, Rose, Tuberose
(c) *Both sexually and asexually propagated:* Lemon grass, Linaloe, Marigold, Palmarosa, Rosemary, Thyme, Vetiver

According to Taxonomy (Botanical Classification)

Family	Plant Species
Compositae(Asteraceae)	*Salidago odora* *Blumea glabra* *Helichrysum angustifolium* *Tagetes spp.* *Anthemis nobilis* *Achillea moschata* *Santolina chamaecyparissus* *Matricaria chamomilla* *Artemisia maritima* *Arnica montana* *Saussurea lappa* *Tanacetum vulgare*
Labiate (Lamiaceae)	*Rosmarinum flexuosus* *Rosmarinum officinalis* *Lavandula hybrida* *Lavandula spica* *Nepeta spicata* *Salvia officinalis* *Monarda punctata*

Family	Plant Species
	Melissa officinalis *Hyssopus officinalis* *Origanum elongatum* *Marjorana hortensis* *Thymus spp.* *Mentha spp.* *Ocimum* spp. *Perilla frutescens* *Pogostemon cablin*
Leguminosae	*Acacia floribunda* *Lupinus luteus* *Myrocarpus frondosus* *Parkinsonia aculeata* *Glycyrrhiza glabra*
Rosaceae	*Spiraea ulmaria* *Rosa spp.* *Prunus amygdalus*
Rutaceae	*Citrus spp.* *Murraya exotica* *Skimmia laureola* *Xanthoxylum piperitum* *Ruta montana* *Cusparia trifoliata*
Cruciferae (Brassicaceae)	*Cochlearia armoracia* *Brassica alba* *Brassica juncea* *Brassica napus* *Brassica nigra* *Raphanus sativus*
Lilliaceae	*Lilium candidum* *Hyacinthus orientalis*
Caryophyllaceae	*Dianthus caryophyllus*
Cupressaceae	*Callitropsis araucarioides* *Thuja plicata* *Cupressus japonica* *Juniperus mexicana*
Ranunculaceae	*Nigella damascena*
Graminae (Poaceae)	*Vetiveria zizanioides* *Cymbopogon citratus* *Cymbopogon flexuosus* *Cymbopogon martinii* *Cymbopogon nardus* *Cymbopogon polyneuros* *Andropogon aciculatus* *Andropogon odoratus*
Malvaceae	*Abelmoschus moschatus*

Family	Plant Species
Cyperaceae	*Cyperus rotundus*
Pandanaceae	*Pandanus fascicularis*
Podocarpaceae	*Dacrydium franklini*
Pinaceae	*Picea alba* *Picea excelsa* *Picea glauca* *Picea nigra* *Picea obovata* *Picea vulgaris* *Tsuga douglasii* *Abies alba* *Abies balsamea* *Cedrus deodara* *Pinus attenuata*
Taxodiaceae	*Cryptomeria japonica*
Acoraceae	*Acorus calamus*
Amaryllidaceae	*Narcissus tagetta* *Polyanthes tuberosa*
Alliaceae	*Allium sativum* *Allium cepa*
Iridaceae	*Crocus sativus* *Iris germanica*
Zingiberaceae	*Kaempferia galanga* *Kaempferia rotunda* *Curcuma amada* *Curcuma aromatica* *Curcuma domestica* *Curcuma longa* *Curcuma xanthorrhiza* *Curcuma zedoaria* *Curcuma zerumbet* *Alpinia galanga* *Zingiber officinale* *Zingiber nigrum* *Zingiber elatum* *Zingiber zerumbet* *Zingiber cassumunar* *Amomum subulatum* *Amomum hirsutum* *Amomum aromaticum* *Elletaria cardamomum*
Piperaceae	*Piper nigrum* *Piper longum* *Piper betle*
Moraceae	*Humulus americanus*

Family	Plant Species
Santalaceae	*Santalum album*
Myristicaceae	*Myristica fragrans*
Lauraceae	*Cinnamomum aromaticum* *Cinnamomum camphora* *Cinnamomum cassia*
Magnoliaceae	*Magnolia grandiflora* *Michelia champak* *Illicium japonicum* *Illicium verum*
Anonaceae	*Cananga odorata* *Cananga latifolia*
Violaceae	*Viola odorata*
Clusiaceae (Guttiferae)	*Calophyllum inophyllum* *Mesua ferra*
Geraniaceae	*Geranium maculatum* *Pelargonium radula* *Pelargonium roseum*
Burseraceae	*Boswellia serrata* *Bursera delpechiana* *Commiphora mukul*
Meliaceae	*Melia azadiracta*
Euphorbiaceae	*Croton eluteria*
Anacardiaceae	*Pistacia lentiscus*
Moringaceae	*Moringa oleifera*
Tiliaceae	*Tilia tomentosa*
Sterculiaceae	*Dombeya viscose*
Myrtaceae	*Myrtus caryophyllata* *Pimenta officinalis* *Eugenia caryophyllata* *Eugenia pimenta* *Eugenia jambolana* *Syzygium cumini* *Eucalyptus grandis,* *Eucalyptus polybractea*
Lythraceae	*Lawsonia alba*
Combretaceae	*Quisqualis indica*
Umbelliferae (Apiaceae)	*Coriandrum sativum* *Cuminum cyminum* *Apium graveolens* *Petroselinum hortense* *Carum carvi* *Pimpinella anisum* *Foeniculum vulgare* *Angelica glabra*

Family	Plant Species
	Ferula asafoetida *Daucus carota*
Oleaceae	*Jasminum officinale* *Jasminum sambac* *Nyctanthes arbortristis*
Apocynaceae	*Nerium odorum* *Tabernaemontana coronaria* *Thevetia nerifolia*
Verbenaceae	*Clerodendron inerme*
Solanaceae	*Capsicum annuum* *Capsicum frutescens*
Bignoniaceae	*Millingtonia hortensis*
Rubiaceae	*Gardenia grandiflora* *Gardenia floribunda* *Gardenia latifolia* *Cinchona calysaya*
Sapotaceae	*Mimusops elengi*
Valerianaceae	*Valeriana officinalis* *Valeriana hardwickii*
Dipterocarpaceae	*Dipterocarpus tuberculatus* *Dipterocarpus turbinatus*

2.9 SPICES AND CONDIMENTS

There are 109 types of spices and condiments in the world out of which 63 are cultivated in India and 52 species come under the purview of Spices board.

Spices are the vegetable crops or parts thereof which are used for imparting colour, flavour and piquancy to taste.

Condiments: A spice can be called as a condiment when used in extra while eating the food, e.g., Sauce, Chutney, Vinegar.

According to International Organisation of Standards there is no clear-cut definition between spices and condiments.

2.9.1 Classification of Spices and Condiments

Based on Completion of Life Cycle

Life Cycle		Examples
Annual	Crop which completes its life cycle in one season	Cumin, Fennel, Coriander, Fenugreek,
Biennial	Crop which completes its life cycle in two seasons	Onion, Garlic
Perennial	Crop which completes its life cycle in more than two seasons	Cardamom (small and large), Cinnamon, Clove, Nutmeg, Saffron, All spice

Based on Their Importance

Type	Examples
Major spice or primary spice	Turmeric, Chilli, Ginger, Black pepper, Cardamom
Minor spice or secondary spice	Cumin, Fennel, Coriander, Fenugreek, Clove, Nutmeg, Cinnamon

Based on Their Utility

Use	Examples
Colour imparting	Turmeric, Chilli, Saffron
Flavour imparting	Garlic, Clove, Curry leaf
Taste imparting	Cinnamon, Asafoetida (Devil's dung/Hing)

Based on Growth Pattern

Growth Pattern	Examples
Herb	Cumin, Coriander, Fenugreek, Onion, Garlic, Turmeric, Ginger, Saffron
Shrub	Black pepper, Cardamom, Vanilla
Tree or woody	Cinnamon, Tejpat, Nutmeg, Clove, Allspice

Based on Cultural Management

Cultural Management	Examples
Horticultural spices	Black pepper, Long pepper
Plantation spices	Clove, Nutmeg, Allspice, Tamarind
Agronomic spices	Dill, Cumin, Fennel, Mustard

Based on Parts Used

Parts Used	Examples
Root spice	Horse radish
Bark spice	Cinnamon, Cassia
Leafy spice	Coriander, Tejpat, Curry leaf, Mentha, Fenugreek, Basil, Tulsi, Rosemary,
Bulb spice	Onion, Garlic
Flower bud	Clove
Floral part(stigma)	Saffron
Leaf and flowering tip	Thyme, Marjoram
Fruit spice	Cumin, Chilli, Capsicum, Cardamom
Seed spice	Coriander, Fenugreek, Dill, Ajowan, Mustard.
Resin	Asafoetida
Unripe fruit	Amchur
Aril	Mace, Mangosteen, Pomegranate
Kernel	Nutmeg
Bean	Vanilla
Pod	Tamarind

Based on Botanical Relationship

1. Monocot

Family	Examples
Alliaceae	Onion, Garlic
Acoraceae	Sweet flag
Zingiberaceae	Ginger, Turmeric, Cardamom

2. Dicot

Family	Examples
Piperaceae	Black pepper
Apiaceae	Cumin, Coriander, Fennel
Solanaceae	Chilli
Lamiaceae	Sacred basil (tulsi)
Myrtaceae	Clove
Myristicaceae	Nutmeg
Lauraceae	Cinnamon, Tejpat
Rutaceae	Curry leaves
Cruciferae	Mustard
Iridaceae	Saffron
Orchidaceae	Vanilla

2.10 SILVICULTURE

Silviculture is the practice of controlling the establishment, growth, composition, health and quality of forests to meet diverse needs and values. It can be defined as "a branch of forestry dealing with the development and care of forests". The word silviculture is derived from Latin word '*silva*' meaning wood and French word '*culture*' meaning cultivation.

This system involves the conscious and deliberate use of land for the concurrent production of agricultural crops including tree, crops and forest crops. Based on the nature of the components this system can be grouped into various forms:

(a) Improved fallow species in shifting cultivation

(b) Taungya system

(c) Multispecies tree gardens

(d) Alley cropping (Hedge row inter-cropping)

(e) Multipurpose trees and shrubs on farmlands

(f) Crops combination with plantation crops

(g) Agroforestry for fuel wood production

(h) Shelter belts

(i) Wind breaks

(j) Soil conservation hedges etc.

2.10.1 List of different Silvicultural Plants

Common English Name	Scientific Name
Teak	*Tectona grandis*
Eucalyptus	*Eucalyptus tereticornis*
Tamarind	*Tamarindus indica*
Ailanthus (Tree of Heaven)	*Ailanthus excelsa*
Neem	*Azadirachta indica*
Pungam	*Pongamia pinnata*
Prosopis	*Prosopis juliflora*
Casuarina	*Casuarina equisetifolia*
Silk Cotton	*Ceiba pentandra*
Acacia	*Acacia auriculiformis*
Bamboos (Solid Bamboo)	*Dendrocalamus strictus*

2.11 SERICULTURE

It is the practice of raising of silk worms for production of raw silk. The word sericulture is derived from Latin word ***sericum*** meaning silk and French word ***culture*** meaning cultivation. Sericulture is a silk producing agro-industry and India is the second largest silk producing country in the world. Different types of silk producing silk worms are as follows.

Name of Silk Worm	Name of Silk Produced	Host Plant for Cultivation
Bombyx mori	Mulberry silk	Mulberry
Antheraea mylitta	Tasar silk	Asan, Arjun, Sal, Oak
Philisamia ricini	Eri or Errandi silk	Castor
Antheraea assamensis	Muga silk	Soalu

2.12 APICULTURE

It is the practice of keeping of bees especially in large commercial scale for sale of honey. 'Apis' is the genus of honey bee and hence the name is apiculture. The word apiculture is derived from Latin word ***'apis' meaning*** bee and English word ***culture*** meaning cultivation. Basically, the honey producing bees belong to genus Apis. They have different species and subspecies according to the locations. In general, four different species of *Apis* are responsible for honey production in the world. They are:

1. *Rock bee: Apis dorsata*
2. *Little bee: Apis florea*
3. *Asian bee: Apis cerana*
4. *European bee: Apis mellifera*

In addition, there are some stink less honey bees which also provide honey. They are:

1. *Melipona* sp.
2. *Trigona* sp.

2.13 MUSHROOM

Mushroom is any fungus having a fleshy fruiting body typically produced above ground on soil or on their food sources from a group of mycelia buried in substratum. The word mushroom is derived from late Latin word '*mussirio*', old French word '*mousseron*', the English of which is mushroom. Common edible mushrooms are:

1. Paddy straw mushroom: *Volvariella spp.*
2. Oyster mushroom: *Pleurotus spp.*
3. Button mushroom: *Agaricus spp.*
4. Milky mushroom: *Calocybe spp.*
5. Shiitake mushroom: *Lentinulla spp.*
6. Jew's ear mushroom: *Auricularia spp.*

2.14 POST HARVEST MANAGEMENT AND PROCESSING

Post-harvest management may be defined as "a branch of horticulture which deals with all operations right from the harvest of the crop, even some preharvest stages, till the commodity reaches at the consumer either in fresh or processed form and efficient utilization of wastes in a profitable manner".

2.14.1 Post Harvest Management

Different post-harvest operations followed to prolong the shelf life are:

1. Proper stage and time of harvesting
2. Precooling
3. Sorting
4. Grading
5. Curing
6. Waxing
7. Transportation
8. Storage
9. Inducing or enhancing ripening
10. Retarding or delaying ripening

2.14.2 Preservation

Value addition is an important aspect of post-harvest processing. Different processed products are:

1. Juice
2. Squash
3. Jam
4. Jelly
5. Ready-to-serve beverage

6. Preserve
7. Toffee
8. Chutney
9. Pickle
10. Canned fruits
11. Canned vegetables
12. Canned mushrooms
13. Amchur powder
14. Fruit concentrate
15. Candy
16. Syrup
17. Nectar
18. Sauce
19. Ketchup
20. Cordial
21. Marmalade
22. Raisin
23. Chips
24. Starch
25. Juice soup
26. Puree
27. Paste

REFERENCES

Pruthi JS 2014. *Spices and Condiments*. National Book Trust India, New Delhi.

Saini GS 2015. *Post-Harvest Management and Preservation of Fruits and Vegetables*. Aman Publishing House, Meerut.

Singh J 2018. *Basic Horticulture*. Kalyani Publisher, Ludhiana.

Skaria BP, Joy PP, Mathew S, Mathew G, Joseph A and Joseph R 2007. *Aromatic plants; Horticulture Science Series-1*. New India Publishing Agency, New Delhi.

e-resources:

e-coursesonline.iasri.res.in

https://en.m.wikipedia.org> wiki

https://www.dictionary.com>browse

https://www.etymonline.com

https://www.merriam-webster.com

https://www.culinarylore.com>food history

https://www.researchgate.net

https://www.yourdictionary.com

https://www.dictionary.com
https://www.learnpick.in>details
https://www.encyclopedia.com>food
https://www.fruitsinfo.com>temperate
https://www.daff.gov.za>portal
https://www.fruitsinfo.com>list of temperate and sub-tropical fruit trees
https://www.quora.com
http://agritech.tnau.ac.in/horticulture/horti_Landscaping_principles.html
nph.gov.in>introduction-and-importance
vikaspedia.in>farmbased enterprisers
www.yourarticlelibrary.com>fruits

OUTCOMES ASSESSMENT

PART A

Answer the following questions. (True or False)

1. Olericulture deals with fruits. (True or False)
2. Post-harvest operations prolong the shelf life. (True or False)
3. Black pepper belongs to Apiaceae. (True or False)
4. Castor is host plant for eri silk. (True or False)
5. Cumin is biennial. (True or False)

PART B

Answer the following questions.

1. What is pomology?
2. Define mushroom.
3. Three/four different species of *Apis* are responsible for honey production in the world.
4. Cite one example of root spice.
5. Keeping of bees in large commercial scale for sale of honey is known as —.

PART C

Write a brief note on each of the following.

1. Classification of medicinal plants.
2. Classification of flowers.
3. Botanical classification of vegetables.
4. Classification of fruits based on climatic requirements.
5. Requirement of classification of horticultural crops.

Chapter 3

Role of Fruits and Vegetables in Human Nutrition

3.1 INTRODUCTION

Life cannot be sustained without adequate nourishment. Human beings need adequate food for growth, development and to lead an active and healthy life. Animals satisfy basic food requirement mainly through natural selection. But human beings have access to a wide range of foods to choose to make up their diet. Since all foods are not of same quality from nutritional point of view, man's ability to meet his nutritional need and maintenance of good health depend upon type and quantity of foods he is able to include in his diet to satisfy hunger. Human beings need a wide range of nutrients to perform various functions in the body and to lead a healthy life. Several reports have shown that adequate intake of fruits and vegetables forms an important part of a healthy diet and low fruit and vegetable intake constitutes a risk factor for chronic diseases such as cancer, coronary heart disease (CHD), stroke and cataract formation. The nutrients include proteins, fats, carbohydrates, vitamins and minerals. The foods containing these nutrients which we consume daily are classified as cereals, legumes, nuts and oil seeds, fruits, vegetables, milk and milk products and flesh foods like fish, meat and poultry. Some foods provide only a single nutrient and some multiples. Scientific evidence indicates that frequent consumption of fruits and vegetables can prevent oesophageal, stomach, pancreatic, bladder and cervical cancers and that a diet high in fruits and vegetables could prevent 20% of most types of cancers. Reports indicate that increasing individual fruit and vegetable consumption by 600 g per day could reduce the global burden of stroke by 19% and decrease the risk of coronary heart diseases by 31% respectively. Realizing the food value of fruits and vegetables in human nutrition the Indian Council of Medical Research (ICMR) recommends per capita use of 120g fruits and 280g vegetables per day.

3.2 ACTIVITIES OF NUTRIENTS

Proteins, fats and carbohydrates are oxidised in the body to yield energy. Proteins are the important constituents of tissues and cells, form mussels, tissues and blood. Fat is an important source of energy, imparts palatability. Carbohydrates are a class of energy yielding substances. Though they do not contribute to nutritive value, the presence of fibre in it is necessary for mechanism of digestion and elimination of waste. Energy is essential for rest, activity and growth. Vitamins are organic substances present in small amount and required for carrying out vital body function. Vitamin A is essential for clear vision. Vitamin D is required for bone growth and calcium metabolism. B vitamins play significant role in metabolism, proper utilization of energy, carbohydrates, proteins and fats. Vitamin C is involved in collagen synthesis, bone and teeth calcification, and many other reactions in the body. Among minerals, calcium is required for formation and maintenance of skeleton and teeth, movement of limbs, contraction of heart, nervous activity and blood clotting. Phosphorus is a part of nucleic acid. Iron is essential for haemoglobin of red blood cells and transport of oxygen.

3.3 ROLE OF FRUITS

Among different sources of nutrients, fruits and vegetables play a significant role because they are considered as food of the future. Fruits and vegetables play an important role in human nutrition and health, particularly as sources of vitamin C, thiamine, niacin, pyridoxine, folic acid, minerals and dietary fibre. In the USA, the consumption of fruits and vegetables as a group is known to contribute to an estimated intake of 91% of vitamin C, 48% of vitamin A, 30% of folate, 27% of vitamin B6, 17% of thiamine and 15% of niacin. It is also known that fruit and vegetable intake supply 16% of magnesium, 19% of iron and 9% of the calories. Other vital nutrients supplied by fruits and vegetables include riboflavin, zinc, calcium, potassium and phosphorus. Some components of fruits and vegetables (phytochemicals) are strong antioxidants and modify the metabolic activation and detoxification/disposition of carcinogens and may even influence processes that may change the course of the tumour cell. Fruits like Aonla and Guava are good sources vitamin 'C'. In addition to vitamin C yellow coloured fruits like Mango and Papaya contain β- Carotene, a precursor of vitamin 'A'. Banana is a good source of carbohydrate and provides energy. Different dried fruits contain appreciable amount of protein. Fresh fruits like Guava and Papaya contain pectin which provide bulk to the diet and help in bowl movements. Different seasonal fruits provide Vitamin C and β- Carotene.

Table 3.1 Nutrient Content of Important Fruits
(Values per 100g of edible portion)

Sl. No.	Name of Fruit	Protein (g)	Fat (g)	Minerals (g)	Crude fibre (g)	Carbo-hydrate (g)	Energy (Kcal)	Ca (mg)	P (mg)	Fe (mg)
1.	Ambada	0.7	3.0	0.5	1.0	4.5	48	36	11	3.9
2.	Aonla	0.5	0.1	0.5	3.4	13.7	58	50	20	1.2
3.	Apple	0.2	0.5	0.3	1.0	13.4	59	10	14	0.66
4.	Apricot (fresh)	1.0	0.3	0.7	1.1	11.6	5663	20	25	2.2
5.	Apricot (dry)	1.6	0.7	2.8	2.1	73.4	306	110	70	4.6
6.	Avocado	1.7	22.8	1.1	-	0.8	215	10	80	0.7
7.	Bael	1.8	0.3	1.7	2.9	31.8	137	85	50	0.6
8.	Banana	1.2	0.3	0.8	0.4	27.2	116	17	36	0.36
9.	Bread fruit	1.5	0.2	0.9	2.1	15.8	71	40	30	0.5

(*Contd.*)

Sl. No.	Name of Fruit	Protein (g)	Fat (g)	Minerals (g)	Crude fibre (g)	Carbo-hydrate (g)	Energy (Kcal)	Ca (mg)	P (mg)	Fe (mg)
10.	Bullock heart	1.4	0.2	0.7	5.2	15.7	70	10	10	0.6
11.	Cape Gooseberry	1.8	0.2	0.8	3.2	11.1	53	10	67	2.0
12.	Cashew apple	0.2	0.1	0.2	0.9	12.3	51	10	10	0.2
13.	Cherry (red)	1.1	0.5	0.8	0.4	13.8	64	24	25	0.57
14.	Currant (black)	2.7	0.5	2.2	1.0	75.2	316	130	110	8.5
15.	Dates (fresh)	1.2	0.4	1.7	3.7	33.8	144	22	38	0.96
16.	Dates (dry)	2.5	0.4	2.1	3.9	75.8	317	120	50	7.3
17.	Fig	1.3	0.2	0.6	2.2	7.6	37	80	30	1.0
18.	Grapes (blue)	0.6	0.4	0.9	2.8	13.1	58	20	23	0.5
19.	Grapes (pale green)	0.5	0.3	0.6	2.9	16.5	71	20	30	0.52
20.	Grape fruit	1.0	0.1	0.4		10.0	45	30	30	0.2
21.	Guava	0.9	0.3	0.7	5.2	11.2	51	10	28	0.27
22.	Jack fruits	1.9	0.1	0.9	1.1	19.8	88	20	41	0.56
23.	Lemon	1.0	0.9	0.3	1.7	11.1	57	70	10	0.26
24.	Sweet lemon	0.7	0.3	0.5	0.7	7.3	35	30	20	0.7
25.	Litchi	1.1	0.2	0.5	0.5	13.6	61	10	35	0.7
26.	Lime	1.5	1.0	0.7	1.3	10.9	59	90	20	0.3
27.	Sweet lime (malta)	0.7	0.2	0.4	0.6	7.8	36	30	20	1.0
28.	Sweet lime (musambi)	0.8	0.3	0.7	0.5	9.3	43	40	30	0.7
29.	Loquat	0.6	0.3	0.5	0.8	9.6	43	30	20	1.3
30.	Mango	0.6	0.4	0.4	0.7	16.9	74	14	16	1.3
31.	Mahua	1.4	1.6	0.7	-	22.7	111	45	22	0.23
32.	Mangosteen	0.5	0.1	0.2	-	14.3	60	10	20	0.2
33.	Mulberry	1.1	0.4	0.6	1.1	10.3	49	70	30	2.3
34.	Orange	0.7	0.2	0.3	0.3	10.9	48	26	20	0.32
35.	Papaya	0.6	0.1	0.5	0.8	7.2	32	17	13	0.5
36.	Passion fruit	0.9	0.1	0.7	9.6	12.4	54	10	60	20
37.	Peach	1.2	0.3	0.8	1.2	10.5	50	15	41	2.4
38.	Pears	0.6	0.2	0.3	1.0	11.9	52	8	15	0.5
39.	Persimmon	0.7	0.2	0.3	0.9	17.9	76	15	10	0.3
40.	Phalsa	1.3	0.9	1.1	1.2	14.7	72	129	39	3.1
41.	Pineapple	0.4	0.1	0.4	0.5	10.8	46	20	9	2.42
42.	Plums	0.7	0.5	0.4	0.4	11.1	52	10	12	0.6
43.	Pomegranate	1.6	0.1	0.7	5.1	14.5	65	10	70	0.79
44.	Prunes	0.5	0.3	0.6	0.5	12.8	56	10	18	---
45.	Pummelo	0.6	0.1	0.5	0.6	10.2	44	30	30	0.3

Sl. No.	Name of Fruit	Protein (g)	Fat (g)	Minerals (g)	Crude fibre (g)	Carbo-hydrate (g)	Energy (Kcal)	Ca (mg)	P (mg)	Fe (mg)
46.	Quince	0.3	0.1	0.3	1.7	11.9	50	10	20	0.4
47.	Raisins	1.8	0.3	2.0	1.1	74.6	308	87	80	7.7
48.	Raspberry	1.0	0.6	0.9	1.0	11.7	56	40	110	2.3
49.	Rose apple	0.7	0.2	0.3	1.2	8.5	39	10	30	0.5
50.	Sapota	0.7	1.1	0.5	2.6	21.4	98	28	27	1.25
51.	Seetaphal	1.6	0.4	0.9	3.1	23.5	104	17	47	4.31
52.	Strawberry	0.7	0.2	0.4	1.1	9.8	44	30	30	1.8
53.	Tomatillo	0.7	0.6	0.6	0.6	5.8	31	7	40	1.4
54.	Wood apple	7.1	3.7	1.9	5.0	18.1	134	130	110	1.48
55.	Ber	0.8	0.3	0.3	-	17.0	74	4	9	0.5

Fruits like Mango and Papaya are rich in vitamin A; Cashew nut, Walnut and Apricot in vitamin B_1; Bael, Papaya and Litchi in vitamin B_2; Barbados cherry, Aonla, Guava, Lime, Lemon and Sweet orange in vitamin C; Apricot, Date, Karonda, Banana, Bael, Custard apple, Cashew nut, Jamun and Jackfruit in carbohydrate; Cashew nut, Almond, and Walnut in protein; Walnut, Almond, Cashew nut and Avocado in fat; Guava, Wood apple, Aonla, Grape, Pomegranate and Walnut in fibres; Litchi, Karonda and Wood apple in calcium; Cashew nut, Walnut, Litchi and Wood apple in phosphorus; Karonda, Date, Cashew nut and Walnut in iron.

3.4 ROLE OF VEGETABLES

Vegetables are considered as protective foods. Vegetables contain appreciable amount of anti-oxidants that can neutralize free radicals produced inside body during metabolic process. On nutritional point of view vegetables are grouped into green leafy vegetables, root and tuber crops and other fruit vegetables.

3.4.1 Green Leafy vegetables

Commonly consumed green leafy vegetables are Amaranthus, Fenugreek, Drumstick, Palak, Mint etc. Green leafy vegetables are rich sources of calcium, iron and β- carotene and vitamins like vitamin C, riboflavin and folic acid. They contain all important nutrients required for growth and maintenance of health. Hence it is necessary to consume green leafy vegetables by all age groups including children, pregnant and nourishing women.

Table 3.2 Nutrient Content of Important Green Leafy Vegetables
(Values per 100g of edible portion)

Sl. No.	Name of Leafy vegetable	Protein (g)	Fat (g)	Minerals (g)	Crude fibre (g)	Carbohy-drate (g)	Energy (Kcal)	Ca (mg)	P (mg)	Fe (mg)
1.	Agasthi	8.4	1.4	3.1	2.2	11.8	93	1130	80	3.9
2.	Amaranthus	4.0	0.5	2.7	1.0	6.1	45	397	83	3.49
3.	Bathua leaves	3.7	0.4	2.6	0.8	2.9	30	150	80	4.2

(*Contd.*)

Sl. No.	Name of Leafy vegetable	Protein (g)	Fat (g)	Minerals (g)	Crude fibre (g)	Carbohydrate (g)	Energy (Kcal)	Ca (mg)	P (mg)	Fe (mg)
4.	Bengal gram leaves	7.0	1.4	2.1	2.0	14.1	97	340	120	23.8
5.	Bottle gourd leaves	2.3	0.7	1.7	1.3	6.1	39	80	59	-
6.	Brussels sprouts	4.7	0.5	1.0	1.2	7.1	52	43	38	1.8
7.	Cabbage	1.8	0.1	0.6	1.0	4.6	27	39	44	0.8
8.	Cauliflower	5.9	1.3	3.2	2.0	7.6	66	626	107	40.0
9.	Celery leaves	6.3	0.6	2.1	1.4	1.6	37	230	140	6.3
10.	Celery stock	0.8	0.1	0.9	1.2	3.5	18	30	38	4.8
11.	Colocasia leaves	6.8	2.0	2.5	1.8	8.1	77	460	125	0.98
12.	Coriander leaves	3.3	0.6	2.3	1.2	6.3	44	184	71	1.42
13.	Curry leaves	6.1	1.0	4.0	6.4	18.7	108	830	57	0.93
14.	Drumstick leaves	6.7	1.7	2.3	0.9	12.5	92	440	70	0.85
15.	Fenugreek leaves	4.4	0.9	1.5	1.1	6.0	49	395	51	1.93
16.	Ipomoea leaves	2.9	0.4	2.1	1.2	3.1	28	110	46	3.9
17.	Knol-khol leaves	3.5	0.4	1.2	1.8	6.4	43	740	50	13.3
18.	Lettuce	2.1	0.3	1.2	0.5	2.5	21	50	28	2.4
19.	Mint	4.8	0.6	12.9	2.0	5.8	48	200	62	15.6
20.	Mustard leaves	4.0	0.6	1.6	0.8	3.2	34	155	26	16.3
21.	Parsley	5.9	1.0	3.2	1.8	13.5	87	390	175	17.9
22.	Pumpkin leaves	4.6	0.8	2.7	2.1	7.9	57	392	112	-
23.	Radish leaves	3.8	0.4	1.6	1.0	2.4	28	265	59	0.09
24.	Spinach	2.0	0.7	1.7	0.6	2.9	26	73	21	1.14
25.	Turnip	4.0	1.5	2.2	1.0	9.4	67	710	60	28.4

Leaves of Mustard, Colocasia, Coriander, Beet, Spinach, Fenugreek, Drumstick and Carrot are rich sources of vitamin A; Colocasia in vitamin B_1; Amaranthus and Fenugreek in vitamin B_2; Drumstick and Coriander in vitamin C; Curry leaves in carbohydrate; Mustard, Beet and Spinach in fibres; Coriander, Curry leaves, Amaranthus, Fenugreek and Radish in calcium; Amaranthus and Coriander in iron.

3.4.2 Root and tuber crops

Some important tuber crops like Tapioca, Potato, Sweet Potato, Carrots, Radish, Yam and Colocasia are rich in carbohydrate and form an important source of energy in our diet. Carrots and yellow varieties of root and tuber crops are rich in carotene. Potato is a significant source of vitamin C. Besides, providing carbohydrate Tapioca and Yams are rich in calcium. Tapioca sometimes replaces cereals during short supply of cereals in Kerala.

Table 3.3 Nutrient Content of Important Root and Tubers
(Values per 100g of edible portion)

Sl. No.	Name of Root and Tuber crops	Protein (g)	Fat (g)	Minerals (g)	Crude fibre (g)	Carbohy-drate (g)	Energy (Kcal)	Ca (mg)	P (mg)	Fe (mg)
1.	Arrow root powder	0.2	0.1	0.1	-	83.1	334	10	20	1.0
2.	Banana rhizome	0.4	0.2	1.4	1.1	11.8	51	25	10	1.1
3.	Beet root	1.7	0.1	0.8	0.9	8.8	43	18.3	55	1.19
4.	Carrot	0.9	0.2	1.1	1.2	10.6	48	80	530	1.03
5.	Colocasia	3.0	0.1	1.7	1.0	1.1	97	40	140	0.42
6.	Mango ginger	1.1	0.7	1.4	1.3	10.5	53	25	90	2.6
7.	Onion (big)	1.2	0.1	0.4	0.6	11.1	50	46.9	50	0.6
8.	Onion (small)	1.8	0.1	0.6	0.6	12.6	59	40	60	1.2
9.	Parsnip	1.3	0.3	1.1	1.7	23.2	101	50	40	0.5
10.	Potato	1.6	0.1	0.6	0.4	22.6	97	10	40	0.48
11.	Radish (pink)	0.6	0.3	0.9	0.6	6.8	32	50	20	0.37
12.	Radish (white)	0.7	0.1	0.6	0.8	3.4	17	35	22	0.4
13.	Radish (table)	0.5	0.1	0.7	0.6	3.2	16	20	20	1.0
14.	Sweet Potato	1.2	0.3	1.0	0.8	28.2	120	46	50	0.21
15.	Tapioca	0.7	0.2	1.0	0.6	38.1	157	50	40	0.9
16.	Turnip	0.5	0.2	0.6	0.9	6.2	29	30	40	0.4
17.	Elephant foot yam	1.2	0.1	0.8	0.8	1.4	79	50	34	0.6
18.	Yam	1.4	0.1	1.6	1.0	26	111	35	20	1.19

Beet and Carrot are rich in vitamin A; Tapioca, Sweet potato, Yam and Potato in carbohydrate.

3.4.3 Other Fruit Vegetables

Vegetables not coming under green leafy vegetables and root and tuber crops are included in these groups. They are Brinjal, different gourds, Tomato, Okra, French beans, Guar beans etc. They provide vitamin C and some minerals. These vegetables are good sources of dietary fibres and provide bulk to the diet. High consumption of tomatoes and tomato products have been associated with reduced carcinogenesis, especially of prostate cancer and is thought to be due to the presence of lycopene, which gives red colour to tomatoes.

Table 3.4 Nutrient Content of Important Other Fruit Vegetables
(Values per 100g of edible portion)

Sl. No.	Name of Fruit vegetable	Protein (g)	Fat (g)	Minerals (g)	Crude fibre (g)	Carbo-hydrate (g)	Energy (Kcal)	Ca (mg)	P (mg)	Fe (mg)
1.	Ash gourd	0.4	0.1	0.3	0.8	1.9	10	30	20	0.8
2.	Beans	7.4	1.0	1.6	1.9	29.8	158	50	160	2.6
3.	Bitter gourd	2.1	1.0	1.4	1.7	10.6	60	23	38	2.0
4.	Bottle gourd	0.2	0.1	0.5	0.6	2.5	12	20	10	0.46

(Contd.)

Sl. No.	Name of Fruit vegetable	Protein (g)	Fat (g)	Minerals (g)	Crude fibre (g)	Carbo-hydrate (g)	Energy (Kcal)	Ca (mg)	P (mg)	Fe (mg)
5.	Brinjal	1.4	0.3	0.3	1.3	4.0	24	18	47	0.38
6.	Broad bean	4.5	0.1	0.8	2.0	7.2	48	50	64	1.4
7.	Cauliflower	2.6	0.4	1.0	1.2	4.0	30	33	57	1.23
8.	Chow-chow	0.7	0.1	0.4	0.6	5.7	27	140	30	0.6
9.	Cluster bean	3.2	0.4	1.4	3.2	10.8	16	113	57	1.08
10.	Cowpea	3.5	0.2	0.9	2.0	8.1	48	72	59	2.5
11.	Cucumber	0.4	0.1	0.3	0.4	2.5	13	10	25	0.6
12.	Drumstick	2.5	0.1	2.0	4.8	3.7	26	30	110	0.18
13.	Drumstick flower	3.6	0.8	1.3	1.3	7.1	50	51	90	-
14.	French bean	1.7	0.1	0.5	1.8	4.5	26	50	28	0.61
15.	Capsicum	1.3	0.3	0.7	1.0	4.3	24	10	30	0.47
16.	Jack fruit	2.6	0.3	0.9	2.8	9.4	51	30	40	1.7
17.	Jack fruit seed	6.6	0.4	1.2	1.5	25.8	133	50	97	1.5
18.	Knol-khol	1.1	0.2	0.7	1.5	3.8	21	20	35	1.54
19.	Okra	1.9	0.2	0.7	1.2	6.4	35	66	56	0.35
20.	Leeks	1.8	0.1	0.7	1.3	17.2	77	50	70	2.3
21.	Onion stalk	0.9	0.2	0.8	1.6	8.9	41	50	50	7.43
22.	Papaya	0.7	0.2	0.5	0.9	5.7	27	28	40	0.9
23.	Pointed gourd	2.0	0.3	0.5	3.0	2.2	20	30	40	1.7
24.	Plantain	1.4	0.2	0.5	0.7	14.0	64	10	29	6.27
25.	Plantain flower	1.7	0.7	1.3	1.3	5.1	34	32	42	1.6
26.	Pumpkin	1.4	0.1	0.6	0.7	4.6	25	10	30	0.44
27.	Pumpkin flower	2.2	0.8	1.4	0.7	5.8	39	120	60	-
28.	Ridge gourd	0.5	0.1	0.3	0.5	3.4	17	18	26	0.39
29.	Snake gourd	0.5	0.3	0.5	0.8	3.3	18	26	20	1.51
30.	Sword bean	2.7	0.2	0.6	1.5	7.8	44	60	40	2.0
31.	Tinda	1.4	0.2	0.5	1.0	3.4	21	25	24	0.9
32.	Tomato (green)	1.9	0.1	0.6	0.7	3.6	23	20	36	1.8
33.	Summer squash	0.5	0.1	0.3	0.8	3.4	17	10	30	0.6
34.	Water chestnut (fresh)	4.7	0.3	1.1	0.6	23.3	115	20	150	1.35
35.	Water chestnut (dry)	13.4	0.8	3.1	-	68.9	330	70	440	2.4

Vegetables like Chilli and red ripe Tomato are rich in vitamin A; Papaya in vitamin B_2; Chilli and Tomato in vitamin C; Peas, Beans, vegetable Soyabean and Cowpea in protein.

3.5 ROLE OF SPICES AND CONDIMENTS

Spices and condiments are accessory foods primarily used for flavouring dishes to improve palatability. They are used in small quantities and their contribution to nutrient intake is very limited. Some spices are rich in iron, potassium, and trace metals. Chilli and Coriander provide β- Carotene. Green chillies provide β- Carotene and vitamin C. Most of the spices contain high level of tannin. Spices contain several pharmacologically active substances like choline, biogenic amines etc. Asafoetida and Garlic have anti-bacterial properties and inhibit putrefying bacteria.

REFERENCES

Gopalan C, Ramasastri BV, Balasubramaniam SC, Narasinga Rao BS, Beosthale YG and Pant KC 1989. *Nutritive Value of Indian Foods*. National Institute of Nutrition. Indian Council of Medical Research, Hyderabad.

Singh J 2018. *Basic Horticulture*. Kalyani Publishers, Ludhiana.

e-resources
https://www.intechopen.com

OUTCOMES ASSESSMENT

PART A

Answer the following questions (True or False).

1. Mango and papaya are rich in vitamin C. (True/False)
2. Tapioca is rich in carbohydrate.(True/False)
3. Aonla and guava are good sources of vitamin A. (True/False)
4. Peas, beans, vegetable soybean and cowpea are rich in protein.(True/False)
5. ICMR recommends per capita use of 120 g fruits per day.(True/False)

PART B

Answer the following questions.

1. Banana is a good source of — and provides energy.
2. Name one fruit rich in vitamin B_1.
3. Green chillies provide —and vitamin C.
4. Spices and condiments are primarily used for — dishes to improve palatability.
5. What are 'other fruit vegetables'?

PART C

Write a brief note on each of the following.

1. What are nutrients and their activities?
2. Role of vegetables in human nutrition.
3. Role of fruits in human nutrition.

Chapter 4

Economic Geography of Horticulture

4.1 INTRODUCTION

India enjoys a wide range of climates and agro-meteorological zones which are suitable for growing various kinds of horticultural crops like fruits, vegetables, flowers, spices and condiments, plantation crops, and medicinal and aromatic plants.

India is the largest producer, consumer and exporter of spices in the world. India stands second both in production of fruits and vegetables next to China. China contributes 21.2% in the world fruit production whereas India produces 12.6%. When we think of vegetables, China produces 49.5% and India produces 14.09%. India is the largest producer of Mango, Banana, Papaya, Coconut, Arecanut, and Cashewnut in the world. In the country, Andhra Pradesh (13.0%) is the leading state in fruit production followed by Maharashtra (11.2%). In case of vegetables, Uttar Pradesh stands first occupying 15.1% share followed by West Bengal and Madhya Pradesh in the range of 14.6% and 10% respectively. In cut flower production West Bengal ranks first.

Total horticultural production for the year 2017–18 was 311.71 million tonnes from cropping area of 25.43 million hectares of land and an estimated production for the crop year 2018-19 is 314.87 million tonnes from an area of 25.6 million hectares.

Table 4.1 Estimated Productions of Horticultural Crops for the Year 2018–19

Sl. No.	Particulars	Year 2017-18	Year 2018-19
1.	Estimated production	311.71 million tonnes	314.87 million tonnes
2.	Area	25.43 million hectares	25.6 million hectares
	Crop year	**July–June**	**July–June**
1.	Production of Fruits	97.36 million tonnes	97.38 million tonnes
2.	Production of Vegetables	171.1 million tonnes	187.36 million tonnes
3.	Production of Onion	23.25 million tonnes	23.28 million tonnes
4.	Production of Potato	16.55 million tonnes	52.96 million tonnes
5.	Production of Tomato	29.49 million tonnes	19.66 million tonnes
6.	Production of Spices	-	8.61 million tonnes

4.2 FRUIT GROWING REGIONS IN INDIA

1. **Temperate Region**: Covering the states of Jammu and Kashmir, part of Uttar Pradesh, Arunachal Pradesh, part of Nagaland, Nilgiris and Palani Hills of Tamil Nadu.

2. **North Western Sub-Tropical Region:** Covering states of Rajasthan, Punjab, Haryana, part of Uttar Pradesh, part of Madhya Pradesh.
3. **North Eastern Sub-Tropical Region:** It includes states of Bihar, Assam, Meghalaya, Tripura, part of West Bengal and part of Arunachal Pradesh.
4. **Central Tropical Region:** It covers part of Madhya Pradesh, part of Maharashtra, Gujarat, part of Odisha, part of West Bengal, part of Andhra Pradesh and part of Karnataka.
5. **Southern Tropical Region:** Part of Karnataka, part of Andhra Pradesh, part of Tamil Nadu, part of Kerala.
6. **Coastal Tropical Humid Region:** It covers Coast of Maharashtra, Kerala, Andhra Pradesh, Tamil Nadu, Odisha, West Bengal, Tripura, Mizoram, part of Gujarat, along the sea and Indian Island.

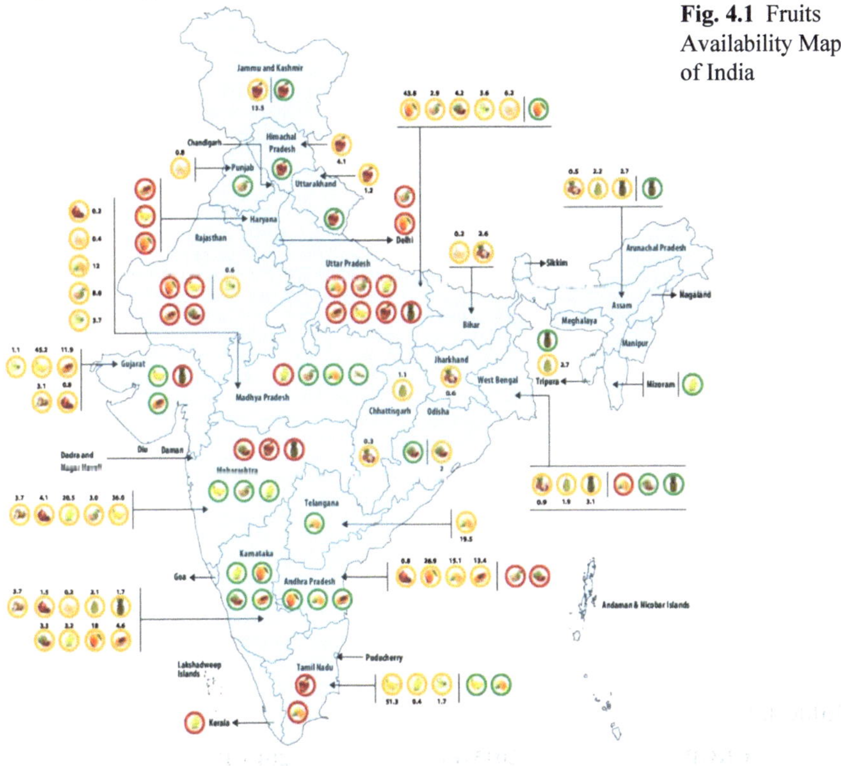

Fig. 4.1 Fruits Availability Map of India

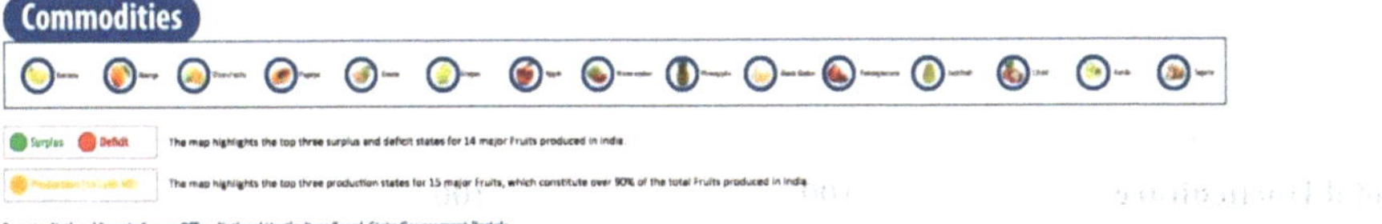

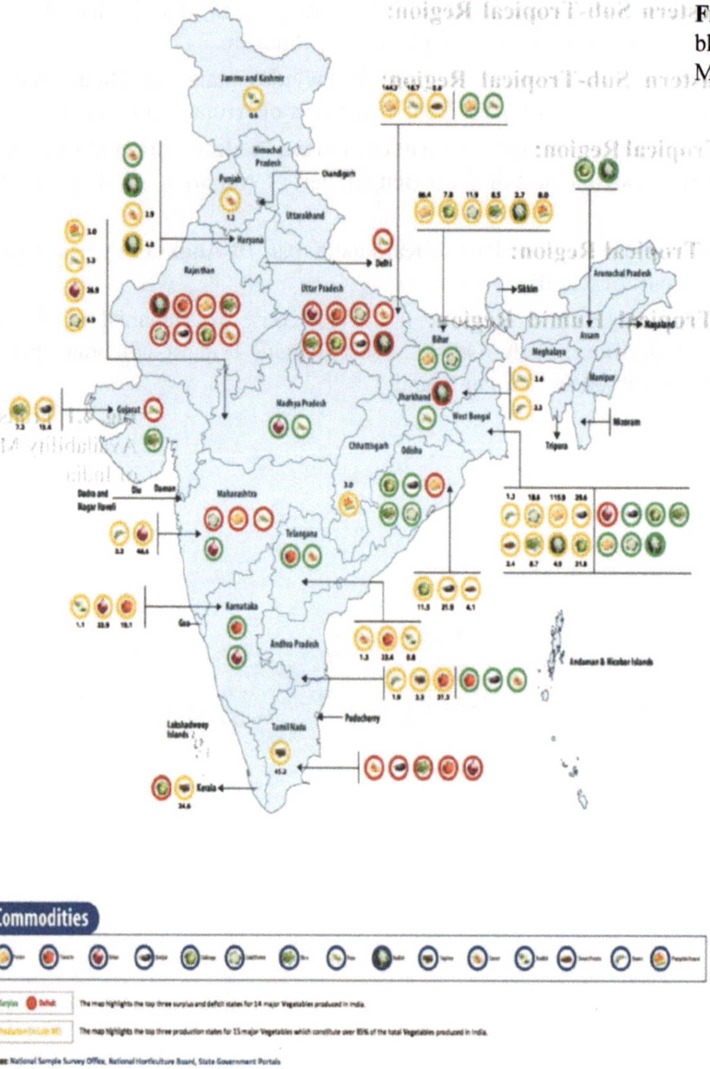

Fig. 4.2 Vegetables Availability Map of India

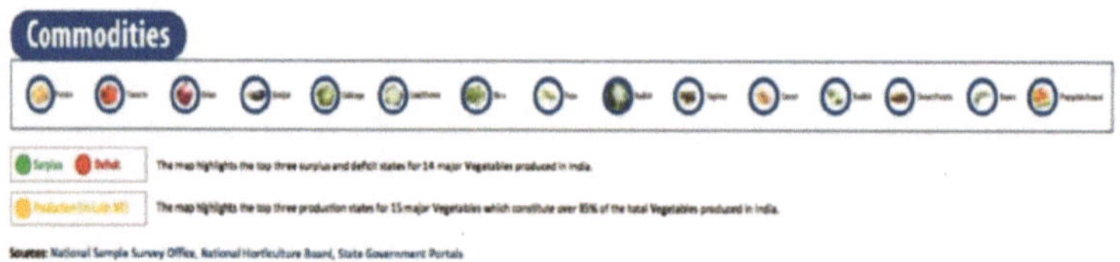

Table 4.2 Percentage Share of Production of various Horticulture Crops

CROP	2015-16	2016-17
Fruits	31.5	31.5
Vegetables	59.1	59.3
Flowers & Aromatics	1.1	1.1
Plantation Crops	5.8	5.7
Spices	2.4	2.4
Total Horticulture	**100**	**100**

Table 4.3 Percentage Share of Horticulture Output in Agriculture Output

	2011-12	2012-13	2013-14
All Agricultural Crops	**100**	**100**	**100**
Total Fruits and Vegetables	22.3	23.3	23
Total Spices and Condiments	3.2	3.1	3.3
Total Floriculture	1.5	1.4	1.4
Total Horticulture	29.2	30	29.8

Table 4.4 Crop wise Area and Production of Fruit Crops for three years

			Area in '000 Ha and Production in '000 MT			
Crop	**2014-15**		**2015-16**		**2016-17**	
Fruits	**Area**	**Production**	**Area**	**Production**	**Area**	**Production**
Almond	21	10	12	8	12	8
Aonla/Gooseberry	95	1173	88	972	91	989
Apple	319	2134	277	2521	277	2242
Banana	822	29221	841	29135	858	29163
Ber	42	401	44	425	49	481
Citrus						
i. Lime/Lemon	268	2950	245	2438	259	2789
ii. Mandarin	299	3699	397	4113	429	4754
iii. Sweet Orange (Mosambi)	275	4229	244	3468	209	3497
iv. Others	111	777	138	1562	157	1706
Citrus Total (i to iv)	953	11655	1024	11581	1055	12746
Custard apple	30	228	37	298	44	367
Grapes	123	2823	122	2590	136	2683
Guava	246	3994	255	4048	262	3648
Jackfruit	118	2088	151	1732	156	1826
Kiwi	5	8	4	11	4	11
Litchi	85	528	90	559	92	583
Mango	2163	18527	2209	18643	2263	19687
Muskmelon	42	863	45	935	47	962
Papaya	115	4913	132	5667	136	6108
Passion Fruit	19	129	13	78	14	79
Peach	19	97	18	107	18	107
Pear	42	303	40	323	40	312
Picanut	1	0	1	1	1	1
Pineapple	116	1984	110	1924	121	2038
Plum	23	72	22	82	22	76
Pomegranate	181	1789	197	2306	209	2442
Sapota	106	1339	107	1294	107	1285
Strawberry	1	8	1	5	1	5
Walnut	115	238	92	229	92	228

(Contd.)

Crop			Area in '000 Ha and Production in '000 MT			
	2014-15		2015-16		2016-17	
Fruits	**Area**	**Production**	**Area**	**Production**	**Area**	**Production**
Watermelon	84	2049	95	2325	101	2480
Others	349	2938	275	2386	272	2289
Total Fruits	**6235**	**89514**	**6301**	**90183**	**6480**	**92846**

Table 4.5 Crop wise Area and Production of Vegetable Crops for three years

Crop			Area in '000 Ha and Production in '000 MT			
	2014-15		2015-16		2016-17	
	Area	**Production**	**Area**	**Production**	**Area**	**Production**
Vegetables						
Beans	218	2204	232	2334	230	2278
Bitter gourd	76	770	93	1046	96	1083
Bottle gourd	108	1826	149	2458	157	2572
Brinjal	673	12589	663	12515	669	12400
Cabbage	386	8585	394	8806	407	8971
Capsicum	32	183	46	288	46	327
Carrot	64	968	82	1338	86	1379
Cauliflower	411	7926	426	8090	452	8499
Cucumber	43	678	71	1202	78	1142
Chillies (Green)	181	1998	292	2955	287	3406
Elephant Foot Yam	24	678	28	733	26	659
Mushroom		51	170	436	183	459
Okra/Lady's finger	504	5709	511	5849	528	6146
Onion	1173	18927	1320	20931	1270	21564
Parwal/Pointed gourd	18	347	18	264	18	252
Peas	476	4652	498	4811	546	5452
Potato	2076	48009	2117	43417	2164	46546
Radish	168	2307	199	2844	206	2927
Pumpkin/Sitaphal/Kaddu	49	1122	68	1509	72	1582
Sweet Potato	107	1228	126	1454	135	1639
Tapioca	208	4373	204	4344	196	4096
Tomato	767	16385	774	18732	809	19697
Others	1654	25053	1625	22707	1628	21932
Total Vegetables	**9417**	**166566**	**10106**	**169064**	**10290**	**175008**

Table 4.6 Crop wise Area and Production of Aromatic, Flowers, Honey, Plantation and Spice Crops for three years

			Area in '000 Ha and Production in '000 MT			
	2014-15		2015-16		2016-17	
Crop	**Area**	**Production**	**Area**	**Production**	**Area**	**Production**
Aromatic	659	1000	634	1022	634	1031
Flowers Cut		484		528		593

Crop	Area in '000 Ha and Production in '000 MT					
	2014-15		2015-16		2016-17	
	Area	Production	Area	Production	Area	Production
Flowers Loose	249	1659	278	1656	309	1653
Total Flowers	249	2143	278	2184	309	2246
Honey		81		88		88
Plantation Crops						
Arecanut	450	747	474	714	466	730
Cashewnut	1030	745	1036	671	1035	779
Cocoa	78	16	81	17	83	19
Coconut	1976	14067	2088	15256	2092	15339
Total Plantation	3534	15575	3680	16658	3677	16867
Spices						
Ajowan	24	16	24	16	24	14
Cardamom	100	24	86	24	84	27
Chillies (Dried)	761	1605	811	1520	831	1872
Cinnamon/Tejpata	3	5	3	5	3	5
Celery,Dill & Poppy	24	21	26	23	36	35
Clove	2	1	2	1	2	1
Coriander	553	462	582	585	663	609
Cumin	890	486	808	503	760	486
Fenugreek	123	131	219	247	218	220
Fennel	39	60	76	129	75	125
Garlic	262	1425	281	1617	274	1271
Ginger	142	760	164	1109	165	1081
Nutmeg	21	14	21	14	23	16
Pepper	129	65	129	55	131	72
Vanilla	6	1	4	0	5	0
Tamarind	54	202	53	194	49	191
Turmeric	184	830	186	943	193	1052
Total Spices	3317	6108	3474	6988	3535	7077
Total	**23410**	**280986**	**24472**	**286188**	**24925**	**295164**

Provisional: 2nd Advance Estimate

Table 4.7 All India Area, Production and Productivity of Fruit Crops over the Years 1991–92 to 2016–17

Year	Fruits		
	Area ('000 ha)	Production ('000 MT)	Productivity (MT/ha)
1991-92	2874	28632	9.96
2001-02	4010	43001	10.72
2002-03	3788	45203	11.93
2003-04	4661	45942	9.86

(Contd.)

Year	Fruits		
	Area ('000 ha)	Production ('000 MT)	Productivity (MT/ha)
2004-05	5049	50867	10.07
2005-06	5324	55356	10.4
2006-07	5554	59563	10.72
2007-08	5857	65587	11.2
2008-09	6101	68466	11.22
2009-10	6329	71516	11.3
2010-11	6383	74878	11.73
2011-12	6705	76424	11.4
2012-13	6982	81285	11.64
2013-14	7216	88977	12.33
2014-15	6110	86602	14.17
2015-16	6301	90183	14.31
2016-17	6480	92846	14.33

India ranks second in fruit production in the World after China due to diverse climate. India is the largest producer of Banana, Papaya and Mangoes.

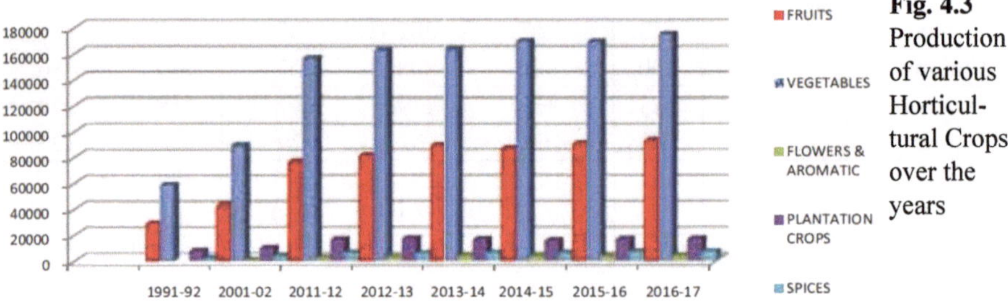

Fig. 4.3 Production of various Horticultural Crops over the years

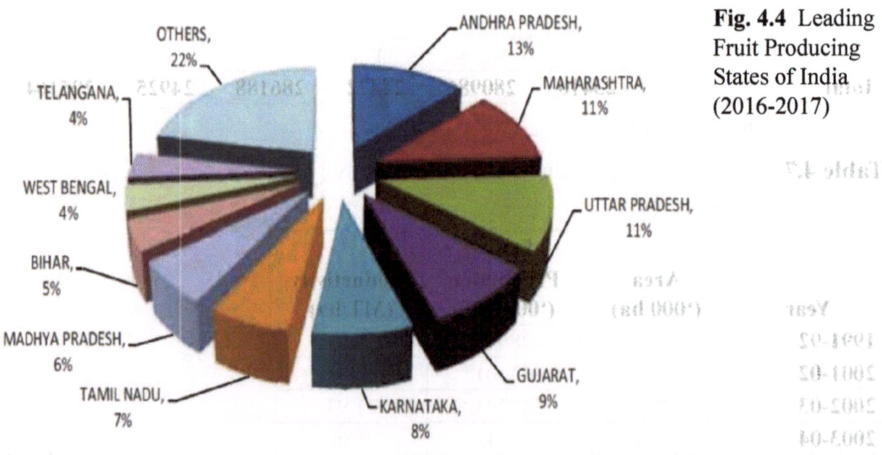

Fig. 4.4 Leading Fruit Producing States of India (2016-2017)

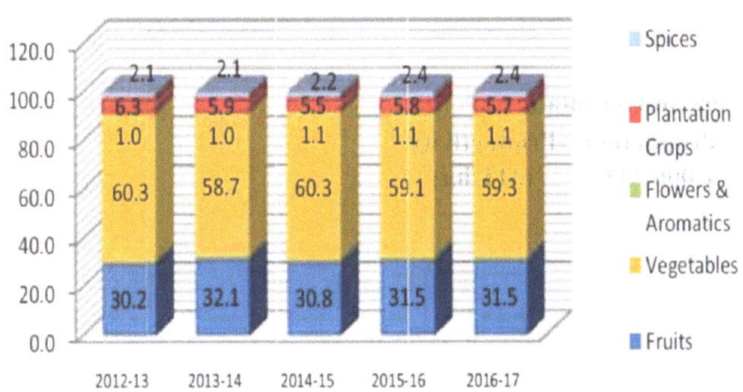

Fig. 4.5 Production share of various Horticultural Crops (% Share in Horticulture)

Table 4.8 All India Area, Production and Productivity of Vegetable Crops over the Years 1991–92 to 2016–17

Year	Vegetables		
	Area ('000 ha)	**Production ('000 MT)**	**Productivity (MT/ha)**
1991-92	5593	58532	10.47
2001-02	6156	88622	14.4
2002-03	6092	84815	13.92
2003-04	6082	88334	14.52
2004-05	6744	101246	15.01
2005-06	7213	111399	15.44
2006-07	7581	114993	15.17
2007-08	7848	128449	16.37
2008-09	7981	129077	16.17
2009-10	7985	133738	16.75
2010-11	8495	146554	17.25
2011-12	8989	156325	17.39
2012-13	9205	162187	17.62
2013-14	9396	162897	17.34
2014-15	9542	169478	17.76
2015-16	10106	169064	16.73
2016-17	10290	175008	17.01

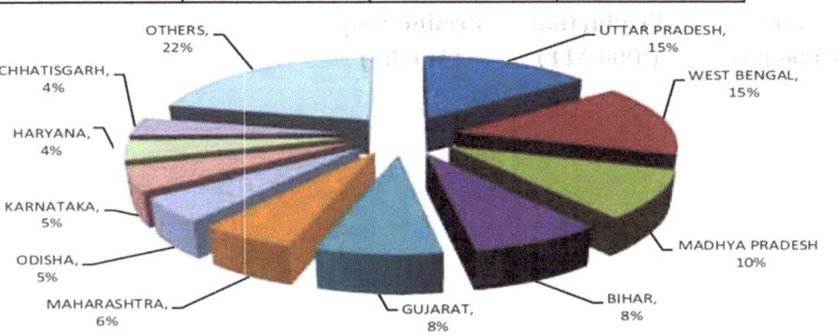

Fig. 4.6 Leading Vegetable Producing States in India (2016–2017)

Table 4.9 All India Area, Production and Productivity of Flower and Aromatic Crops over the Years 1991–92 to 2016–17

Year	Flowers and Aromatic		
	Area ('000 ha)	Production ('000 MT)	Productivity (MT/ha)
1991–92			
2001–02	106	535	5.05
2002–03	70	735	10.5
2003–04	101	580	5.74
2004–05	118	659	5.58
2005–06	129	654	5.07
2006–07	144	880	6.11
2007–08	166	868	5.23
2008–09	167	987	5.91
2009–10	183	1021	5.58
2010–11	191	1031	5.4
2011–12	760	2218	2.92
2012–13	790	2647	3.35
2013–14	748	3192	4.27
2014–15	908	3143	3.46
2015–16	912	3206	3.52
2016–17	943	3277	3.48

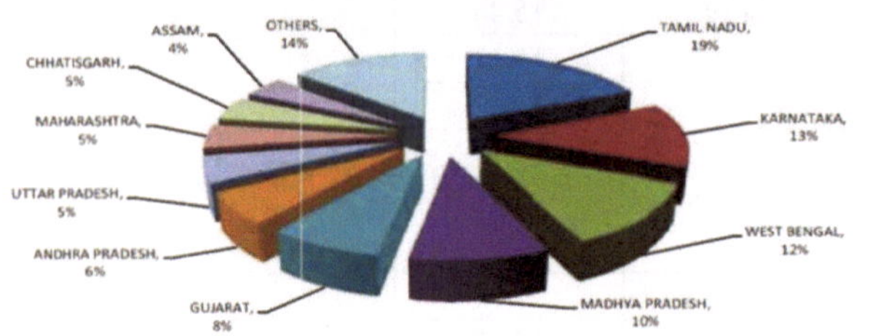

Fig. 4.7 Leading Flower Producing States in India (2016-2017)

Table 4.10 All India Area, Production and Productivity of Plantation Crops over the Years 1991–92 to 2016–17

Year	Plantation Crops		
	Area ('000 ha)	Production ('000 MT)	Productivity (MT/ha)
1991–92	2298	7498	3.26
2001–02	2984	9697	3.25
2002–03	2984	9697	3.25
2003–04	3102	13161	4.24
2004–05	3147	9835	3.13
2005–06	3283	11263	3.43
2006–07	3207	12007	3.74

Year	Plantation Crops		
	Area ('000 ha)	Production ('000 MT)	Productivity (MT/ha)
2007–08	3190	11300	3.54
2008–09	3217	11336	3.52
2009–10	3265	11928	3.65
2010–11	3306	12007	3.63
2011–12	3577	16359	4.57
2012–13	3641	16985	4.66
2013–14	3675	16301	4.44
2014–15	3534	15575	4.41
2015–16	3680	16658	4.53
2016–17	3677	16867	4.59

Table 4.11 All India Area, Production and Productivity of Spices Crops over the Years 1991-92 to 2016-17

Year	Spices		
	Area ('000 ha)	Production ('000 MT)	Productivity (MT/ha)
1991–92	2005	1900	0.95
2001–02	3220	3765	1.17
2002–03	3220	3765	1.17
2003–04	5155	5113	0.99
2004–05	3150	4001	1.27
2005–06	2366	3705	1.57
2006–07	2448	3953	1.61
2007–08	2617	4357	1.66
2008–09	2629	4145	1.58
2009–10	2464	4016	1.63
2010–11	2940	5350	1.82
2011–12	3212	5951	1.85
2012–13	3076	5744	1.87
2013–14	3163	5908	1.87
2014–15	3317	6108	1.84
2015–16	3474	6988	2.01
2016–17	3535	7077	2.002

Table 4.12 All India Area, Production and Productivity of Horticultural Crops over the Years 1991-92 to 2016-17

Year	Total		
	Area ('000 ha)	Production ('000 MT)	Productivity (MT/ha)
1991–92	12770	96562	7.56
2001–02	16592	145785	8.79

(Contd.)

Year	Total		
	Area ('000 ha)	Production ('000 MT)	Productivity (MT/ha)
2002–03	16270	144380	8.87
2003–04	19208	153302	7.98
2004–05	18445	166939	9.05
2005–06	18707	182816	9.77
2006–07	19389	191813	9.89
2007–08	20207	211235	10.45
2008–09	20662	214716	10.39
2009–10	20876	223089	10.69
2010–11	21825	240531	11.02
2011–12	23243	257277	11.07
2012–13	23694	268848	11.35
2013–14	24198	277352	11.46
2014–15	23410	280986	12.00
2015–16	24472	286188	11.69
2016–17	24925	295164	11.84

Table 4.13 India's Position in World Agriculture 2014

Items	India	World	India's		Next to
			% Share	Rank	
1. Fruits & Vegetables (million tonnes)					
Vegetables & Melons	127	1169	10.8	Second	China
Okra	6.3	9.5	66.3	First	
Potatoes	46	382	12	Second	China
Tomato	19	171	11.1	Second	China
Onion (dry)	19	88	21.5	Second	China
Cabbages & other Brassicas	9	72	12.5	Second	China
Cauliflower & Broccoli	8.5	24	35.2	Second	China
Fruits excluding Melons	88	689	13	Second	China
Banana	28	114	24.5	First	
Mango, Mangosteen and Guava	18	45	40	First	
Lemon & Lime	2.8	16.2	17.2	First	
Papaya	5.6	12.6	44.4	First	
2. Commercial Crops (million tonnes)					
Sugarcane	352	1884	18.6	Second	Brazil
Tea	1.2	5.5	21.7	Second	China

Items	India	World	India's		
			% Share	Rank	Next to
Coffee (Green)	0.3	8.8	3.5	Sixth	Brazil, Vietnam, Colombia, Indonesia, Ethiopia
Tobacco leaves	0.7	7.18	10.4	Third	China, Brazil

Source:

1. Agricultural Statistics at a Glance, 2014, Directorate of Economics and Statistics, Ministry of Agriculture, Govt. of India (Website: http://www.dacnet.nic.in/eands).
2. FAOSTAT Website (http://faostat3.fao.org/home/E) accessed on 3 July 2015.

Table 4.14 Area and Production of Almond for Major Producing Districts

A = Area in '000 Ha
P = Production in '000 MT

States	Sl. No.	Districts	2014–15		2015–16	
			A	P	A	P
1. Himachal Pradesh	1.1	Mandi	6.47	4.15	4.67	3.55
	1.2	Shimla	7.13	3.19	1.47	2.47
	1.3	Kinnaur	0.26	0.95	0.26	0.95
2. Jammu and Kashmir	2.1	Pulwama	1.53	0.17	1.53	0.29
	2.2	Badgam	1.48	0.32	1.48	0.27
	2.3	Baramulla	1.01	0.15	1.00	0.14

Source: Horticulture Statistics Division, Deptt. of Agriculture & Coopn.

Table 4.15 Area and Production of Apple for Major Producing Districts

A – Area in '000 Ha
P = Production in '000 MT

States	Sl. No.	Districts	2014-15		2015-16	
			A	p	A	P
1. Jammu and Kashmir	1.1	Baramulla	24.95	523.99	25.20	529.27
	1.2	Kupwara	19.02	130.37	19.02	278.66
	1.3	Shopian	21.60	183.41	21.61	248.04
	1.4	Kulgam	17.15	72.52	18.19	209.21
2. Himachal Pradesh	2.1	Shimla	38.78	407.75	39.73	482.39
	2.2	Kullu	25.81	104.59	26.03	143.48
3. Uttarakhand	3.1	Uttar	8.35	18.25	8.94	19.53
	3.2	Almora	1.58	14.14	1.58	14.14
	3.3	Nainital	7.81	30.48	1.24	9.07
	3.4	Dehradun	4.80	12.40	4.80	7.34
4. Arunachal Pradesh	4.1	West Kameng	3.43	6.48	3.46	6.61

Source: Horticulture Statistics Division, Deptt. of Agriculture & Coopn.

Table 4.16 Area and Production of Aonla for Major Producing Districts

A = Area in '000 Ha
P = Production in '000 MT

States	Sl. No.	Districts	2014-15		2015-16	
			A	p	A	P
1. Uttar Pradesh	1.1	Pratapgarh	16.83	183.44	17.33	185.82
	1.2	Allahabad	3.00	32.66	3.09	33.08
	1.3	Mathura	1.83	19.96	1.89	20.22
	1.4	Azamgarh	1.63	17.73	1.68	17.97
	1.5	Basti	1.43	15.62	1.48	15.82
2. Madhya Pradesh	2.1	Singrauli	1.35	35.44	1.50	28.20
	2.2	Shahdol	0.98	21.56	1.00	21.71
	2.3	Chhindwara	0.74	11.06	0.80	12.00
	2.4	Katni	0.71	5.81	0.87	10.30
	2.5	Satna	0.80	7.17	0.87	8.76
	2.6	Chhatarpur	0.42	8.28	0.43	8.34
	2.7	Betul	0.76	6.94	0.76	7.61
	2.8	Vidisha	0.49	4.74	0.50	7.40
	2.9	Rewa	0.52	5.32	0.57	5.89
	2.1	Shajapur	0.22	1.42	0.43	5.59
	2.11	Jhabua	0.27	4.63	0.26	5.05
	2.12	Sagar	0.59	4.38	0.63	5.04
	2.13	Shivpuri	0.33	4.97	0.33	4.97
	2.14	Sidhi	1.68	12.60	0.84	4.60
	2.15	Jabalpur	0.75	4.25	0.75	4.29
3. Tamil Nadu	3.1	Tirunelveli	2.55	53.84	2.39	50.47
	3.2	Dindigul	1.22	25.69	1.35	28.15
	3.3	Sivaganga	0.50	10.63	0.59	12.99
	3.4	Tiruppur	0.61	12.93	0.57	11.46
	3.5	Theni	0.43	9.04	0.47	9.94
	3.6	Coimbatore	0.37	7.75	0.37	7.81
4. Gujarat	4.1	Kheda	2.16	27.08	2.05	24.68
	4.2	Anand	1.21	12.40	1.22	12.39
	4.3	Mahesana	1.90	13.59	1.30	9.08
	4.4	Vadodara	0.86	10.81	0.72	8.94
	4.5	Gandhinagar	0.46	5.57	0.46	5.57
5. Chhattisgarh	5.1	Kabirdham	0.60	7.20	0.65	8.40
	5.2	Koriya	0.58	7.95	0.59	8.07
	5.3	Kondagaon	0.19	3.57	0.21	3.90
	5.4	Dhamtari	0.51	3.10	0.52	3.19
	5.5	Raigarh	0.21	2.84	0.21	2.84
	5.6	Raipur	0.20	2.73	0.21	2.75

States	Sl. No.	Districts	2014-15		2015-16	
			A	p	A	P
6. Andhra Pradesh	6.1	Anantapur	NA	NA	0.29	8.64
	6.2	Visakhapatnam	0.24	2.40	0.24	4.80
	6.3	Chittoor	NA	NA	0.14	2.88

Source: Horticulture Statistics Division, Deptt. of Agriculture & Coopn.

Table 4.17 Area and Production of Banana for Major Producing Districts

A = Area in '000 Ha
P = Production in '000 MT

States	Sl. No.	Districts	2014-15		2015-16	
			A	p	A	P
1. Tamil Nadu	1.1	Theni	6.18	430.33	6.79	473.36
	1.2	Tiruchirappalli	7.36	405.95	7.00	386.33
	1.3	Coimbatore	7.63	350.47	8.20	381.16
	1.4	Tuticorin	9.42	436.19	9.38	337.67
	1.5	Pudukkottai	2.37	102.27	2.21	331.35
	1.6	Erode	10.74	339.25	9.92	324.73
	1.7	Tiruvannamalai	3.06	183.03	4.12	251.13
	1.8	Tirunelveli	7.41	228.72	7.18	221.67
	1.9	Kannyakumari	6.74	214.03	6.64	210.75
2. Gujarat	2.1	Bharuch	15.42	1083.26	12.51	891.96
	2.2	Anand	12.56	775.46	12.56	778.09
	2.3	Narmada	8.60	595.12	8.15	565.20
	2.4	Surat	7.66	526.63	7.80	546.00
	2.5	Vadodara	10.40	719.23	5.84	397.08
3. Andhra Pradesh	3.1	Kadapa	15.89	953.28	12.80	832.00
	3.2	Anantapur	9.76	585.60	11.71	784.70
	3.3	East Godavari	17.05	596.89	14.74	589.72
	3.4	West Godavari	9.46	378.60	9.46	378.60
4. Uttar Pradesh	4.1	Gorakhpur	11.47	529.10	18.04	813.67
	4.2	Kushi Nagar	9.15	423.02	14.39	650.55
	4.3	Fatehpur	4.78	230.56	7.51	354.57
	4.4	Kaushambi	4.03	195.59	6.34	300.79
	4.5	Maharajganj	3.41	155.24	5.37	238.75
5. Maharashtara	5.1	Jalgaon	46.00	2378.20	41.37	2126.11
	5.2	Nanded	8.68	303.63	6.48	231.75
6. Karnataka	6.1	Chamarajanagar	12.81	209.98	13.11	300.05
	6.2	Mysore	7.82	213.15	7.20	187.37
	6.3	Shimoga	5.32	141.13	5.20	138.13
	6.4	Bellary	5.51	216.55	5.84	125.07
	6.5	Chitradurga	5.90	126.55	5.84	125.07

(Contd.)

States	Sl. No.	Districts	2014-15		2015-16	
			A	p	A	P
	6.6	Kolar	3.72	91.49	3.80	121.99
	6.7	Gulbarga	4.06	111.53	3.80	116.28
	6.8	Haveri	3.95	107.64	3.95	107.57
	6.9	Chikmagalur	7.49	179.82	4.95	106.90
	6.10	Hassan	3.56	98.98	3.34	104.80
	6.11	Belgaum	3.12	96.22	3.28	102.89
	6.12	Ramanagara	4.30	106.52	4.19	99.53
	6.13	Uttar Kannad	2.91	90.30	3.00	99.27

Source: Horticulture Statistics Division, Deptt. of Agriculture & Coopn.

Table 4.18 Area and Production of Citrus for Major Producing Districts

A = Area in '000 Ha
P = Production in '000 MT

States	Sl. No.	Districts	2014-15		2015-16	
			A	p	A	P
1. Andhra Pradesh	1.1	Anantapur	41.19	654.54	47.53	730.80
	1.2	Spsr Nellore	20.28	337.49	18.92	314.11
	1.3	Kadapa	14.32	275.59	15.40	308.00
2. Telangana	2.1	Nalgonda	76.83	1843.87	90.39	1355.82
	2.2	Mahbubnagar	47.62	655.36	12.99	194.83
3. Maharashtra	3.1	Amravati	73.19	657.87	73.50	530.13
	3.2	Aurangabad	21.13	293.07	21.24	295.37
	3.3	Jalna	21.53	218.20	14.38	193.87
	3.4	Nagpur	26.34	157.39	26.34	157.40
4. Madhya Pradesh	4.1	Chhindwara	23.72	468.31	24.11	475.92
	4.2	Agar Malwa	38.04	294.30	39.00	346.74
	4.3	Rajgarh	13.90	179.66	15.62	198.44
	4.4	Mandsaur	9.60	114.32	12.02	120.32
5. Punjab	5.1	Fazilka	28.49	648.91	29.50	613.16
	5.2	Hoshiarpur	6.51	153.23	6.25	140.20
	5.3	Muktsar	5.91	187.88	5.93	130.18
6. Gujarat	6.1	Mahesana	10.76	122.13	11.86	168.10
	6.2	Bhavnagar	7.17	73.13	7.00	91.28
	6.3	Anand	4.98	63.43	5.10	65.54
	6.4	Vadodara	2.83	34.40	2.57	34.00
	6.5	Gandhinagar	2.06	26.10	2.08	27.04
	6.6	Kheda	2.08	27.30	1.96	26.18

States	Sl. No.	Districts	2014-15		2015-16	
			A	p	A	P
7. Rajasthan	7.1	Jhalawar	0.04	0.34	12.04	261.36
	7.2	Ganganagar	8.75	255.08	7.61	135.69
8.Bihar	8.1	Pashchim Champaran	2.80	25.30	2.86	25.81
	8.2	Purbi Champaran	2.80	23.80	2.86	24.28
	8.3	Darbhanga	2.10	18.00	2.14	18.36
	8.4	Bhagalpur	2.00	17.50	2.04	17.85
	8.5	Muzaffarpur	2.00	17.50	2.04	17.85
	8.6	Madhubani	1.90	17.10	1.94	17.44
	8.7	Samastipur	1.90	17.10	1.94	17.44
	8.8	Vaishali	1.70	16.90	1.73	17.24
	8.9	Siwan	1.80	16.70	1.84	17.04
	8.1	Saran	1.80	16.60	1.84	16.93
	8.11	Nalanda	1.60	15.70	1.63	16.02
	8.12	Begusarai	1.50	13.80	1.53	14.08
	8.13	Saharsa	1.50	13.50	1.53	13.77
	8.14	Buxar	0.80	7.70	0.82	7.85
	8.15	Bhojpur	1.40	12.90	1.43	13.16
	8.16	Gaya	1.40	12.80	1.43	13.06
	8.17	Gopalganj	1.40	12.80	1.43	13.06
	8.18	Aurangabad	1.40	12.40	1.43	12.65
	8.19	Purnia	1.30	12.40	1.33	12.65
	8.2	Nawada	1.30	11.70	1.33	11.94
9. Karnataka	9.1	Bijapur	7.12	168.35	6.82	170.38
	9.2	Kodagu	1.98	49.58	2.12	52.88
	9.3	Chikmagalur	1.47	27.36	1.37	27.91
	9.4	Hassan	1.17	27.50	1.15	26.06
	9.5	Gulbarga	1.12	24.07	1.06	25.97
10. Assam	10.1	Dima Hasao	4.84	62.54	4.87	68.09
	10.2	Kamrup	2.74	30.76	2.71	30.39
	10.3	Karbi Anglong	1.62	23.41	1.69	29.37
	10.4	Tinsukia	1.64	27.47	1.77	27.97
	10.5	Kamrup Metro	2.17	21.25	2.26	22.13

(Contd.)

States	Sl. No.	Districts	2014-15		2015-16	
			A	p	A	P
	10.6	Chirang	1.68	19.11	1.50	17.27
	10.7	Udalguri	1.21	12.93	1.51	15.74
	10.8	Goalpara	0.75	6.73	1.22	11.55
	10.9	Sonitpur	0.70	7.19	1.19	11.21
11. Haryana	11.1	Sirsa	9.45	198.33	9.66	191.79
	11.2	Hisar	1.89	32.09	1.86	33.26
12. Odisha	12.1	Mayurbhanj	3.79	43.59	3.78	42.13
	12.2	Ganjam	2.52	23.52	2.52	23.52
	12.3	Sundargarh	1.66	19.20	1.66	19.01
	12.4	Gajapati	1.84	3.14	1.85	16.85
	12.5	Anugul	1.67	17.16	2.06	16.72
	12.6	Kendujhar	1.48	2.12	1.48	14.07
	12.7	Kandhamal	0.03	0.26	1.22	12.12
	12.8	Kalahandi	1.18	3.95	1.18	11.12
	12.9	Rayagada	1.04	4.35	1.04	9.87
	12.1	Bhadrak	0.00	0.00	0.88	8.74
	12.11	Dhenkanal	0.89	8.70	0.89	8.69
	12.12	Baleshwar	0.75	7.50	0.84	8.11

Source: Horticulture Statistics Division, Deptt. of Agriculture & Coopn.

Table 4.19 Area and Production of Grapes for Major Producing Districts

A = Area in '000 Ha
P = Production in '000 MT

States	Sl. No.	Districts	2014-15		2015-16	
			A	p	A	P
1. Maharashtra	1.1	Nashik	52.32	1123.41	56.23	1237.06
	1.2	Sangli	23.07	657.49	21.45	611.27
	1.3	Solapur	5.63	87.18	5.06	61.03
	1.4	Pune	2.48	54.37	2.50	56.54
	1.5	Osmanabad	2.54	38.95	2.40	38.25
2. Karnataka	2.1	Bijapur	9.96	201.22	10.58	190.86
	2.2	Chikballapur	2.55	56.31	2.97	72.08
	2.3	Belgaum	2.44	37.98	2.70	43.08
	2.4	Bagalkot	1.73	40.22	2.14	42.11
	2.5	Bangalore	2.14	35.62	2.16	35.85
3. Tamil Nadu	3.1	Theni	1.79	23.70	1.96	26.07

Source: Horticulture Statistics Division, Deptt. of Agriculture & Coopn.

Table 4.20 Area and Production of Strawberry for Major Producing Districts

A = Area in '000 Ha
P = Production in '000 MT

States	Sl. No.	Districts	2014-15		2015-16	
			A	p	A	P
Haryana	1.1	Hisar	0.097	0.860	0.085	0.900
	1.2	Bhiwani	0.035	0.320	0.036	0.650
Mizoram	2.1	Aizawl	NA	NA	0.000	0.000
	2.2	Aizawl	NA	NA	0.000	0.000
	2.3	Saiha	NA	NA	0.000	0.000
	2.4	Kolasib	NA	NA	0.000	0.000
Meghalaya	3.1	Ri Bhoi	0.025	0.460	0.047	0.493
	3.2	East Khasi Hills	0.008	0.175	0.009	0.197
Maharashtra	4.1	Nashik	0.070	0.700	0.070	0.690
Himachal Pradesh	5.1	Sirmaur	0.038	0.548	0.038	0.463

Source: Horticulture Statistics Division, Deptt. of Agriculture & Coopn.

Table 4.21 Area and Production of Guava for Major Producing Districts

A = Area in '000 Ha
P = Production in '000 MT

States	Sl. No.	Districts	2014-15		2015-16	
			A	p	A	P
1. Madhya Pradesh	1.1	Indore	1.26	28.20	1.30	28.20
	1.2	Khargone	1.19	16.20	1.21	26.33
	1.3	Vidisha	1.28	19.14	1.43	25.66
	1.4	Katni	0.70	11.89	0.81	22.18
	1.5	Singrauli	1.20	42.60	1.32	21.12
	1.6	Sheopur	NA	NA	0.76	19.76
	1.7	Morena	0.86	18.12	0.90	19.00
	1.8	Ratlam	0.83	17.76	1.06	15.55
	1.9	Jabalpur	1.03	13.02	1.22	15.33
	1.10	Shajapur	0.48	5.55	0.49	12.25
	1.11	Tikamgarh	0.28	10.76	0.32	11.31
	1.12	Chhatarpur	0.42	10.58	0.43	10.48
	1.13	Harda	0.34	10.03	0.36	10.44
	1.14	Shahdol	0.57	22.72	0.57	10.39
	1.15	Sagar	0.67	10.43	0.69	10.38
	1.16	Ashoknagar	0.34	10.20	0.38	10.20
	1.17	Ujjain	0.62	15.46	0.63	9.93
	1.18	Chhindwara	0.80	9.58	0.81	9.68
	1.19	Bhopal	0.59	1.67	0.69	9.40
	1.20	Jhabua	0.46	8.76	0.46	9.17

(*Contd.*)

States	Sl. No.	Districts	2014-15		2015-16	
			A	p	A	P
2. Uttar Pradesh	2.1	Budaun	4.37	68.81	4.76	74.32
	2.2	Kasganj	3.72	65.60	4.06	70.85
	2.3	Aligarh	3.64	64.20	3.97	69.34
	2.4	Farrukhabad	2.43	56.66	2.65	61.20
	2.5	Etah	2.84	50.05	3.10	54.05
	2.6	Hathras	2.66	46.94	2.90	50.69
	2.7	Kanpur Nagar	1.94	34.76	2.11	37.54
	2.8	Moradabad	1.07	32.04	1.17	34.60
	2.9	Kaushambi	1.82	30.76	1.98	33.22
	2.10	Bulandshahr	1.56	30.13	1.70	32.54
	2.11	Rampur	1.28	28.25	1.40	30.51
	2.12	Agra	1.06	27.65	1.15	29.87
	2.13	Bareilly	0.91	27.46	0.99	29.65
	2.14	Unnao	1.03	22.60	1.12	24.41
	2.15	Sambhal	0.86	22.00	0.94	23.76
3. Bihar	3.1	Nalanda	1.50	146.20	1.50	146.20
	3.2	Rohtas	3.30	27.00	3.30	27.00
	3.3	Bhojpur	1.90	16.50	1.93	16.50
	3.4	Purbi Champaran	1.70	14.00	1.70	14.00
	3.5	Pashchim Champaran	1.60	13.80	1.60	13.80
	3.6	Buxar	1.60	12.50	1.60	12.50
	3.7	Muzaffarpur	1.50	11.80	1.50	11.80
	3.8	Vaishali	1.50	11.20	1.50	11.20
	3.9	Kaimur (Bhabua)	1.40	10.60	1.40	10.60
4. West Bengal	4.1	Paraganas South	2.48	37.90	2.55	39.31
	4.2	Birbhum	1.35	18.13	1.37	18.38
	4.3	Paraganas North	1.07	15.36	1.11	15.98
	4.4	Medinipur West	1.44	15.45	1.48	15.92
	4.5	Nadia	1.28	13.25	1.31	13.52
	4.6	Murshidabad	1.25	11.50	1.50	13.43
	4.7	Purulia	0.82	10.20	0.85	10.56
	4.8	Bardhaman	0.74	9.48	0.77	9.71
	4.9	Medinipur East	0.76	9.45	0.78	9.65
5. Punjab	5.1	Patiala	0.94	21.50	0.95	21.60
	5.2	Ludhiana	0.91	20.92	0.90	20.70
	5.3	S.A.S Nagar	0.71	15.67	0.72	15.85
	5.4	Sangrur	0.70	15.91	0.69	15.71
	5.5	Jalandhar	0.55	12.46	0.56	12.55
	5.6	Muktsar	0.52	9.28	0.55	12.16

States	Sl. No.	Districts	2014-15		2015-16	
			A	p	A	P
	5.7	Rupnagar	0.43	9.72	0.43	9.72
	5.8	Bathinda	0.41	7.82	0.43	9.64
	5.9	Tarn Taran	0.33	7.17	0.34	7.34
6. Chhattisgarh	6.1	Korba	2.10	22.41	2.10	22.62
	6.2	Bilaspur	2.38	16.34	2.46	16.92
	6.3	Kanker	1.66	13.84	1.71	13.68
	6.4	Mahasamund	0.94	11.24	1.12	12.50
	6.5	Kabirdham	0.83	12.38	0.85	12.38
	6.6	Janjgir-Champa	1.31	10.36	1.31	11.11
	6.7	Surguja	0.90	8.55	1.01	9.00
	6.8	Raigarh	1.14	8.86	1.14	8.86
	6.9	Raipur	0.94	7.40	0.98	7.71
	6.10	Baloda Bazar	0.77	6.20	0.90	7.26
	6.11	Korea	0.78	6.47	0.79	6.80
7. Gujarat	7.1	Bhavnagar	3.96	47.16	4.00	47.60
	7.2	Vadodara	2.11	35.83	1.67	27.82
	7.3	Kachchh	0.51	8.18	0.56	9.51
	7.4	Kheda	0.63	9.63	0.57	8.84
	7.5	Gandhinagar	0.57	8.18	0.59	8.57
	7.6	Chhotaudepur	NA	NA	0.51	8.49
8. Haryana	8.1	Sonipat	0.82	16.18	0.88	16.48
	8.2	Karnal	0.67	10.35	0.65	14.01
	8.3	Hisar	0.98	12.20	0.98	13.13
	8.4	Yamunanagar	0.79	7.00	0.82	12.80
	8.5	Jind	0.51	10.01	0.52	11.01
	8.6	Rohtak	0.49	4.64	0.53	10.43
	8.7	Jhajjar	0.69	12.48	0.74	9.80
	8.8	Panipat	0.42	7.82	0.44	8.73
	8.9	Fatehabad	0.51	5.60	0.52	7.82
	8.10	Gurgaon	0.53	7.48	0.53	7.58
9. Maharashtra	9.1	Solapur	4.03	46.47	3.63	32.53
	9.2	Nashik	1.11	14.37	1.16	15.07
	9.3	Ahmednagar	0.92	13.16	1.00	14.23
	9.4	Pune	1.12	15.57	1.15	13.68
	9.5	Aurangabad	0.51	7.07	0.58	8.05
	9.6	Jalgaon	0.50	6.50	0.51	7.11
	9.7	Osmanabad	0.64	6.73	0.65	6.17

Source: Horticulture Statistics Division, Deptt. of Agriculture & Coopn.

Table 4.22 Area and Production of Mango for Major Producing Districts

A = Area in '000 Ha
P = Production in '000 MT

States	Sl. No.	Districts	2014-15		2015-16	
			A	p	A	P
1. Uttar Pradesh	1.1	Lucknow	28.07	563.78	29.47	585.20
	1.2	Saharanpur	28.14	554.05	29.55	575.10
	1.3	Unnao	16.18	342.93	16.99	355.97
	1.4	Bulandshahr	14.50	251.91	15.22	261.48
	1.5	Amroha	8.80	189.11	9.24	196.29
	1.6	Sitapur	15.12	178.70	15.89	185.49
	1.7	Faizabad	7.24	145.45	7.60	150.97
	1.8	Sultanpur	8.81	136.58	9.25	141.77
	1.9	Meerut	7.63	123.32	8.01	128.01
	1.10	Bijnor	5.59	113.08	5.87	117.38
	1.11	Muzaffarnagar	4.93	91.05	5.18	94.51
	1.12	Hardoi	5.34	88.79	5.60	92.17
	1.13	Kasganj	4.95	87.93	5.60	92.17
	1.14	Ambedkar Nagar	4.77	80.99	5.01	84.07
	1.15	Aligarh	4.68	78.54	4.91	81.53
	1.16	Kheri	4.03	73.13	4.23	75.91
2. Andhra Pradesh	2.1	Chittoor	73.53	588.22	73.98	665.79
	2.2	Vizianagaram	42.29	507.46	44.40	532.84
	2.3	Krishna	63.50	508.00	63.46	380.73
	2.4	Anantapur	39.74	317.90	47.69	286.12
	2.5	Kadapa	23.86	238.60	26.40	211.20
3. Telangana	3.1	Mahbubnagar	32.58	326.96	32.91	329.08
	3.2	Karimnagar	31.50	283.50	31.50	283.50
	3.3	Khammam	45.56	410.05	29.43	264.85
	3.4	Warangal	24.72	222.48	25.72	231.48
	3.5	Adilabad	24.92	224.31	25.23	145.06
4. Karnataka	4.1	Kolar	48.82	390.96	49.64	421.19
	4.2	Ramanagara	23.46	234.59	22.72	228.01
	4.3	Tumkur	15.15	151.52	15.67	167.91
	4.4	Chikballapur	13.95	109.70	16.00	128.97
	4.5	Dharwad	11.55	55.79	12.50	107.43
	4.6	Mandya	6.54	73.28	6.50	74.43
	4.7	Bangalore Rural	6.97	66.06	6.98	66.12
	4.8	Belgaum	5.13	63.98	5.08	62.95
5. Bihar	5.1	Darbhanga	13.50	120.50	13.69	141.91
	5.2	Samastipur	10.60	89.30	10.75	100.29
	5.3	Muzaffarpur	9.80	82.50	9.94	97.47
	5.4	Purbi Champaran	9.30	82.30	9.42	93.22

States	Sl. No.	Districts	2014-15		2015-16	
			A	p	A	P
	5.5	Vaishali	8.40	76.00	8.55	86.86
	5.6	Pashchim Champaran	7.30	72.50	7.39	73.23
	5.7	Bhagalpur	7.50	60.50	7.39	73.23
	5.8	Madhubani	6.10	51.00	6.16	61.61
	5.9	Rohtas	5.70	52.80	5.76	59.39
	5.10	Sitamarhi	5.30	49.50	5.45	53.03
	5.11	Banka	6.30	48.50	6.36	51.01
	5.12	Saran	5.10	50.00	5.23	50.50
	5.13	Bhojpur	4.60	42.80	4.70	47.27
	5.14	Begusarai	4.10	36.50	4.14	40.91
	5.15	Patna	4.00	32.50	4.08	38.89
6. Gujarat	6.1	Bhavnagar	3.96	47.16	-	-
	6.2	Vadodara	2.11	35.83	-	-
	6.3	Mehsana	0.81	7.92	-	-
	6.4	Kheda	0.63	9.63	-	-
	6.5	Gandhinagar	0.57	8.18	-	-
	6.6	Kutch	0.51	8.18	-	-
7. Tamil Nadu	7.1	Vellore	11.95	165.57	15.50	201.50
	7.2	Dharmapuri	10.67	62.67	17.67	176.65
	7.3	Dindigul	16.77	98.18	18.42	108.50
	7.4	Thiruvallur	10.91	88.91	10.60	84.82
	7.5	Tirunelveli	6.41	84.04	6.26	82.08
	7.6	Sivaganga	2.29	27.62	3.82	45.99
	8.1	Rayagada	11.51	51.89	11.53	54.38
8. Odisha	8.2	Anugul	9.39	51.52	9.40	51.55
	8.3	Mayurbhanj	14.11	50.83	15.79	51.03
	8.4	Dhenkanal	9.29	41.43	9.30	44.45
	8.5	Koraput	12.93	43.70	12.94	44.12
	8.6	Kalahandi	11.39	43.96	11.41	43.98
	8.7	Kendujhar	10.32	42.65	10.48	42.66
	8.8	Sundargarh	8.85	39.19	8.88	39.31
	8.9	Ganjam	10.56	38.27	10.67	38.29
	8.10	Kandhamal	10.58	27.03	10.56	30.67
	8.11	Sambalpur	2.57	11.52	6.26	28.07
	8.12	Nabarangpur	7.14	27.87	7.16	27.91
	8.13	Malkangiri	7.98	27.58	8.00	27.59
	8.14	Balangir	9.35	24.36	8.36	25.82

(Contd.)

States	Sl. No.	Districts	2014-15		2015-16	
			A	p	A	P
9. West Bengal	9.1	Maldah	30.00	375.00	30.48	270.00
	9.2	Murshidabad	17.85	145.00	18.04	146.08
	9.3	24 Paraganas North	7.51	62.80	7.62	65.42
	9.4	Nadia	5.42	54.00	5.53	54.28
	10.1	Ratnagiri	59.30	207.56	60.11	190.13
10. Maharashtra	10.2	Sindhudurg	31.36	40.44	31.52	53.28
	10.3	Raigad	12.33	54.89	12.33	36.98
	10.4	Beed	4.13	19.55	4.26	20.44
	10.5	Pune	6.80	23.79	6.83	18.15
	10.6	Solapur	3.73	11.55	3.35	15.71
	11.1	Surguja	5.30	48.92	6.01	57.10
11. Chhattisgarh	11.2	Jashpur	4.46	44.55	4.47	46.15
	11.3	Surajpur	4.20	35.28	4.41	37.55
	11.4	Raigarh	7.10	35.14	7.11	35.19
	11.5	Korba	5.70	23.81	5.70	23.88
	11.6	Balrampur	2.85	19.06	3.35	22.41
	11.7	Bilaspur	5.84	19.65	6.02	20.24
	11.8	Baloda Bazar	2.10	13.76	2.78	17.75
	11.9	Korea	2.70	17.55	2.73	17.75
	11.10	Kabirdham	2.77	16.60	2.87	16.60

Source: Horticulture Statistics Division, Deptt. of Agriculture & Coopn.

Table 4.23 Area and Production of Papaya for Major Producing Districts

A = Area in '000 Ha
P = Production in '000 MT

States	Sl. No.	Districts	2014-15		2015-16	
			A	p	A	P
1. Gujarat	1.1	Kachchh	3.54	299.83	3.15	266.65
	1.2	Tapi	2.02	121.20	2.07	126.60
	1.3	Vadodara	2.44	126.86	1.89	102.01
	1.4	Aravalli			1.48	82.73
	1.5	Chhotaudaipur			0.88	62.00
	1.6	Banas Kantha	0.66	38.84	1.05	61.53
	1.7	Sabar Kantha	2.43	134.87	1.10	61.21
	1.8	Mahesana	0.94	41.36	1.17	56.35
	1.9	Bharuch	0.79	45.76	0.80	46.59
	1.10	Anand	0.59	45.86	0.56	43.89
2. Andhra Pradesh	2.1	Chittoor	1.26	95.46	2.10	315.00
	2.2	Anantapur	2.53	253.10	3.04	273.33
	2.3	Kadapa	1.53	91.68	2.40	180.00

States	Sl. No.	Districts	2014-15		2015-16	
			A	p	A	P
3. Karnataka	3.1	Chitradurga	0.90	70.87	0.83	65.37
	3.2	Koppal	0.87	55.00	0.53	41.96
	3.3	Bellary	0.61	48.29	0.49	40.81
	3.4	Mandya	0.39	29.51	0.51	35.97
	3.5	Gulbarga	0.45	32.69	0.45	32.97
	3.6	Kolar	0.37	27.63	0.39	28.39
	3.7	Chamarajanagar	0.30	23.66	0.33	24.67
	3.8	Raichur	0.26	20.56	0.30	23.60
	3.9	Bidar	0.34	18.72	0.38	21.55
	3.10	Mysore	0.27	16.42	0.34	20.78
	3.11	Tumkur	0.34	20.97	0.31	20.78
4. Madhya Pradesh	4.1	Khargone	1.06	80.51	1.19	92.47
	4.2	Dhar	0.06	14.72	0.88	52.85
	4.3	Sidhi	0.14	37.50	0.15	43.18
	4.4	Barwani	0.24	7.18	0.32	28.67
	4.5	Shivpuri	0.41	20.60	0.41	20.60
	4.6	Alirajpur	0.02	0.05	0.17	16.43
	4.7	Satna	0.43	15.95	0.43	16.26
	4.8	Jhabua	0.16	12.80	0.17	13.20
5. Tamil Nadu	5.1	Erode	0.24	56.86	0.50	122.70
	5.2	Dharmapuri	0.02	38.88	0.22	50.07
	5.3	Vellore	0.19	45.68	0.16	37.76
	5.4	Salem	0.03	7.05	0.14	32.20
	5.5	Theni	0.12	28.67	0.13	31.54
	5.6	Tiruppur	0.08	18.47	0.10	24.58
	5.7	Coimbatore	0.08	18.47	0.10	24.58
6. Telangana	6.1	Khammam	0.76	61.04	1.46	116.82
	6.2	Medak	0.55	43.76	1.18	108.25
	6.3	Mahbubnagar	0.71	57.75	0.96	76.18
7. West Bengal	7.1	Paraganas	1.90	55.87	1.92	56.35
	7.2	Paraganas South	1.10	34.77	1.11	36.22
	7.3	Nadia	0.86	30.50	0.88	31.11
	7.4	Murshidabad	0.75	28.03	0.71	25.43
	7.5	Birbhum	0.71	19.70	0.73	20.21
	7.6	Hooghly	0.76	19.59	0.77	19.68
	7.7	Bankura	0.70	18.85	0.70	19.00
	7.8	Bardhaman	0.63	18.30	0.63	17.97
	7.9	Medinipur East	0.81	17.20	0.82	17.40
	7.10	Jalpaiguri	0.31	16.17	0.31	16.17

(Contd.)

States	Sl. No.	Districts	2014-15		2015-16	
			A	p	A	P
8. Chhattisgarh	8.1	Durg	1.27	49.94	1.29	51.20
	8.2	Bilaspur	1.81	4.79	1.88	46.41
	8.3	Raipur	0.99	39.72	1.09	43.76
	8.4	Bemetara	0.63	25.13	0.65	26.13
	8.5	Baloda Bazar	0.76	16.58	0.90	19.68
	8.6	Mahasamund	1.33	19.77	1.44	19.04
	8.7	Janjgir-Champa	0.75	14.93	0.76	17.26
	8.8	Balrampur	0.80	15.11	0.90	17.01
9. Maharashtra	9.1	Nandurbar	0.62	26.23	2.13	74.66
	9.2	Dhule	1.87	89.90	1.87	33.17
	9.3	Jalgaon	0.83	28.87	0.85	28.80
	9.4	Hingoli	0.58	13.95	0.53	18.13
	9.5	Akola	0.40	16.48	0.47	15.65
	9.6	Washim	0.26	11.96	0.29	13.68
	9.7	Solapur	0.91	17.61	0.81	12.32
	9.8	Nanded	0.37	28.41	0.39	10.34
10. Assam	10.1	Sonitpur	0.52	12.42	0.51	12.09
	10.2	Dima Hasao	0.61	11.34	0.61	11.34
	10.3	Nagaon	0.76	11.22	0.74	10.92
	10.4	Barpeta	0.45	11.00	0.43	10.37
	10.5	Karbi Anglong	0.58	10.22	0.58	10.22
	10.6	Goalpara	0.31	8.28	0.30	8.07
	10.7	Kamrup	0.35	7.76	0.33	7.44
	10.8	Kokrajhar	0.29	7.66	0.27	7.10
	10.9	Baksa	0.33	6.46	0.35	6.98
	10.1	Nalbari	0.18	4.67	0.24	6.23
	10.11	Cachar	0.31	6.03	0.29	5.70
	10.12	Bongaigaon	0.21	5.27	0.22	5.50

Source: Horticulture Statistics Division, Deptt. of Agriculture & Coopn.

Table 4.24 Area and Production of Pine Apple for Major Producing Districts

A = Area in '000 Ha
P = Production in '000 MT

States	Sl. No.	Districts	2014-15		2015-16	
			A	p	A	P
1. West Bengal	1.1	Darjeeling	4.50	140.60	4.53	141.50
	1.2	Dinajpur Uttar	3.14	87.91	3.27	96.79
2. Kerala	2.1	Kottayam	0.00	0.10	0.00	0.15
	2.2	Ernakulam	0.00	0.19	0.00	0.12
3. Assam	3.1	Dima Hasao	3.51	67.24	3.60	67.16
	3.2	Karbi Anglong	2.03	36.15	2.03	36.69

States	Sl. No.	Districts	2014-15		2015-16	
			A	p	A	P
	3.3	Kamrup	1.61	30.26	1.59	31.75
	3.4	Cachar	1.25	23.38	1.34	25.15
	3.5	Hailakandi	1.41	20.41	1.45	21.00
	3.6	Chirang	0.71	13.44	0.77	15.18
	3.7	Sonitpur	0.84	14.84	0.83	13.22
4. Tripura	4.1	Dhalai	2.97	42.66	3.48	49.94
	4.2	North Tripura	2.07	29.33	2.19	30.62
	4.3	West Tripura	1.18	16.93	1.23	18.50
	4.4	Unakoti	1.18	16.96	1.24	16.35
	4.5	Sepahijala	0.98	13.95	1.02	14.59
5. Karnataka	5.1	Shimoga	1.38	82.62	1.41	84.66
	5.2	Uttar Kannad	0.44	32.82	0.42	31.26
6. Manipur	6.1	Senapati	3.43	35.99	3.85	40.23
	6.2	Churachandpur	2.25	20.25	2.25	22.50
	6.3	Thoubal	2.62	30.70	1.63	20.32
	6.4	Imphal East	1.44	17.95	1.13	14.04
	6.5	Imphal West	0.69	8.96	0.80	10.09
7. Nagaland	7.1	Dimapur	2.80	32.98	2.75	32.50
	7.2	Peren	1.56	20.53	1.59	20.13
	7.3	Mokokchung	1.13	15.21	1.13	15.21
	7.4	Wokha	0.91	13.46	0.91	13.50
	7.5	Phek	0.49	9.42	0.49	9.42
8. Meghalaya	8.1	Ri Bhoi	3.87	46.77	3.85	45.71
	8.2	West Garo Hills	2.81	24.89	2.79	24.57
	8.3	East Garo Hills	0.87	16.30	0.87	16.18
9. Bihar	9.1	Kishanganj	2.10	59.60	2.12	60.20
	9.2	Purnia	1.70	43.70	1.72	44.14

Source: Horticulture Statistics Division, Deptt. of Agriculture & Coopn.

Table 4.25 Area and Production of Pomegranate for Major Producing Districts

A = Area in '000 Ha
P = Production in '000 MT

States	Sl. No.	Districts	2014-15		2015-16	
			A	p	A	P
1. Maharashtra	1.1	Nashik	42.36	724.71	42.95	729.06
	1.2	Ahmednagar	16.11	162.10	16.41	165.06
	1.3	Solapur	19.03	169.80	17.13	118.86
2. Karnataka	2.1	Chitradurga	10.86	145.39	10.87	145.93
	2.2	Bellary	2.06	18.20	3.23	36.08
	2.3	Tumkur	1.87	15.04	3.33	35.40
	2.4	Bijapur	2.28	22.79	2.61	26.06

(*Contd.*)

States	Sl. No.	Districts	2014-15		2015-16	
			A	p	A	P
3. Gujarat	3.1	Banas Kantha	4.88	53.68	6.36	101.60
	3.2	Kachchh	3.34	46.72	4.55	72.49
	3.3	Mahesana	0.59	6.74	1.02	16.13
	3.4	Sabar Kantha	0.85	12.88	0.60	9.56

Source: Horticulture Statistics Division, Deptt. of Agriculture & Coopn.

Table 4.26 Area and Production of Sapota for Major Producing Districts

A = Area in '000 Ha
P = Production in '000 MT

States	Sl. No.	Districts	2014-15		2015-16	
			A	p	A	P
1. Karnataka	1.1	Kolar	3.60	55.32	3.40	52.32
	1.2	Mandya	2.27	37.86	2.20	34.91
	1.3	Bellary	2.43	45.68	2.17	27.34
	1.4	Chikballapur	1.83	20.96	1.78	21.70
	1.5	Haveri	1.90	19.94	1.88	19.42
	1.6	Belgaum	1.75	18.24	1.67	17.66
	1.7	Chikmagalur	1.31	15.28	1.16	14.37
	1.8	Mysore	1.47	13.12	1.52	13.58
	1.9	Chitradurga	1.44	13.92	1.39	13.47
	1.10	Dharwad	1.39	15.74	1.13	12.36
	1.11	Tumkur	1.03	12.34	1.06	10.59
2. Gujarat	2.1	Navsari	7.66	96.91	7.75	98.14
	2.2	Junagadh	4.93	49.99	3.69	37.65
	2.3	Valsad	2.91	30.15	2.92	31.87
	2.4	Bhavnagar	2.35	24.68	2.36	25.02
	2.5	Surat	2.13	23.45	2.15	23.87
	2.6	Kachchh	1.52	17.15	1.48	16.67
3. Tamil Nadu	3.1	Dindigul	1.68	59.26	1.85	65.02
	3.2	Vellore	0.52	18.36	0.52	18.20
	3.3	Theni	0.45	16.00	0.50	17.60
	3.4	Namakkal	0.36	12.67	0.46	16.61
	3.5	Tirunelveli	0.39	13.79	0.39	13.58
	3.6	Virudhunagar	0.60	20.94	0.54	13.43
	3.7	Madurai	0.31	10.74	0.29	10.16
	3.8	Salem	0.25	8.67	0.27	9.86
4. Maharashtra	4.1	Palghar	3.02	36.26	3.04	34.09
	4.2	Pune	2.46	28.66	2.48	24.64
	4.3	Solapur	2.91	29.04	2.62	20.33
	4.4	Ahmednagar	1.83	17.91	1.90	18.57
	4.5	Osmanabad	1.23	11.65	1.25	11.91

States	Sl. No.	Districts	2014-15		2015-16	
			A	p	A	P
5. Andhra Pradesh	5.1	Anantapur	3.43	41.17	4.12	57.64
	5.2	Prakasam	2.39	28.69	2.54	25.44
	5.3	Guntur	NA	NA	1.55	15.51
	5.4	Kurnool	0.68	12.17	0.69	12.47

Source: Horticulture Statistics Division, Deptt. of Agriculture & Coopn.

Table 4.27 Area and Production of Walnut for Major Producing Districts

A = Area in '000 Ha
P = Production in '000 MT

States	Sl. No.	Districts	2014-15		2015-16	
			A	p	A	P
1. Jammu And Kashmir	1.1	Kupwara	8.67	20.79	8.67	35.82
	1.2	Kulgam	6.12	34.45	5.77	23.09
	1.3	Doda	6.42	10.59	6.48	19.78
	1.4	Pulwama	4.38	7.73	4.81	16.84
	1.5	Badgam	7.44	18.72	3.99	14.53
	1.6	Poonch	8.07	11.51	7.01	14.48
	1.7	Baramulla	3.11	14.01	3.11	14.01
	1.8	Kishtwar	4.59	6.49	4.64	11.84
2. Uttarakhand	2.1	Almora	2.81	8.47	2.81	8.47
	2.2	Dehradun	2.69	3.02	2.69	3.82
	2.3	Pauri Garhwal	2.00	2.01	1.99	2.03
3. Himachal Pradesh	3.1	Sirmaur	1.11	0.65	1.10	0.94
	3.2	Chamba	1.28	0.52	1.26	0.60
	3.3	Shimla	0.32	0.03	0.32	0.30
4. Arunachal Pradesh	4.1	West Kameng	0.36	0.48	0.39	0.39
	4.2	Tawang	0.02	0.04	0.02	0.04

Source: Horticulture Statistics Division, Deptt. of Agriculture & Coopn.

Table 4.28 Area and Production of Beans for Major Producing Districts

A = Area in '000 Ha
P = Production in '000 MT

States	Sl. No.	Districts	2014-15		2015-16	
			A	p	A	P
1. Gujurat	1.1	Banas Kantha	11.63	132.45	11.64	134.06
	1.2	Anand	7.23	79.02	7.28	79.11
	1.3	Kheda	6.38	71.53	5.63	63.18
	1.4	Ahmadabad	5.87	58.96	5.86	58.96
	1.5	Mahesana	3.81	41.20	4.06	43.93
	1.6	Vadodara	5.09	53.10	3.66	38.00

(Contd.)

States	Sl. No.	Districts	2014-15		2015-16	
			A	p	A	P
	1.7	Surat	3.03	29.74	3.06	30.13
	1.8	Chhotaudepur	NA	NA	2.24	23.41
	1.9	Junagadh	3.54	41.93	1.82	21.01
	1.1	Gir Somnath	NA	NA	1.80	20.76
2. Telangana	2.1	Medak	0.96	11.52	1.09	18.91
	2.2	Adilabad	0.08	9.24	0.92	13.76
	2.3	Rangareddi	1.25	14.96	0.69	13.79
3. Tamil Nadu	3.1	Krishnagiri	8.43	235.82	2.20	49.57
	3.2	Vellore	1.48	39.16	1.51	41.04
	3.3	Theni	1.11	22.30	0.98	22.56
	3.4	Dindigul	0.82	16.24	0.77	14.12
4. Karnataka	4.1	Kolar	3.40	35.31	3.33	35.44
	4.2	Belgaum	1.53	14.72	1.58	15.03
	4.3	Bangalore Rural	0.70	13.32	0.69	13.16
	4.4	Chamarajanagar	1.15	12.35	1.32	13.15
	4.5	Chikballapur	0.55	5.49	0.63	7.30
	4.6	Haveri	0.98	7.93	0.85	7.15
5. Andhra Pradesh	5.1	Chittoor	3.35	20.09	4.00	60.00
	5.2	Kurnool	0.92	11.45	5.95	47.57
	5.3	Anantapur	0.63	7.50	0.75	11.25
6. West Bengal	6.1	Paraganas North	2.62	20.84	2.85	22.71
	6.2	Murshidabad	2.39	10.24	2.68	12.56
	6.3	Nadia	2.67	11.79	2.68	11.87
	6.4	Hooghly	1.26	10.94	1.26	10.99
	6.5	24 Paraganas South	2.29	10.21	2.31	10.32
	6.6	Birbhum	2.30	8.44	2.29	8.41
	6.7	Medinipur West	1.42	7.47	1.42	7.49
	6.8	Bankura	1.24	6.89	1.25	6.91
	6.9	Coochbehar	0.73	6.55	0.74	6.62
7. Jharkhand	7.1	Ranchi	0.50	0.50	2.97	46.04
	7.2	Khunti	0.17	0.17	1.60	24.80
	7.3	Lohardaga	0.24	0.25	1.37	21.27
8. Bihar	8.1	Katihar	1.00	8.20	1.01	8.28
	8.2	Nalanda	0.90	7.50	0.91	7.58
	8.3	Siwan	1.00	7.20	1.01	7.27
	8.4	Vaishali	0.80	6.50	0.81	6.57
	8.5	Darbhanga	0.80	6.20	0.81	6.26
	8.6	Muzaffarpur	0.70	5.50	0.71	5.56
	8.7	Saharsa	0.70	5.50	0.71	5.56

States	Sl. No.	Districts	2014-15		2015-16	
			A	p	A	P
	8.8	Madhepura	0.70	5.20	0.71	5.25
	8.9	Pashchim	0.40	3.80	0.40	3.84
	8.1	Purnia	0.50	3.70	0.51	3.74
	8.11	Patna	0.40	3.60	0.40	3.64
	8.12	Samastipur	0.40	3.60	0.40	3.64
	8.13	Purbi Champaran	0.40	3.20	0.40	3.23
	8.14	Begusarai	0.40	3.00	0.40	3.03
	8.15	Bhojpur	0.30	2.70	0.30	2.73
	8.16	Bhagalpur	0.30	2.60	0.30	2.63

Source: Horticulture Statistics Division, Deptt. of Agriculture & Coopn.

Table 4.29 Area and Production of Bootle gourd for Major Producing Districts

A = Area in '000 Ha
P = Production in '000 MT

States	Sl. No.	Districts	2014-15		2015-16	
			A	p	A	P
1. Bihar	1.1	Vaishali	1.70	40.00	1.72	40.40
	1.2	Muzaffarpur	1.70	37.00	1.72	37.37
	1.3	Pashchim Champaran	1.50	33.40	1.52	33.73
	1.4	Purbi Champaran	1.40	31.00	1.41	31.31
	1.5	Katihar	1.40	27.60	1.41	27.88
	1.6	Bhagalpur	1.20	26.20	1.21	26.46
	1.7	Samastipur	1.00	23.70	1.01	23.94
	1.8	Gopalganj	1.20	23.50	1.21	23.74
	1.9	Saharsa	1.10	23.50	1.11	23.74
	1.1	Sitamarhi	1.20	23.00	1.21	23.23
	1.11	Begusarai	1.10	22.90	1.11	23.13
	1.12	Darbhanga	1.00	22.20	1.01	22.42
	1.13	Patna	9.70	22.00	9.80	22.22
	1.14	Purnia	1.10	20.90	1.11	21.11
	1.15	Saran	1.00	20.00	1.01	20.20
	1.16	Nalanda	0.90	19.60	0.91	19.80
	1.17	Madhepura	0.90	17.50	0.91	17.68
	1.18	Siwan	0.90	17.40	0.91	17.57
2. Uttar Pradesh	2.1	Agra	0.95	27.78	1.43	41.79
	2.2	Mainpuri	0.58	16.72	0.87	25.14
	2.3	Faizabad	0.50	15.84	0.74	23.81
	2.4	Hapur	0.42	12.27	0.63	18.46
	2.5	Etah	0.41	12.55	0.62	18.13

(*Contd.*)

States	Sl. No.	Districts	2014-15		2015-16	
			A	p	A	P
	2.6	Aligarh	0.36	10.63	0.55	15.99
	2.7	Lucknow	0.34	10.00	0.51	15.04
	2.8	Meerut	0.32	9.36	0.48	14.08
	2.9	Bareilly	0.29	8.51	0.44	12.81
	2.1	Unnao	0.27	8.16	0.40	12.27
	2.11	Kushi Nagar	0.28	8.15	0.42	12.26
	2.12	Kannauj	0.28	8.11	0.42	12.19
	2.13	Hathras	0.27	7.93	0.41	11.93
	2.14	Bulandshahr	0.26	7.61	0.39	11.45
	2.15	Gonda	0.22	6.70	0.33	10.06
	2.16	Mathura	0.21	6.09	0.31	9.16
	2.17	Amroha	0.15	5.60	0.23	8.42
	2.18	Baghpat	0.19	5.60	0.29	8.42
	2.19	Firozabad	0.19	5.47	0.28	8.22
	2.2	Moradabad	0.17	4.88	0.25	7.34
	2.21	Kanpur Nagar	0.16	4.57	0.24	6.87
3. Haryana	3.1	Mewat	NA	NA	4.10	54.00
	3.2	Yamunanagar	NA	NA	2.85	42.20
	3.3	Gurgaon	NA	NA	2.21	33.21
	3.4	Faridabad	NA	NA	2.30	32.00
	3.5	Sonipat	NA	NA	1.37	18.60
	3.6	Ambala	1.34	15.11	1.54	17.08
	3.7	Palwal	NA	NA	1.10	15.80
	3.8	Rohtak	NA	NA	1.35	13.36
4. Madhya Pradesh	4.1	Jabalpur	1.59	63.68	1.37	54.96
	4.2	Satna	1.15	19.49	1.32	22.42
	4.3	Chhindwara	0.61	12.10	0.68	13.17
	4.4	Shivpuri	0.68	11.56	0.68	11.56
	4.5	Chhatarpur	0.53	11.63	0.51	11.26
	4.6	Gwalior	0.40	7.32	0.48	10.60
	4.7	Indore	0.35	7.70	0.39	8.47
	4.8	Shahdol	0.27	7.16	0.29	7.90
	4.9	Dhar	0.14	1.00	0.37	7.45
	4.1	Barwani	0.13	2.68	0.33	7.13
5. Chhattisgarh	5.1	Durg	1.49	38.61	1.52	38.65
	5.2	Bemetara	1.02	27.65	1.02	27.65
	5.3	Korba	1.41	24.83	1.41	24.07
	5.4	Mahasamund	1.09	14.75	1.28	17.15
	5.5	Balod	0.46	11.66	0.51	12.82
	5.6	Kondagaon	0.72	11.94	0.77	12.77
	5.7	Raipur	0.50	9.35	0.65	12.15

States	Sl. No.	Districts	2014-15		2015-16	
			A	p	A	P
	5.8	Surguja	0.68	11.46	0.68	11.46
	5.9	Raigarh	0.63	10.71	0.63	10.71
6. Odisha	6.1	Sundargarh	0.63	9.20	0.65	9.49
	6.2	Mayurbhanj	0.64	9.18	0.61	8.78
	6.3	Kendujhar	0.54	7.94	0.52	7.64
	6.4	Ganjam	0.57	7.79	0.55	7.52
	6.5	Cuttack	0.48	6.64	0.46	6.36
	6.6	Dhenkanal	0.49	6.15	0.46	6.17
	6.7	Kalahandi	0.39	5.51	0.42	5.86
	6.8	Anugul	0.38	4.54	0.40	5.82
	6.9	Sonepur	0.45	5.50	0.47	5.75
	6.1	Kandhamal	0.35	5.12	0.37	5.42
	6.11	Sambalpur	0.16	2.30	0.37	5.23
	6.12	Nayagarh	0.37	5.48	0.35	5.18
	6.13	Gajapati	0.41	4.92	0.43	5.16
	6.14	Koraput	0.34	4.74	0.36	5.02
	6.15	Bargarh	0.33	4.46	0.36	4.80
7. Punjab	7.1	Sangrur	0.49	8.02	0.43	6.62
	7.2	Kapurthala	0.24	4.53	0.24	4.61
	7.3	Patiala	0.25	3.99	0.25	4.08
	7.4	Nawanshahr	0.25	3.03	0.25	3.10
	7.5	Faridkot	0.06	2.33	0.08	3.07
	7.6	Gurdaspur	0.10	1.63	0.19	2.89
	7.7	Amritsar	0.05	2.44	0.04	2.15
8. Telangana	8.1	Medak	0.73	7.98	1.09	18.48
	8.2	Warangal	1.15	12.63	1.08	11.86
	8.3	Adilabad	0.03	0.37	0.30	9.09
	8.4	Mahbubnagar	0.13	2.21	0.64	6.41
	8.5	Khammam	0.87	5.19	1.06	6.38
9. Assam	9.1	Dhubri	NA	NA	0.35	5.17
	9.2	Nagaon	NA	NA	0.16	5.12
	9.3	Kamrup	NA	NA	0.23	4.96
	9.4	Darrang	NA	NA	0.15	4.90
	9.5	Dima Hasao	NA	NA	0.15	3.54
	9.6	Karbi Anglong	NA	NA	0.13	3.42
	9.7	Nalbari	NA	NA	0.10	3.28
	9.8	Dhemaji	NA	NA	0.12	2.47
	9.9	Dibrugarh	NA	NA	0.10	2.08
	9.10	Cachar	NA	NA	0.15	1.54

Source: Horticulture Statistics Division, Deptt. of Agriculture & Coopn.

Table 4.30 Area and Production of Brinjal for Major Producing Districts

A = Area in '000 Ha
P = Production in '000 MT

States	Sl. No.	Districts	2014-15		2015-16	
			A	p	A	P
1. West Bengal	1.1	Murshidabad	17.44	329.19	17.54	352.02
	1.2	Nadia	12.41	277.33	12.46	297.81
	1.3	Maldah	12.55	211.50	12.72	212.76
	1.4	Bankura	10.16	200.30	10.14	200.17
	1.5	24 Paraganas South	10.33	190.73	10.33	191.19
	1.6	24 Paraganas North	9.76	199.79	9.75	176.28
	1.7	Medinipur West	9.36	175.08	9.35	174.85
	1.8	Birbhum	10.54	169.77	10.53	169.73
	1.9	Purulia	8.27	165.16	8.26	164.94
	1.1	Medinipur East	9.52	164.96	9.52	164.83
2. Odisha	2.1	Kendujhar	11.25	197.48	10.67	187.28
	2.2	Ganjam	5.02	129.16	4.97	127.85
	2.3	Kalahandi	7.98	133.58	7.53	126.95
	2.4	Mayurbhanj	7.23	123.74	7.03	119.86
	2.5	Bhadrak	5.64	94.70	7.20	118.14
	2.6	Baleshwar	4.95	82.96	5.42	87.35
	2.7	Cuttack	5.44	92.46	4.70	79.52
	2.8	Dhenkanal	4.92	83.33	4.63	78.47
	2.9	Balangir	4.52	73.45	4.71	76.58
	2.1	Jajapur	4.67	78.58	4.47	74.94
	2.11	Anugul	4.47	75.43	4.26	71.80
	2.12	Boudh	4.36	73.71	4.26	71.57
	2.13	Rayagada	4.20	73.28	4.15	71.50
	2.14	Kendrapara	4.14	69.39	4.07	68.21
	2.15	Koraput	4.07	62.10	4.03	61.46
3. Gujarat	3.1	Anand	7.55	165.72	7.56	167.45
	3.2	Vadodara	10.05	211.05	8.21	165.49
	3.3	Banas Kantha	4.74	102.69	4.76	104.68
	3.4	Surat	5.16	99.33	5.27	101.96
	3.5	Kheda	4.55	93.73	4.31	88.32
	3.6	Ahmadabad	3.78	72.01	3.59	71.86
	3.7	Tapi	3.72	66.68	3.72	67.24
	3.8	Gandhinagar	2.95	65.37	2.95	65.87
	3.9	Navsari	2.77	54.25	2.84	55.61
	3.1	Kachchh	2.91	52.00	2.94	52.43
	3.11	Surendranagar	3.45	62.72	2.85	51.87
	3.12	Valsad	2.61	48.86	2.75	51.33

States	Sl. No.	Districts	2014-15		2015-16	
			A	p	A	P
4. Madhya Pradesh	4.1	Sagar	3.69	95.48	3.79	113.61
	4.2	Raisen	3.43	60.86	3.54	85.87
	4.3	Chhindwara	2.86	74.44	3.14	69.72
	4.4	Jabalpur	2.48	43.45	2.70	47.17
	4.5	Tikamgarh	1.65	36.34	1.74	38.35
	4.6	Shahdol	1.29	33.41	1.25	34.36
	4.7	Chhatarpur	3.24	48.53	2.16	26.51
	4.8	Dewas	0.95	21.90	1.05	24.18
	4.9	Vidisha	1.34	10.07	1.54	23.15
	4.1	Satna	1.40	20.66	1.56	23.13
	4.11	Narsinghpur	1.06	17.52	1.33	22.71
	4.12	Ujjain	1.18	18.82	1.17	22.50
	4.13	Seoni	0.81	18.90	0.84	19.10
	4.14	Shivpuri	0.99	17.77	0.99	17.77
	4.15	Bhind	NA	NA	0.70	17.29
	4.16	Katni	2.02	33.87	2.08	15.56
5. Bihar	5.1	Nalanda	7.20	148.50	7.27	149.99
	5.2	Vaishali	3.20	69.00	3.23	69.69
	5.3	Begusarai	2.80	58.90	2.83	59.49
	5.4	Samastipur	2.30	58.50	2.32	59.09
	5.5	Darbhanga	2.50	51.70	2.53	52.22
	5.6	Patna	1.90	49.00	1.92	49.49
	5.7	Pashchim Champaran	2.00	45.00	2.02	45.45
	5.8	Madhubani	2.10	43.20	2.12	43.63
	5.9	Saran	1.80	38.50	1.82	38.89
	5.1	Muzaffarpur	2.90	36.60	2.93	36.97
	5.11	Purbi Champaran	1.60	36.30	1.62	36.66
	5.12	Bhagalpur	1.70	36.00	1.72	36.36
	5.13	Siwan	1.60	35.00	1.62	35.35
	5.14	Madhepura	1.60	33.90	1.62	34.24
	5.15	Gopalganj	1.40	30.40	1.41	30.70
	5.16	Khagaria	1.20	29.60	1.21	29.90
6. Chhattisgarh	6.1	Durg	4.06	103.40	4.34	109.42
	6.2	Kondagaon	4.41	64.68	4.61	67.61
	6.3	Raipur	2.95	52.40	3.10	55.06
	6.4	Bemetara	1.45	36.25	1.45	36.25
	6.5	Balod	1.10	28.35	1.21	31.19
	6.6	Rajnandgaon	1.90	28.53	1.92	28.77
	6.7	Janjgir-Champa	1.37	27.40	1.37	27.58

(Contd.)

States	Sl. No.	Districts	2014-15		2015-16	
			A	p	A	P
	6.8	Kabirdham	1.00	20.00	1.20	24.00
	6.9	Surguja	1.45	23.98	1.45	23.98
	6.1	Bilaspur	1.75	22.27	1.86	23.72
	6.11	Kanker	1.56	23.03	1.61	22.54
7. Karnataka	7.1	Belgaum	4.01	109.15	3.87	104.17
	7.2	Haveri	1.40	38.73	1.38	37.76
	7.3	Kolar	1.29	41.52	1.05	32.79
	7.4	Mandya	1.18	29.59	1.19	28.45
	7.5	Bijapur	1.15	28.75	0.93	23.13
	7.6	Koppal	0.98	23.38	0.85	19.94
	7.7	Mysore	0.68	19.69	0.67	18.71
	7.8	Chamarajanagar	0.51	0.08	0.56	16.78
	7.9	Bellary	0.74	22.60	0.56	16.71
8. Andhra Pradesh	8.1	East Godavari	3.63	73.42	2.79	83.70
	8.2	Chittoor	3.60	71.96	3.60	71.96
	8.3	Kadapa	0.87	43.45	1.30	58.50
	8.4	Krishna	1.23	24.50	1.48	29.58
	8.5	Guntur	NA	NA	1.08	27.08

Source: Horticulture Statistics Division, Deptt. of Agriculture & Coopn.

Table 4.31 Area and Production of Cabbage for Major Producing Districts

A = Area in '000 Ha
P = Production in '000 MT

States	Sl. No.	Districts	2014-15		2015-16	
			A	p	A	P
1. West Bengal	1.1	Murshidabad	10.55	328.13	10.40	371.65
	1.2	Nadia	8013.00	209.87	8.19	212.07
	1.3	24 Paraganas South	6.51	206.03	6.52	207.33
	1.4	Bankura	5.34	169.85	5.34	170.12
	1.5	24 Paraganas North	5.15	149.56	5.09	144.59
	1.6	Coochbehar	4.63	140.67	4.46	140.75
	1.7	Medinipur West	6.20	139.41	6.22	139.90
	1.8	Maldah	4.52	126.49	4.83	128.93
	1.9	Dinajpur Dakshin	4.13	106.34	4.13	106.40
2. Odisha	2.1	Jagatsinghapur	2.35	66.70	2.35	65.73
	2.2	Kandhamal	2.79	78.32	2.25	63.23
	2.3	Khordha	2.26	64.77	2.20	63.16
	2.4	Kendujhar	2.84	80.06	2.11	59.47
	2.5	Koraput	2.17	59.70	2.03	55.61
	2.6	Sundargarh	2.21	64.75	1.85	53.90

States	Sl. No.	Districts	2014-15		2015-16	
			A	p	A	P
	2.7	Dhenkanal	1.98	54.56	1.93	53.34
	2.8	Balangir	1.82	49.99	1.77	48.73
	2.9	Mayurbhanj	1.80	49.69	1.74	48.22
	2.1	Anugul	1.80	49.61	1.73	47.81
	2.11	Kendrapara	1.71	46.28	1.67	45.31
	2.12	Cuttack	2.01	56.54	1.55	43.47
	2.13	Ganjam	1.51	42.05	1.47	41.07
	2.14	Boudh	1.43	41.07	1.40	40.32
	2.15	Jajapur	1.33	36.59	1.30	35.71
3. Assam	3.1	Barpeta	2.96	52.05	2.96	65.88
	3.2	Nagaon	2.78	62.92	2.88	58.93
	3.3	Dhubri	1.97	32.80	2.07	40.21
	3.4	Lakhimpur	1.28	46.38	2.05	38.56
	3.5	Darrang	1.80	38.27	1.90	38.27
	3.6	Golaghat	1.35	47.69	1.35	37.69
	3.7	Cachar	1.80	33.86	1.80	36.58
	3.8	Kamrup	1.60	30.43	1.70	35.85
	3.9	Nalbari	1.55	26.36	1.55	31.79
	3.1	Tinsukia	1.25	38.27	1.25	28.27
	3.11	Dibrugarh	0.96	27.71	0.96	27.71
	3.12	Kokrajhar	1.10	17.39	1.10	23.99
	3.13	Jorhat	1.15	33.69	1.15	23.69
4. Bihar	4.1	Vaishali	2.50	55.00	2.53	55.55
	4.2	Muzaffarpur	2.80	51.00	2.83	51.51
	4.3	Katihar	2.30	38.90	2.32	39.29
	4.4	Samastipur	1.90	38.50	1.92	38.89
	4.5	Patna	2.50	38.00	2.53	38.38
	4.6	Nalanda	1.60	33.00	1.62	33.33
	4.7	Darbhanga	1.60	30.30	1.62	30.60
	4.8	Purbi Champaran	1.60	30.30	1.62	30.60
	4.9	Pashchim Champaran	1.40	30.00	1.41	30.30
	4.10	Madhubani	1.50	27.10	1.52	27.37
	4.11	Madhepura	1.50	26.00	1.52	26.26
	4.12	Purnia	1.20	24.60	1.21	24.85
	4.13	Bhagalpur	1.20	23.20	1.21	23.43
	4.14	Begusarai	1.90	20.90	1.92	21.11
	4.15	Gopalganj	1.00	19.40	1.01	19.59
	4.16	Saran	0.90	18.10	0.91	18.28

(*Contd.*)

States	Sl. No.	Districts	2014-15		2015-16	
			A	p	A	P
5. Gujarat	5.1	Sabar Kantha	2.56	94.04	2.60	97.50
	5.2	Banas Kantha	3.87	73.97	3.57	68.08
	5.3	Kheda	2.37	54.27	2.29	51.92
	5.4	Gandhinagar	1.80	48.34	1.83	49.28
	5.5	Anand	2.03	42.63	2.13	44.67
	5.6	Rajkot	3.39	77.46	1.53	33.55
	5.7	Vadodara	1.67	34.67	1.40	27.34
	5.8	Ahmadabad	1.21	21.73	1.17	21.00
	5.9	Bhavnagar	1.09	24.70	0.99	20.39
	5.1	Junagadh	2.00	36.40	1.03	18.77
6. Jharkhand	6.1	Giridih	1.23	1.17	3.71	26.75
	6.2	Ranchi	10.91	10.36	6.54	10.22
7. Madhya Pradesh	7.1	Chhindwara	3.64	101.98	3.80	106.46
	7.2	Indore	1.51	33.11	1.61	35.31
	7.3	Betul	2.45	29.42	2.46	31.05
	7.4	Shahdol	NA	NA	0.99	26.78
	7.5	Jabalpur	0.59	14.68	0.85	21.33
	7.6	Satna	0.77	19.26	0.84	21.19
	7.7	Sidhi	0.52	15.60	0.58	17.25
	7.8	Dewas	0.57	13.40	0.62	14.75
	7.9	Tikamgarh	0.61	11.57	0.69	12.37
	7.1	Katni	0.85	12.81	0.90	11.85
	7.11	Singrauli	NA	NA	0.43	11.65
	7.12	Rajgarh	0.66	10.51	0.69	11.07
	7.13	Sagar	NA	NA	0.41	10.33
	7.14	Shajapur	0.24	4.96	0.50	10.18
	7.15	Dhar	0.42	9.09	0.56	9.58
8. Chhattisgarh	8.1	Murshidabad	10.55	328.13	13.10	371.65
	8.2	Durg	3.11	63.76	3.19	64.40
	8.3	Kondagaon	1.80	28.78	2.00	31.98
	8.4	Bemetara	1.38	27.53	1.46	29.23
	8.5	Balod	1.21	25.83	1.33	28.41
9. Haryana	9.1	Sonipat	3.69	54.55	2.83	45.30
	9.2	Karnal	1.91	29.91	2.10	32.00
	9.3	Panipat	1.30	14.20	2.86	29.03
	9.4	Rohtak	0.42	6.71	1.12	22.35
	9.5	Gurgaon	1.15	19.75	1.82	20.17
	9.6	Jind	0.85	15.17	0.70	16.55
	9.7	Yamunanagar	0.76	11.75	0.85	15.60

States	Sl. No.	Districts	2014-15		2015-16	
			A	p	A	P
	9.8	Fatehabad	0.66	11.06	1.00	13.57
	9.9	Faridabad	0.86	14.08	0.80	13.20
	9.1	Hisar	0.57	9.05	0.84	12.80
10. Uttar Pradesh	10.1	Unnao	0.26	9.23	0.94	32.86
	10.2	Mainpuri	0.23	8.04	0.83	28.61
	10.3	Mathura	0.22	7.76	0.80	27.62
	10.4	Agra	0.22	7.67	0.79	27.29
	10.5	Faizabad	0.16	5.59	0.56	19.90
	10.6	Kannauj	0.13	4.46	0.46	15.88
	10.7	Gonda	0.11	3.73	0.38	13.26
	10.8	Hapur	0.08	2.69	0.27	9.58
	10.9	Firozabad	0.06	2.09	0.22	7.45
	10.1	Kheri	0.06	2.03	0.21	7.21
	10.11	Lucknow	0.05	1.82	0.18	6.46
	10.12	Kasganj	0.05	1.77	0.18	6.29
	10.13	Basti	0.05	1.68	0.17	5.98

Source: Horticulture Statistics Division, Deptt. of Agriculture & Coopn.

Table 4.32 Area and Production of Capsicum for Major Producing Districts

A = Area in '000 Ha
P = Production in '000 MT

States	Sl. No.	Districts	2014-15		2015-16	
			A	P	A	P
1. Himachal Pradesh	1.1	Solan	1.12	33.69	1.13	34.02
	1.2	Sirmaur	0.31	5.16	0.32	5.42
	1.3	Mandi	0.28	4.27	0.29	4.49
2. Karnataka	2.1	Mandya	0.43	6.69	0.43	10.37
	2.2	Belgaum	0.6	9.71	0.6	10.24
	2.3	Kolar	0.45	9	0.48	9.95
	2.4	Haveri	0.62	6.08	0.52	8.67
	2.5	Chikballapur	0.08	1.91	0.2	7.54
	2.6	Bangalore Rural	0.08	0.61	0.17	7.33
3. Madhya Pradesh	3.1	Jhabua	0.63	26.25	0.65	27.3
	3.2	Tikamgarh	NA	NA	0.15	1.38
	3.3	Chhatarpur	0.12	1.29	0.12	1.29
	3.4	Khargone	0.01	0.48	0.03	1.12
4. Maharashtra	4.1	Kolhapur	NA	NA	1.75	10.95
	4.2	Amravati	0.78	6.04	0.78	6.1
	4.3	Solapur	0.46	2.54	0.41	3.82
	4.4	Nashik	0.25	2.76	0.25	2.76

(Contd.)

States	Sl. No.	Districts	2014-15		2015-16	
			A	P	A	P
5. Haryana	5.1	Karnal	NA	NA	1.05	10.1
	5.2	Kurukshetra	NA	NA	0.49	4.95
	5.3	Panipat	NA	NA	0.33	2.1
	5.4	Yamunanagar	NA	NA	0.2	1.9
	5.5	Ambala	0.06	0.08	0.4	1.82
6. Jharkhand	6.1	Pakur	NA	NA	0.02	0.1
	6.2	Bokaro	0	0.01	0	0.01
	6.3	Dumka	0.01	0.01	NA	NA
7 Uttarakhand	7.1	Pithoragarh	0.38	4.62	0.38	4.62
	7.2	Haridwar	0.13	1.51	0.14	1.58
	7.3	Tehri Garhwal	0.31	1.38	0.32	1.42
	7.4	Almora	0.42	1.23	0.42	1.23
	7.5	Nainital	0.23	1.12	0.19	1.19
8. Odisha	8.1	Sundargarh	0.02	0.27	0.04	0.54
	8.2	Khordha	0.02	0.3	0.03	0.51
	8.3	Kendrapara	0.01	0.16	0.04	0.5
	8.4	Sambalpur	0.01	0.11	0.04	0.48
	8.5	Mayurbhanj	0.02	0.22	0.03	0.44
	8.6	Bhadrak	NA	NA	0.03	0.44
	8.7	Kandhamal	0.02	0.21	0.03	0.42
	8.8	Sonepur	0.01	0.12	0.03	0.42
	8.9	Kalahandi	0.01	0.02	0.03	0.38
	8.10	Nabarangpur	0.01	0.16	0.03	0.37
	8.11	Jagatsinghapur	0.01	0.18	0.02	0.35

Source: Horticulture Statistics Division, Deptt. of Agriculture & Coopn.

Table 4.33 Area and Production of Carrot for Major Producing Districts

A = Area in '000 Ha
P = Production in '000 MT

States	Sl. No.	Districts	2014-15		2015-16	
			A	p	A	P
1. Haryana	1.1	Sonipat	2.86	30.40	3.15	51.30
	1.2	Rohtak	0.63	9.58	2.98	40.66
	1.3	Panipat	2.24	25.60	2.55	37.78
	1.4	Ambala	1.95	24.00	2.28	33.11
	1.5	Hisar	1.45	16.50	1.65	24.50
	1.6	Karnal	1.30	20.22	1.51	22.00
	1.7	Jhajjar	1.21	17.77	1.34	20.50
	1.8	Kurukshetra	0.80	11.33	1.93	18.38
	1.9	Fatehabad	0.88	12.26	0.96	16.13

States	Sl. No.	Districts	2014-15		2015-16	
			A	p	A	P
2. Uttar Pradesh	2.1	Bulandshahr	0.92	22.87	1.21	30.01
	2.2	Agra	0.47	11.87	0.62	15.59
	2.3	Aligarh	0.34	8.33	0.44	10.93
	2.4	Meerut	0.27	6.59	0.35	8.65
	2.5	Mainpuri	0.26	6.58	0.34	8.63
	2.6	Unnao	0.23	5.85	0.31	7.68
	2.7	Sambhal	0.14	3.51	0.19	4.61
	2.8	Firozabad	0.13	3.27	0.17	4.30
	2.9	Moradabad	0.12	2.91	0.15	3.82
	2.10	Bareilly	0.11	2.82	0.15	3.69
	2.11	Shamli	0.10	2.39	0.13	3.14
	2.12	Kannauj	0.08	2.05	0.11	2.70
	2.13	Shahjahanpur	0.08	1.97	0.10	2.59
	2.14	Kaushambi	0.06	1.60	0.08	2.10
	2.15	Hapur	0.06	1.56	0.08	2.04
3. Punjab	3.1	Hoshiarpur	3.90	88.69	3.79	90.90
	3.1	Ludhiana	0.85	20.04	0.72	18.10
	3.3	Kapurthala	0.68	14.36	0.67	15.01
	3.4	Nawanshahr	0.65	6.97	0.66	12.98
4. Tamil Nadu	4.1	TheNilgiris	2.60	96.81	3.19	83.04
	4.2	Dindigul	0.77	27.16	0.93	33.69
5. Karnataka	5.1	Kolar	1.39	27.05	1.39	27.12
	5.2	Chikballapur	0.62	12.11	0.84	15.57
	5.3	Belgaum	0.74	11.41	0.72	11.03
	5.4	BangaloreRural	0.30	5.64	0.29	5.42
	5.5	Gulbarga	0.22	4.42	0.20	3.57
	5.6	Bidar	0.19	3.24	0.20	3.39
6. Assam	6.1	Darrang	0.30	6.37	0.53	9.51
	6.2	Goalpara	0.25	2.17	0.37	7.61
	6.3	Nagaon	0.36	7.57	0.28	5.70
	6.4	Barpeta	0.31	4.74	0.31	5.46
	6.5	Bongaigaon	0.24	5.03	0.24	4.97
	6.6	Dhubri	0.24	3.57	0.23	4.51
	6.7	Chirang	0.22	4.37	0.21	4.38
	6.8	Kokrajhar	0.20	4.22	0.19	4.16
	6.9	Udalguri	0.20	1.51	0.19	3.95
7. Bihar	7.1	Vaishali	0.30	3.50	0.30	3.54
	7.2	Nalanda	0.30	3.40	0.30	3.43
	7.3	Muzaffarpur	0.30	3.20	0.30	3.23
	7.4	Patna	0.30	3.20	0.30	3.23

(Contd.)

States	Sl. No.	Districts	2014-15		2015-16	
			A	p	A	P
	7.5	Pashchim Champaran	0.20	2.80	0.20	2.83
	7.6	Samastipur	0.20	2.60	0.20	2.63
	7.7	Bhagalpur	0.10	1.80	0.10	1.82
	7.8	Begusarai	0.10	1.70	0.10	1.72
	7.8	Katihar	1.50	1.70	1.52	1.72
	7.10	PurbiChamparan	1.40	1.70	1.41	1.72
	7.11	Madhubani	0.10	1.60	0.10	1.62
	7.12	Purnia	0.10	1.60	0.10	1.62
	7.13	Saharsa	0.10	1.50	0.10	1.52
	7.14	Siwan	1.20	1.50	1.21	1.52
	7.15	Gopalganj	0.10	1.40	0.10	1.41
	7.16	Madhepura	0.10	1.40	0.10	1.41
	7.17	Saran	0.10	1.40	0.10	1.41
	7.18	Sitamarhi	0.10	1.40	0.10	1.41
8. Madhya Pradesh	8.1	Indore	0.83	16.50	0.40	8.00
	8.2	Ujjain	0.62	6.36	0.65	5.84
	8.3	Sagar	NA	NA	0.17	4.35
	8.4	Datia	NA	NA	0.19	3.80
	8.5	Sheopur	NA	NA	0.20	3.43
	8.6	Mandsaur	0.19	3.18	0.20	3.32
	8.7	Tikamgarh	0.13	2.79	0.17	3.30
	8.8	Bhind	NA	NA	0.16	3.10
	8.9	Barwani	0.05	1.86	0.15	3.02
	8.10	Betul	0.23	2.76	0.24	2.88
	8.11	Neemuch	0.12	2.44	0.14	2.72
9. Andhra Pradesh	9.1	Kurnool	0.82	16.48	0.85	17.00
	9.2	Chittoor	0.77	7.71	0.78	15.50
10. Telangana	10.1	Rangareddi	3.83	38.34	2.47	27.92
	10.2	Medak	0.16	1.64	0.54	6.02

Source: Horticulture Statistics Division, Deptt. of Agriculture & Coopn.

Table 4.34 Area and Production of Cauliflower for Major Producing Districts

A = Area in '000 Ha
P = Production in '000 MT

States	Sl. No.	Districts	2014-15		2015-16	
			A	p	A	P
1. West Bengal	1.1	Murshidabad	10.29	304.73	10.03	290.92
	1.2	Nadia	7.55	214.28	7.60	216.56
	1.3	Bankura	5.75	154.29	5075.00	154.48
	1.4	24 Paraganas North	5.29	143.40	5.39	145.60

States	Sl. No.	Districts	2014-15		2015-16	
			A	p	A	P
	1.5	Coochbehar	4.61	140.12	4.62	140.10
	1.6	24 Paraganas South	5.59	116.55	5.61	117.73
	1.7	Maldah	3.86	112.07	3.96	114.05
	1.8	Medinipur West	6.02	102.80	6.02	109.50
	1.9	Hooghly	3.84	99.28	3.85	99.54
2. Bihar	2.1	Vaishali	6.00	100.00	6.06	101.00
	2.2	Muzaffarpur	3.50	68.00	3.54	68.68
	2.3	Nalanda	3.10	60.00	3.13	60.60
	2.4	Samastipur	3.00	59.50	3.03	60.10
	2.5	Pashchim Champaran	2.90	53.40	2.93	53.93
	2.6	Madhubani	2.60	45.80	2.63	46.26
	2.7	Purbi Champaran	2.90	39.30	2.93	38.38
	2.8	Saran	2.90	39.30	2.93	39.69
	2.9	Purnia	2.10	35.20	2.12	35.55
	2.10	Begusarai	1.80	34.90	1.82	35.25
	2.11	Gopalganj	1.90	33.90	1.92	34.24
	2.12	Saharsa	1.80	31.50	1.82	31.82
	2.13	Madhepura	1.80	30.90	1.82	31.21
	2.14	Bhagalpur	1.70	29.20	1.72	29.49
	2.15	Darbhanga	1.60	28.50	1.62	28.79
	2.16	Siwan	1.60	28.40	1.62	28.68
3. Madhya Pradesh	3.1	Chhindwara	6.67	186.84	7.03	196.84
	3.2	Indore	2.26	51.07	2.34	51.37
	3.3	Sagar	1.80	45.08	1.97	47.23
	3.4	Satna	1.10	32.95	1.21	36.25
	3.5	Jabalpur	1.17	29.25	1.59	32.75
	3.6	Dewas	0.99	21.96	1.09	23.44
	3.7	Shahdol	0.74	22.94	0.72	21.45
	3.8	Katni	1.32	18.21	1.67	19.59
	3.9	Shajapur	0.85	16.60	0.85	15.95
	3.10	Tikamgarh	0.92	16.47	0.93	15.84
	3.11	Sidhi	0.71	14.20	0.79	15.70
	3.12	Rajgarh	0.97	15.49	0.98	15.70
	3.13	Khandwa	0.42	12.93	0.46	14.11
4. Odisha	4.1	Kendujhar	4.43	72.33	3.82	62.12
	4.2	Mayurbhanj	3.28	49.59	3.24	48.59
	4.3	Kandhamal	3.46	53.09	3.13	47.73
	4.4	Sundargarh	3.10	45.54	2.67	39.09
	4.5	Ganjam	2.41	35.56	2.37	34.81
	4.6	Puri	2.18	34.24	2.16	33.77

(*Contd.*)

States	Sl. No.	Districts	2014-15		2015-16	
			A	p	A	P
	4.7	Koraput	1.80	29.25	1.78	28.85
	4.8	Dhenkanal	2.11	32.14	1.88	28.52
	4.9	Kendrapara	1.77	25.67	1.80	25.97
	4.1	Khordha	1.25	19.11	1.24	18.89
	4.11	Sambalpur	0.58	8.81	1.23	18.71
	4.12	Balangir	1.36	20.02	1.28	18.71
	4.13	Jagatsinghapur	1.35	19.69	1.27	18.54
	4.14	Jajapur	1.44	21.10	1.24	18.18
5. Haryana	5.1	Sonipat	4.90	82.33	5.41	95.10
	5.2	Panipat	3.80	59.00	5.25	86.36
	5.3	Karnal	3.24	50.72	3.53	64.02
	5.4	Yamunanagar	2.70	27.90	3.14	51.50
	5.5	Gurgaon	2.27	36.35	2.47	38.97
	5.6	Fatehabad	1.28	22.42	2.32	31.83
	5.7	Ambala	1.74	24.05	2.09	31.21
	5.8	Kurukshetra	1.35	24.56	1.30	24.16
6. Gujarat	6.1	Sabar Kantha	4.71	140.05	4.52	134.85
	6.2	Vadodara	2.68	77.58	2.85	81.18
	6.3	Banas Kantha	3.84	77.82	3.82	72.64
	6.4	Kheda	2.78	56.99	2.56	52.38
	6.5	Ahmadabad	1.80	27.90	1.87	28.81
	6.6	Surat	1.34	27.14	1.35	27.08
7. Assam	7.1	Darrang	1.81	57.60	1.81	44.83
	7.2	Kamrup Metro	1.40	40.59	1.40	40.59
	7.3	Lakhimpur	0.76	11.92	1.33	36.76
	7.4	Barpeta	1.29	48.33	1.29	35.97
	7.5	Nagaon	1.43	35.48	1.43	35.48
	7.6	Dhubri	1.18	30.04	1.18	30.25
	7.7	Golaghat	1.08	25.67	1.08	25.67
	7.8	Kamrup	1.12	27.45	1.12	25.22
	7.9	Nalbari	0.91	24.23	1.13	24.23
	7.1	Karimganj	0.89	13.84	0.81	15.66
	7.11	Kokrajhar	0.76	13.51	0.76	13.51
	7.12	Tinsukia	0.78	11.37	0.78	12.37
8. Chhattisgarh	8.1	Kondagaon	2.71	70.90	3.01	78.76
	8.2	Durg	3.05	64.05	3.21	66.30
	8.3	Bemetara	1.32	26.40	1.42	28.30
	8.4	Raipur	1.33	24.54	1.40	25.83
	8.5	Balod	1.03	21.00	1.13	23.10
	8.6	Surajpur	1.25	18.63	1.30	19.66
	8.7	Korba	1.09	19.15	1.09	19.19

States	Sl. No.	Districts	2014-15		2015-16	
			A	p	A	P
	8.8	Bilaspur	1.18	16.80	1.25	17.82
	8.9	Surguja	1.02	16.45	1.02	16.80
	8.1	Kanker	0.97	15.55	1.02	16.32
9. Uttar Pradesh	9.1	Agra	0.78	17.78	1.20	27.44
	9.2	Unnao	0.51	11.56	0.78	17.83
	9.3	Bareilly	0.46	10.52	0.71	16.23
	9.4	Bulandshahr	0.44	10.07	0.68	15.54
	9.5	Maharajganj	0.41	9.26	0.62	14.28
	9.6	Saharanpur	0.35	8.04	0.54	12.41
	9.7	Sambhal	0.33	7.59	0.51	11.72
	9.8	Kushi Nagar	0.33	7.54	0.51	11.63
	9.9	Lucknow	0.31	6.96	0.47	10.75
	9.10	Gorakhpur	0.28	6.35	0.43	9.79
	9.11	Shamli	0.27	6.27	0.42	9.67
	9.12	Mainpuri	0.27	6.16	0.42	9.51
	9.13	Faizabad	0.26	5.85	0.40	9.02
	9.14	Gonda	0.25	5.73	0.38	8.84
	9.15	Aligarh	0.25	5.61	0.38	8.66
	9.16	Budaun	0.24	5.41	0.37	8.34
	9.17	Meerut	0.23	5.30	0.36	8.17
	9.18	Bahraich	0.23	5.15	0.35	7.94
	9.19	Kanpur Nagar	0.20	4.45	0.30	6.87
	9.20	Kasganj	0.18	4.10	0.28	6.32
	9.21	Muzaffarnagar	0.17	3.93	0.27	6.07
	9.22	Fatehpur	0.17	3.90	0.26	6.01
	9.23	Moradabad	0.16	3.67	0.25	5.66
	9.24	Kannauj	0.16	3.55	0.24	5.48
	9.25	Kheri	0.15	3.41	0.23	5.26
	9.26	Basti	0.15	3.40	0.23	5.25
	9.27	Varanasi	0.15	3.36	0.23	5.18
	9.28	Sultanpur	0.15	3.35	0.23	5.15
	9.29	Rampur	0.15	3.32	0.23	5.12
10. Jharkhand	10.1	Lohardaga	0.23	0.22	1.07	16.89
	10.2	Giridih	0.11	0.11	2.42	15.79
	10.3	Ranchi	0.99	0.94	7.97	12.56
11. Punjab	11.1	Hoshiarpur	1.71	31.82	2.21	41.35
	11.2	Gurdaspur	1.84	33.17	1.84	34.80
	11.3	Sangrur	1.50	27.02	1.50	27.86
	11.4	Kapurthala	1.39	25.36	1.40	25.57
	11.5	Ludhiana	1.01	18.43	1.35	25.44

(Contd.)

States	Sl. No.	Districts	2014-15		2015-16	
			A	p	A	P
	11.6	Fatehgarh Sahib	1.16	21.42	1.17	21.58
	11.7	Patiala	1.01	18.31	1.11	20.17

Source: Horticulture Statistics Division, Deptt. of Agriculture & Coopn.

Table 4.35 Area and Production of Cucumber for Major Producing Districts

A = Area in '000 Ha
P = Production in '000 MT

States	Sl. No.	Districts	2014-15		2015-16	
			A	p	A	P
1. Haryana	1.1	Karnal	NA	NA	2.24	25.1
	1.2	Sonipat	NA	NA	1.9	24.2
	1.3	Panipat	NA	NA	1.3	20.5
	1.4	Gurgaon	NA	NA	1.01	17.01
	1.5	Rohtak	NA	NA	1.77	14.77
	1.6	Yamunanagar	NA	NA	0.97	14
	1.7	Kurukshetra	NA	NA	1.31	12.09
2. Karnataka	2.1	Haveri	1.02	13.91	1	15.32
	2.2	Belgaum	0.74	13.19	0.75	13.64
	2.3	Mandya	0.77	11.59	0.73	11.18
	2.4	Hassan	0.56	10.11	0.75	10.67
	2.5	Chikballapur	0.3	4.64	0.37	7.21
	2.6	Bellary	0.37	5.42	0.41	6.53
	2.7	Dharwad	0.21	3.67	0.15	6.36
	2.8	Koppal	0.5	7.55	0.36	5.68
	2.9	Ramanagara	0.18	2.48	0.2	5.61
	2.1	Bagalkot	0.28	3.84	0.35	4.97
	2.11	Bidar	0.31	4.63	0.3	4.57
	2.12	Bangalore Rural	0.22	3.5	0.24	4.37
3. Madhya Pradesh	3.1	Shahdol	NA	NA	0.54	23.5
	3.2	Barwani	0.33	8.13	0.66	10.74
	3.3	Satna	0.63	8.84	0.69	9.72
	3.4	Sagar	NA	NA	0.15	6.25
	3.5	Betul	0.39	3.72	0.39	4.18
	3.6	Khandwa	0.15	3.55	0.16	3.72
	3.7	Khargone	0.02	0.94	0.01	3.51
	3.8	Jabalpur	0.66	4.99	0.51	3.4
4. Tamil Nadu	4.1	Dharmapuri	NA	NA	0.01	104.52
	4.2	Krishnagiri	0.03	0.27	0.49	5.68
5. Andhra Pradesh	5.1	Krishna	1.15	23	1.28	25.56
	5.2	Guntur	NA	NA	1	23.98
	5.3	Prakasam	0.88	17.5	1.07	12.85

States	Sl. No.	Districts	2014-15		2015-16	
			A	p	A	P
6. Telangana	6.1	Medak	0.35	7	1.53	30.82
	6.2	Nalgonda	0.7	14	0.98	19.73
	6.3	Mahbubnagar	0	0.03	0.52	9.45
7. Assam	7.1	Kokrajhar	0.34	3.69	0.56	6.47
	7.2	Dhubri	0.73	7.48	0.53	5.73
	7.3	Goalpara	0.44	4.84	0.44	4.83
	7.4	Tinsukia	0.4	3.79	0.4	4.7
	7.5	Darrang	0.35	4.45	0.35	4.41
	7.6	Nagaon	0.33	3.29	0.33	4.25
	7.7	Bongaigaon	0.36	4.25	0.36	4.21
	7.8	Baksa	0.29	3.83	0.28	3.79
	7.9	Sonitpur	0.35	4.5	0.28	3.79
	7.1	Kamrup	0.25	3.59	0.25	3.54
	7.11	Barpeta	0.35	3.56	0.34	3.51
	7.12	Kamrup Metro	0.32	3.3	0.32	3.26
8. Bihar	8.1	Nalanda	0.1	46.5	0.1	46.97
	8.2	Vaishali	0.1	1.6	0.1	1.62
9. Maharashtra	9.1	Ahmednagar	NA	NA	0.68	8.37
	9.2	Pune	0.54	6.11	0.54	6.3
	9.3	Gadchiroli	0.17	3.42	0.17	4.8
	9.4	Raigad	0.27	3.45	0.27	3.45
	9.5	Nashik	0.28	2.91	0.29	3.38
	9.6	Thane	0.05	0.61	0.24	3.17
	9.7	Osmanabad	0.13	0.96	0.15	3.15
	9.8	Kolhapur	0.02	0.34	0.16	2.36
10. Kerala	10.1	Kannur	0	0.01	0	0.01
	10.2	Kasaragod	0	0.01	0	0.01
	10.3	Malappuram	0	0.01	0	0.01
	10.4	Thiruvananthapuram	0	NA	0	0
	10.5	Palakkad	0	0	0	0
11. Punjab	11.1	Amritsar	0.77	11.51	0.91	15.44
	11.2	Sangrur	0.5	8.75	0.52	9.52
	11.3	Ludhiana	0.05	0.72	0.51	7.38
	11.4	Kapurthala	0.27	5.09	0.27	5.18
	11.5	Nawanshahr	0.35	4.24	0.4	4.92

Source: Horticulture Statistics Division, Deptt. of Agriculture & Coopn.

Table 4.36 Area and Production of Chilli for Major Producing Districts

A = Area in '000 Ha
P = Production in '000 MT

States	S. No.	Districts	2014-15		2015-16	
			A	P	A	P
1. Karnataka	1.1	Belgaum	6.45	93.99	6.30	94.29
	1.2	Haveri	7.03	136.59	7.46	90.42
	1.3	Bellary	3.00	49.52	3.10	52.53
	1.4	Mysore	1.96	30.31	2.07	32.10
	1.5	Hassan	1.73	21.28	1.94	30.74
	1.6	Kolar	1.44	29.46	1.41	28.88
	1.7	Bagalkot	2.38	23.69	2.58	27.20
	1.8	Chitradurga	1.89	19.04	2.12	23.69
	1.9	Gadag	0.59	8.31	1.04	20.04
	1.10	Tumkur	0.91	13.20	1.32	19.53
2. Madhya Pradesh	2.1	Khandwa	3.07	110.34	3.17	113.94
	2.2	Jhabua	1.09	65.64	1.16	69.78
	2.3	Alirajpur	0.70	15.67	1.19	59.50
	2.4	Dewas	1.45	41.03	1.76	49.67
	2.5	Shivpuri	1.70	33.96	1.89	37.80
	2.6	Shahdol	0.74	26.71	0.72	25.48
	2.7	Khargone	0.16	10.96	0.27	18.83
3. Bihar	3.1	Patna	2.70	39.00	2.73	39.39
	3.2	Katihar	2.00	22.90	2.02	23.13
	3.3	Muzaffarpur	1.80	22.00	1.82	22.22
	3.4	Vaishali	1.80	21.00	1.82	21.21
	3.5	Begusarai	1.70	19.30	1.72	19.49
	3.6	Pashchim Champaran	1.60	18.00	1.62	18.18
	3.7	Samastipur	1.50	17.50	1.52	17.68
	3.8	Darbhanga	1.40	16.00	1.41	16.16
	3.9	Purnia	1.40	15.90	1.41	16.06
	3.10	Bhagalpur	1.10	14.40	1.11	14.54
	3.11	Madhubani	1.20	14.20	1.21	14.34
	3.12	Siwan	1.10	13.60	1.11	13.74
	3.13	Munger	1.10	13.50	1.11	13.64
	3.14	Purbi Champaran	0.10	13.30	0.10	13.43
	3.15	Saran	1.20	13.30	1.21	13.43
	3.16	Gopalganj	1.00	11.40	1.01	11.51
	3.17	Khagaria	1.00	10.90	1.01	11.01
4. Andhra Pradesh	4.1	Anantapur	3.50	87.50	4.20	126.00
	4.2	Kurnool	2.35	23.50	3.48	69.64
	4.3	Chittoor	1.26	31.40	1.26	31.40
	4.4	Kadapa	1.77	17.70	1.70	25.50

States	S. No.	Districts	2014-15		2015-16	
			A	p	A	P
5. Maharashtra	5.1	Nagpur		9.84	67.21	
	5.2	Jalgaon	2.00	23.00	2.08	25.64
	5.3	Nashik	2.06	22.74	2.06	22.74
	5.4	Nanded	2.11	22.23	1.48	17.62
	5.5	Nandurbar	0.93	14.70	0.93	14.92
	5.6	Palghar	0.67	13.62	0.67	14.22
	5.7	Pune	1.80	8.79	2.06	13.82
	5.8	Jalna	0.84	8.35	1.23	12.03
	5.9	Aurangabad	NA	NA	0.47	9.30
	5.10	Amravati	0.79	8.55	0.79	9.25
6. Chhattisgarh	6.1	Kabirdham	3.40	34.00	3.50	35.00
	6.2	Raigarh	4.56	29.15	4.56	29.15
	6.3	Surguja	2.30	20.31	2.30	22.95
	6.4	Bilaspur	3.03	19.13	3.09	19.51
	6.5	Korba	2.45	16.67	2.45	16.72
	6.6	Balrampur	1.59	13.97	1.70	14.88
	6.7	Kondagaon	2.28	14.68	2.48	14.88
	6.8	Rajnandgaon	2.00	12.98	2.03	13.16
	6.9	Bastar	1.09	11.91	1.16	12.18
7. Telangana	7.1	Mahbubnagar	4.85	93.43	5.49	71.33
	7.2	Medak	1.01	10.08	2.76	30.23
	7.3	Adilabad	0.05	0.96	0.47	14.16
8. Jharkhand	8.1	Ranchi	4.95	5.00	1.83	24.71
	8.2	Pakur	5.15	5.20	1.25	15.64
	8.3	Lohardaga	2.41	2.44	0.97	13.05
	8.4	Giridih	1.61	1.63	11.65	12.66
	8.5	Dhanbad	6.67	6.73	0.89	12.01

Source: Horticulture Statistics Division, Deptt. of Agriculture & Coopn.

Table 4.37 Area and Production of Muskmelon for Major Producing Districts

A = Area in '000 Ha
P = Production in '000 MT

States	S. No.	Districts	2014-15		2015-16	
			A	p	A	P
1. Uttar Pradesh	1.1	Farrukhabad	1.82	47.37	1.84	47.84
	1.2	Etah	1.50	38.90	1.51	39.29
	1.3	Aligarh	1.12	29.22	1.14	29.51
	1.4	Firozabad	1.12	29.07	1.13	29.35
	1.5	Agra	1.05	27.35	1.06	27.64
	1.6	Kannauj	0.84	21.89	0.85	22.10

(Contd.)

States	S. No.	Districts	2014-15		2015-16	
			A	p	A	P
	1.7	Bareilly	0.82	21.42	0.83	21.63
	1.8	Unnao	0.79	20.57	0.80	20.77
	1.9	Allahabad	0.79	20.44	0.79	20.64
	1.10	Mainpuri	0.72	18.75	0.73	18.93
	1.11	Budaun	0.72	18.69	0.73	18.88
	1.12	Hathras	0.71	18.33	0.71	18.51
	1.13	Hardoi	0.49	12.71	0.49	12.84
	1.14	Kanpur Nagar	0.46	11.88	0.46	12.01
	1.15	Kanpur Dehat	0.43	11.23	0.44	11.34
	1.16	Amroha	0.42	11.02	0.43	11.13
	1.17	Kasganj	0.40	10.43	0.41	10.53
2. Andhra Pradesh	2.1	Anantapur	3.00	90.00	3.00	90.00
	2.2	Chittoor	0.95	28.41	1.00	37.50
3. Punjab	3.1	Ludhiana	0.33	5.80	0.33	5.97
	3.2	S.A.S Nagar	0.23	3.71	0.23	3.85
4. Madhya Pradesh	4.1	Mandsaur	0.83	12.45	0.65	9.68
	4.2	Neemuch	0.61	6.69	0.61	9.13
	4.3	Burhanpur	0.16	4.80	0.19	4.50
	4.4	Khargone	0.07	1.31	0.13	3.50
	4.5	Barwani	0.07	1.30	0.15	2.70
5. Haryana	5.1	Mewat	Na	Na	1.05	13.50
	5.2	Sonipat	Na	Na	0.16	3.10
	5.3	Gurgaon	Na	Na	0.12	2.80
	5.4	Karnal	Na	Na	0.30	1.68
	5.5	Yamunanagar	Na	Na	0.22	1.50

Source: Horticulture Statistics Division, Deptt. of Agriculture & Coopn.

Table 4.38 Area and Production of Okra for Major Producing Districts

A = Area in '000 Ha
P = Production in '000 MT

States	S. No.	Districts	2014-15		2015-16	
			A	p	A	P
1. West Bengal	1.1	Nadia	9.35	121.33	9.37	121.93
	1.2	24 Paraganas South	7.56	92.72	7.59	93.15
	1.3	24 Paraganas North	4.67	63.02	4.77	65.91
	1.4	Bankura	5.89	65.52	5.89	65.56
	1.5	Murshidabad	4.96	52.86	5.50	60.60
	1.6	Bardhaman	5.28	59.69	5.41	59.85
	1.7	Medinipur West	4.22	52.01	4.22	52.03
	1.8	Medinipur East	4.75	50.73	4.75	50.74

States	S. No.	Districts	2014-15		2015-16	
			A	p	A	P
	1.9	Hooghly	3.97	45.50	3.97	45.60
	1.10	Jalpaiguri	3.22	41.89	3.22	41.97
2. Gujarat	2.1	Surat	10.98	145.71	11.63	158.71
	2.2	Tapi	9.82	133.55	9.93	134.99
	2.3	Navsari	5.69	72.00	6.09	76.98
	2.4	Banas Kantha	4.53	66.09	4.54	61.25
	2.5	Vadodara	5.61	77.70	3.16	41.05
	2.6	Chhotaudepur	NA	NA	2.62	32.79
	2.7	Gandhinagar	3.39	30.19	3.39	30.19
	2.8	Mahesana	2.17	29.86	2.05	28.25
	2.9	Anand	3.23	27.92	3.25	28.12
	2.10	Bharuch	2.75	26.53	2.81	27.89
3. Bihar	3.1	Vaishali	3.50	50.00	3.50	50.00
	3.2	Nalanda	2.90	41.10	2.90	41.10
	3.3	Patna	3.00	41.00	3.00	41.00
	3.4	Muzaffarpur	2.50	36.90	2.50	36.90
	3.5	Purbi Champaran	2.70	35.20	2.70	35.20
	3.6	Begusarai	2.40	32.80	2.40	32.80
	3.7	Bhagalpur	1.40	30.00	1.40	30.00
	3.8	Pashchim Champaran	2.50	30.00	2.50	30.00
	3.9	Gopalganj	2.00	29.20	2.00	29.20
	3.10	Siwan	2.00	26.00	2.00	26.00
	3.11	Saran	2.10	25.00	2.10	25.00
	3.12	Katihar	1.80	24.50	1.80	24.50
	3.13	Samastipur	1.70	24.50	1.70	24.50
	3.14	Nawada	1.90	23.90	1.90	23.90
	3.15	Aurangabad	1.80	22.70	1.80	22.70
	3.16	Madhubani	1.60	22.00	1.60	22.00
	3.17	Bhojpur	1.60	21.90	1.60	21.90
	3.18	Khagaria	1.30	21.20	1.30	21.20
4. Odisha	4.1	Jajapur	5.13	44.96	4.94	43.32
	4.2	Baleshwar	4.57	40.92	4.54	40.68
	4.3	Sundargarh	4.22	37.47	4.25	37.74
	4.4	Kendujhar	4.16	37.36	4.03	36.23
	4.5	Ganjam	3.46	30.33	3.33	29.13
	4.6	Mayurbhanj	3.09	27.13	3.07	26.94
	4.7	Puri	2.68	24.00	2.66	23.82
	4.8	Nayagarh	2.71	24.33	2.63	23.64
	4.9	Khordha	2.50	22.84	2.50	22.80

(Contd.)

States	S. No.	Districts	2014-15		2015-16	
			A	p	A	P
	4.10	Bhadrak	4.57	40.92	2.48	22.07
	4.11	Dhenkanal	2.51	21.88	2.43	21.21
	4.12	Balangir	2.32	20.18	2.30	20.01
	4.13	Anugul	2.20	19.36	2.15	18.90
	4.14	Boudh	2.10	18.79	2.09	18.68
	4.15	Koraput	2.05	17.80	2.07	18.01
5. Jharkhand	5.1	Hazaribagh	0.08	13.18	0.10	16.47
	5.2	Ramgarh	0.09	12.50	0.10	14.71
	5.3	Bokaro	0.29	4.92	0.30	5.02
6. Madhya Pradesh	6.1	Chhindwara	2.74	60.30	2.90	63.71
	6.2	Jabalpur	2.10	30.00	2.58	38.64
	6.3	Sagar	1.90	33.97	1.93	38.60
	6.4	Hoshangabad	NA	NA	2.00	24.02
	6.5	Tikamgarh	0.95	16.15	1.17	17.52
	6.6	Ratlam	1.19	15.82	1.57	16.80
	6.7	Dewas	0.94	15.00	1.03	16.54
	6.8	Katni	1.45	17.55	1.49	15.57
	6.9	Barwani	0.46	8.23	0.87	13.89
	6.10	Gwalior	0.88	8.84	0.97	13.61
	6.11	Datia	0.24	2.36	0.70	11.20
	6.12	Alirajpur	0.48	0.99	0.89	10.68
	6.13	Bhind	NA	NA	0.81	9.72
	6.14	Dhar	0.55	5.46	1.18	9.41
	6.15	Shivpuri	0.90	8.98	0.90	8.98
	6.16	Chhatarpur	1.24	11.18	1.00	8.96
7. Uttar Pradesh	7.1	Agra	1.22	17.36	1.76	25.00
	7.2	Meerut	1.09	16.16	1.57	23.43
	7.3	Lucknow	0.74	9.52	1.06	13.80
	7.4	Unnao	0.72	9.05	1.04	13.12
	7.5	Mainpuri	0.62	8.84	0.90	12.77
	7.6	Gonda	0.60	7.06	0.86	10.24
	7.7	Hapur	0.47	7.04	0.68	10.20
	7.8	Fatehpur	0.41	6.36	0.60	9.22
	7.9	Kushi Nagar	0.43	5.63	0.62	8.16
	7.10	Aligarh	0.41	5.41	0.59	7.85
	7.11	Kanpur Nagar	0.42	5.36	0.60	7.77
	7.12	Bahraich	0.36	4.61	0.51	6.69
	7.13	Etah	0.34	4.51	0.49	6.54
	7.14	Kannauj	0.31	4.18	0.44	6.07
	7.15	Farrukhabad	0.32	4.02	0.46	5.82
	7.16	Hathras	0.29	3.92	0.42	5.68

States	S. No.	Districts	2014-15		2015-16	
			A	p	A	P
	7.17	Varanasi	0.29	3.89	0.42	5.64
	7.18	Mathura	0.27	3.78	0.39	5.45
	7.19	Bulandshahr	0.25	3.69	0.36	5.35
	7.2	Firozabad	0.25	3.59	0.36	5.16
	7.21	Sitapur	0.30	3.47	0.43	5.02
	7.22	Gorakhpur	0.25	3.41	0.36	4.95
	7.23	Saharanpur	0.24	3.43	0.35	4.94
	7.24	Hardoi	0.26	2.99	0.37	4.34
8. Chhattisgarh	8.1	Kondagaon	3.07	27.21	3.22	28.54
	8.2	Raipur	2.36	24.87	2.45	25.81
	8.3	Durg	3.05	31.97	2.20	22.60
	8.4	Raigarh	1.36	18.57	1.36	18.57
	8.5	Korba	1.79	17.52	1.79	17.55
	8.6	Bilaspur	1.99	15.46	2.11	16.39
	8.7	Surguja	1.40	13.97	1.40	14.97
	8.8	Rajnandgaon	1.20	10.80	1.22	13.43
	8.9	Bemetara	1.30	13.00	1.30	13.00
	8.10	Korea	1.15	10.93	1.16	12.00
	8.11	Mungeli	0.44	10.57	0.47	11.37
	8.12	Surajpur	0.70	10.71	0.73	11.36
9. Andhra Pradesh	9.1	Chittoor	4.35	52.25	4.00	60.00
	9.2	Kurnool	3.10	23.25	2.73	20.48
	9.3	Anantapur	0.94	14.03	1.12	20.20
	9.4	Krishna	1.53	15.25	1.61	16.12
	9.5	East Godavari	1.50	19.55	0.92	13.79
	9.6	Guntur	NA	NA	1.15	13.74
10. Haryana	10.1	Sonipat	2.35	21.65	2.77	31.25
	10.2	Gurgaon	1.81	18.22	1.70	22.40
	10.3	Mewat	3.79	44.50	3.62	18.77
	10.4	Karnal	1.75	20.64	1.87	16.72
	10.5	Panipat	1.40	14.70	1.40	14.50
	10.6	Faridabad	0.93	9.11	0.82	11.00
	10.7	Rohtak	0.17	1.84	1.01	10.81
	10.8	Yamunanagar	1.48	10.20	1.60	10.50
11. Assam	11.1	Nagaon	0.74	16.91	0.94	24.77
	11.2	Chirang	0.67	9.51	0.67	17.37
	11.3	Dhubri	0.90	16.47	0.90	16.47
	11.4	Cachar	0.82	8.38	0.82	16.23
	11.5	Darrang	1.16	29.81	0.96	11.88
	11.6	Sivasagar	0.41	6.50	0.75	11.43

(*Contd.*)

States	S. No.	Districts	2014-15		2015-16	
			A	p	A	P
	11.7	Kamrup	0.67	8.87	0.67	8.87
	11.8	Barpeta	0.53	7.81	0.54	7.81
	11.9	Kamrup Metro	0.44	9.35	0.44	6.41
	11.10	Baksa	0.44	6.37	0.44	6.37
	11.11	Udalguri	0.38	9.14	0.38	6.20
12. Tamil Nadu	12.1	Warangal	6.07	91.08	3.54	53.13
	12.2	Khammam	3.45	44.90	2.19	28.49
	12.3	Nalgonda	1.20	18.00	1.20	18.00
	12.4	Rangareddi	1.57	23.55	1.12	13.79
	12.5	Medak	1.63	24.38	1.40	12.26
	12.6	Coimbatore	0.28	3.35	0.37	4.48
	12.7	Tiruvannamalai	0.35	2.47	0.44	3.62

Source: Horticulture Statistics Division, Deptt. of Agriculture & Coopn.

Table 4.39 Area and Production of Peas for Major Producing Districts

A = Area in '000 Ha
P = Production in '000 MT

States	S. No.	Districts	2014-15		2015-16	
			A	p	A	P
1. Uttar Pradesh	1.1	Jalaun	29.10	353.27	29.39	357.16
	1.2	Jhansi	27.96	329.98	28.24	333.61
	1.3	Mahoba	12.40	109.50	12.53	110.71
	1.4	Kasganj	3.98	62.55	4.02	63.24
	1.5	Etah	3.26	51.19	3.29	51.75
	1.6	Sultanpur	4.52	45.45	4.57	45.95
	1.7	Azamgarh	4.66	44.00	4.71	44.49
	1.8	Hamirpur	4.56	40.29	4.61	40.74
	1.9	Amethi	3.82	38.33	3.85	38.75
2. Madhya Pradesh	2.1	Jabalpur	19.30	114.54	17.72	105.05
	2.2	Chhindwara	5.83	81.66	5.93	82.98
	2.3	Ratlam	7.03	65.45	8.62	80.19
	2.4	Ujjain	6.99	100.64	7.34	71.18
	2.5	Narsinghpur	2.79	38.60	2.98	41.27
	2.6	Dewas	2.47	34.92	5.72	38.52
	2.7	Tikamgarh	6.38	38.25	5.53	27.65
	2.8	Gwalior	1.97	15.73	2.17	21.66
	2.9	Seoni	3.05	21.36	3.01	20.78
	2.1	Shajapur	2.42	42.59	2.20	20.46
3. Punjab	3.1	Amritsar	7.14	74.67	14.81	158.67
	3.2	Hoshiarpur	5.02	51.08	5.29	50.97
	3.3	Nawanshahr	2.01	20.39	2.31	23.13

States	S. No.	Districts	2014-15		2015-16	
			A	p	A	P
4. Himachal Pradesh	4.1	Shimla	6.50	71.67	6.51	71.70
	4.2	Mandi	3.69	48.99	3.71	49.24
	4.3	Lahul And Spiti	3.28	32.94	3.28	32.27
	4.4	Kullu	1.86	27.90	1.87	28.05
	4.5	Kinnaur	2.36	24.21	2.36	24.21
5. Jharkhand	5.1	Ranchi	0.05	0.05	2.27	33.87
	5.2	Lohardaga	0.03	0.03	1.43	21.27
	5.3	Khunti	NA	NA	1.29	19.26
	5.4	Simdega	NA	NA	1.08	16.12
	5.5	Ramgarh	0.12	11.80	0.12	11.80
6. West Bengal	6.1	Jalpaiguri	3.50	28.57	3.50	28.58
	6.2	Nadia	2.72	25.40	2.74	25.38
	6.3	24 Paraganas South	3.62	13.45	3.62	13.45
	6.4	24 Paraganas North	1.37	12.40	1.36	13.16
	6.5	Darjeeling	2.68	12.30	2.68	12.32
	6.6	Hooghly	0.73	7.10	0.73	7.10
7. Chhattisgarh	7.1	Rajnandgaon	0.91	40.95	0.93	41.63
	7.2	Surguja	1.33	12.38	1.33	13.30
	7.3	Baloda Bazar	0.05	0.22	0.65	10.07
	7.4	Surajpur	0.60	8.34	0.62	8.85
	7.5	Bemetara	0.60	6.60	0.60	6.60
	7.6	Raigarh	0.60	5.61	0.60	5.61
8. Haryana	8.1	Sonipat	2.63	18.70	2.65	17.15
	8.2	Karnal	2.41	16.00	2.70	17.05
	8.3	Kurukshetra	2.51	16.77	1.54	16.77
	8.4	Panipat	1.30	8.00	1.35	9.01
	8.5	Ambala	0.71	6.77	0.86	8.73
	8.6	Jind	0.81	7.56	0.66	6.69
	8.7	Gurgaon	0.65	5.10	0.60	5.10

Source: Horticulture Statistics Division, Deptt. of Agriculture & Coopn.

Table 4.40 Area and Production of Potato for Major Producing Districts

A = Area in '000 Ha
P = Production in '000 MT

States	S. No.	Districts	2014-15		2015-16	
			A	p	A	P
1. Uttar Pradesh	1.1	Agra	64.07	1584.95	64.07	1656.28
	1.2	Hathras	49.69	1370.82	49.69	1432.51
	1.3	Firozabad	51.24	1314.57	51.24	1373.73
	1.4	1kannauj	52.50	1126.47	52.50	1177.16

(*Contd.*)

States	S. No.	Districts	2014-15		2015-16	
			A	p	A	P
	1.5	Farrukhabad	36.24	863.87	36.24	902.75
	1.6	Aligarh	24.84	560.93	24.84	586.17
	1.7	Shahjahanpur	13.17	460.75	13.17	481.48
	1.8	Mainpuri	15.14	419.66	15.14	438.55
	1.9	Etawah	17.38	397.18	17.38	415.05
	1.10	Mathura	12.43	338.01	12.43	353.22
	1.11	Barabanki	15.62	284.11	15.62	296.89
	1.12	Budaun	14.37	261.41	14.37	273.18
	1.13	Hardoi	11.96	224.67	11.96	234.78
	1.14	Sambhal	10.74	197.08	10.74	205.95
2. Bihar	2.1	Nalanda	20.80	600.00	21.01	606.00
	2.2	Patna	12.50	360.00	12.63	363.60
	2.3	Vaishali	9.90	290.50	10.00	293.41
	2.4	Saran	10.80	280.50	10.91	283.31
	2.5	Muzaffarpur	85.50	254.00	86.36	256.54
	2.6	Pashchim	9.20	250.00	9.29	252.50
	2.7	Samastipur	9.20	250.00	9.29	252.50
	2.8	Gopalganj	9.10	242.70	9.19	245.13
	2.9	Purbi Champaran	8.90	240.90	8.99	243.31
	2.10	Madhubani	7.70	206.90	7.78	208.97
	2.11	Rohtas	8.00	206.90	8.08	208.97
	2.12	Siwan	7.70	205.90	7.78	207.96
	2.13	Bhojpur	6.30	172.70	6.36	174.43
	2.14	Bhagalpur	6.30	168.30	6.36	169.98
	2.15	Begusarai	6.00	168.00	6.06	169.68
	2.16	Darbhanga	5.80	154.90	5.86	156.45
	2.17	Saharsa	5.70	151.60	5.76	153.12
	2.18	Katihar	5.60	147.80	5.66	149.28
	2.19	Munger	5.30	142.10	5.35	143.52
3. Gujarat	3.1	Banas Kantha	52.80	1710.72	55.70	1804.68
	3.2	Aravalli	NA	NA	21.90	715.18
4. Madhya Pradesh	4.1	Indore	34.25	753.50	39.00	858.00
	4.2	Shajapur	11.08	205.20	12.10	266.20
	4.3	Ujjain	6.06	158.88	12.12	233.31
	4.4	Chhindwara	7.10	177.48	7.31	182.80
	4.5	Dewas	7.51	146.98	7.88	154.38
	4.6	Sagar	4.79	143.75	5.18	129.53
	4.7	Morena	3.92	96.20	3.93	96.34
	4.8	Gwalior	2.29	46.22	2.86	71.50
5. Punjab	5.1	Jalandhar	20.44	528.40	20.58	541.90
	5.2	Hoshiarpur	12.61	310.31	12.91	326.43

States	S. No.	Districts	2014-15		2015-16	
			A	p	A	P
	5.3	Ludhiana	10.02	256.44	10.46	274.75
	5.4	Kapurthala	9.26	235.71	9.26	240.48
	5.5	Amritsar	6.79	173.06	7.70	200.09
	5.6	Moga	6.18	157.30	6.38	167.60
6. Assam	6.1	Kokrajhar	6.06	63.09	6.51	68.83
	6.2	Sonitpur	8.86	62.14	8.29	62.99
	6.3	Barpeta	NA	NA	7.47	61.78
	6.4	Dhubri	3.81	47.14	4.19	49.38
	6.5	Nagaon	6.24	58.40	7.10	47.05
	6.6	Lakhimpur	9.62	42.57	10.43	46.50
	6.7	Dhemaji	7.59	32.48	6.07	43.14
	6.8	Udalguri	7.54	43.94	7.19	37.53
	6.9	Nalbari	4.50	29.18	4.55	28.02
	6.1	Kamrup	3.81	46.04	3.68	26.51
	6.11	Darrang	5.33	48.44	5.27	25.78
7. Haryana	7.1	Kurukshetra	8.51	206.65	8.65	216.85
	7.2	Yamunanagar	5.40	136.72	5.90	150.55
	7.3	Karnal	3.90	95.47	5.20	134.94
	7.4	Sonipat	2.45	56.46	2.71	70.60
	7.5	Ambala	2.32	54.43	2.34	54.45

Source: Horticulture Statistics Division, Deptt. of Agriculture & Coopn.

Table 4.41 Area and Production of Radish for Major Producing Districts

A = Area in '000 Ha
P = Production in '000 MT

States	S. No.	Districts	2014-15		2015-16	
			A	p	A	P
1. West Bengal	1.1	Nadia	5.53	69.17	5.33	67.26
	1.2	Murshidabad	4.64	55.65	4.45	61.30
	1.3	24 Paraganas South	3.89	49.56	3.89	49.60
	1.4	24 Paraganas North	2.08	32.52	1.97	31.00
	1.5	Coochbehar	2.32	30.50	2.32	30.54
	1.6	Medinipur West	2.04	27.86	2.05	28.01
	1.7	Maldah	2.51	27.77	2.57	27.97
	1.8	Dinajpur Uttar	2.28	24.16	2.29	24.28
	1.9	Bankura	1.74	22.87	1.75	22.95
	1.10	Birbhum	1.80	22.35	1.81	22.44

(*Contd.*)

States	S. No.	Districts	2014-15		2015-16	
			A	p	A	P
2. Haryana	2.1	Sonipat	3.40	61.80	3.75	59.35
	2.2	Yamunanagar	2.75	37.10	3.03	41.45
	2.3	Fatehabad	2.26	23.41	2.50	39.72
	2.4	Karnal	2.91	43.01	2.45	38.30
	2.5	Panipat	2.03	20.54	2.20	35.47
	2.6	Ambala	1.84	21.86	2.20	34.47
	2.7	Kurukshetra	1.71	24.80	2.46	32.47
	2.8	Gurgaon	2.04	29.40	2.18	28.31
	2.9	Faridabad	1.52	23.05	1.63	27.60
3. Punjab	3.1	Hoshiarpur	0.95	21.60	0.98	22.76
	3.2	Tarn Taran	0.74	14.97	0.78	16.61
	3.3	Amritsar	0.50	11.09	0.60	13.89
	3.4	Nawanshahr	0.60	6.93	0.61	12.09
	3.5	Gurdaspur	0.45	9.45	0.45	11.78
	3.6	Kapurthala	0.48	10.06	0.47	10.51
4. Bihar	4.1	Nalanda	9.30	15.50	9.39	15.66
	4.2	Vaishali	1.00	15.50	1.01	15.66
	4.3	Patna	0.90	14.50	0.91	14.65
	4.4	Pashchim Champaran	0.70	11.50	0.71	11.62
	4.5	Muzaffarpur	0.70	11.30	0.71	11.41
	4.6	Purbi Champaran	0.70	10.90	0.71	11.01
	4.7	Samastipur	0.70	10.80	0.71	10.91
	4.8	Katihar	0.70	10.00	0.71	10.10
	4.9	Darbhanga	0.60	9.20	0.61	9.29
	4.10	Begusarai	0.60	8.90	0.61	8.99
	4.11	Purnia	0.50	8.00	0.51	8.08
	4.12	Saran	0.50	7.50	0.51	7.58
	4.13	Bhagalpur	0.40	7.10	0.40	7.17
	4.14	Madhubani	0.40	7.00	0.40	7.07
	4.15	Saharsa	0.50	6.80	0.51	6.87
	4.16	Sitamarhi	0.40	6.50	0.40	6.57
	4.17	Siwan	0.40	6.50	0.40	6.57
	4.18	Gopalganj	0.40	6.40	0.40	6.46

States	S. No.	Districts	2014-15		2015-16	
			A	p	A	P
5. Assam	5.1	Sonitpur	0.46	8.58	1.91	21.93
	5.2	Nagaon	0.80	12.74	1.79	18.50
	5.3	Lakhimpur	0.46	6.52	1.29	17.37
	5.4	Chirang	0.68	7.44	1.67	17.02
	5.5	Kamrup	1.72	26.08	1.70	16.47
	5.6	Dhubri	1.61	25.20	1.60	16.17
	5.7	Darrang	0.55	10.23	1.51	14.49
	5.8	Barpeta	1.10	16.79	1.50	13.88
	5.9	Goalpara	0.42	7.07	1.41	13.63
6. Chhattisgarh	6.1	Kondagaon	3.67	62.33	3.87	65.73
	6.2	Korba	0.78	18.43	0.79	18.49
	6.3	Surguja	0.45	11.69	0.46	11.69
	6.4	Kanker	0.63	10.35	0.70	11.55
	6.5	Durg	0.43	5.62	0.81	9.88
	6.6	Rajnandgaon	0.56	9.44	0.58	9.78
	6.7	Raipur	0.36	8.56	0.38	9.16
	6.8	Korea	0.28	7.00	0.29	7.50
7. Madhya Pradesh	7.1	Datia	NA	NA	6.61	99.17
	7.2	Ujjain	0.86	10.63	0.90	9.67
	7.3	Indore	0.31	6.00	0.32	6.40
	7.4	Sheopur	NA	NA	0.29	4.93
	7.5	Betul	0.21	2.16	0.21	4.71
8. Uttar Pradesh	8.1	Mainpuri	0.51	13.93	0.98	26.70
	8.2	Agra	0.18	4.68	0.34	8.97
	8.3	Kanpur Dehat	0.13	3.31	0.25	6.34
	8.4	Shahjahanpur	0.13	3.10	0.26	5.95
	8.5	Kannauj	0.11	3.05	0.22	5.85
	8.6	Faizabad	0.10	2.66	0.20	5.10
	8.7	Kaushambi	0.10	2.64	0.19	5.06
	8.8	Gonda	0.10	2.43	0.20	4.66
	8.9	Kanpur Nagar	0.08	2.14	0.16	4.10
	8.1	Firozabad	0.09	2.06	0.18	3.96

(Contd.)

States	S. No.	Districts	2014-15		2015-16	
			A	p	A	P
	8.11	Varanasi	0.08	1.82	0.16	3.49
	8.12	Bulandshahr	0.07	1.81	0.14	3.48
	8.13	Chitrakoot	0.07	1.77	0.13	3.40
	8.14	Basti	0.06	1.50	0.11	2.87
	8.15	Lucknow	0.05	1.35	0.09	2.58
	8.16	Bahraich	0.06	1.27	0.11	2.43
	8.17	Kheri	0.05	1.21	0.09	2.32
	8.18	Allahabad	0.04	1.15	0.08	2.20
	8.19	Ghaziabad	0.05	1.10	0.09	2.11
	8.20	Ballia	0.04	1.04	0.07	2.00
9. Odisha	9.1	Koraput	0.74	7.68	0.75	7.84
	9.2	Mayurbhanj	0.57	7.34	0.59	7.53
	9.3	Jajapur	0.64	7.71	0.62	7.47
	9.4	Cuttack	0.67	7.61	0.65	7.38
	9.5	Bargarh	0.57	6.22	0.59	6.44
	9.6	Sambalpur	0.28	2.91	0.61	6.40
	9.7	Dhenkanal	0.52	6.50	0.50	6.29
	9.8	Sundargarh	0.57	6.10	0.59	6.26
	9.9	Ganjam	0.54	6.36	0.52	6.15
	9.1	Baleshwar	0.51	5.60	0.49	5.41
	9.11	Rayagada	0.42	4.91	0.40	4.66
	9.12	Kendujhar	0.42	4.73	0.40	4.50
	9.13	Kalahandi	0.39	4.23	0.42	4.46
	9.14	Balangir	NA	NA	0.39	4.39
	9.15	Anugul	0.38	4.03	0.40	4.23
	9.16	Kandhamal	0.31	3.93	0.32	4.12
	9.17	Nayagarh	0.37	4.10	0.35	3.88
10. Karnataka	10.1	Kolar	1.13	13.35	0.92	11.32
	10.2	Belgaum	0.56	5.88	0.56	6.26
	10.3	Chikballapur	0.50	5.52	0.37	5.99
	10.4	Mandya	0.57	7.12	0.57	5.65
	10.5	Chamarajanagar	0.29	2.90	0.34	3.40
	10.6	Chikmagalur	0.27	2.95	0.27	3.26

States	S. No.	Districts	2014-15		2015-16	
			A	p	A	P
	10.7	Haveri	0.28	2.76	0.28	2.73
	10.8	Mysore	0.22	2.07	0.24	2.31
	10.9	Gulbarga	0.24	2.45	0.24	1.93
11. Uttarakhand	11.1	Almora	0.83	15.76	0.84	15.76
	11.2	Pithoragarh	0.80	9.62	0.80	9.62
	11.3	Pauri Garhwal	0.56	3.37	0.59	4.75
	11.4	Udam Singh Nagar	0.23	4.22	0.24	4.33
	11.5	Uttar Kashi	0.60	1.84	0.46	4.25
	11.6	Chamoli	0.38	3.72	0.39	3.77

Source: Horticulture Statistics Division, Deptt. of Agriculture & Coopn.

Table 4.42 Area and Production of Sweet Potato for Major Producing Districts

A = Area in '000 Ha
P = Production in '000 MT

States	S. No.	Districts	2014-15		2015-16	
			A	p	A	P
1. Odisha	1.1	Ganjam	7.52	73.55	7.50	73.35
	1.2	Sundargarh	3.54	35.31	3.26	32.52
	1.3	Koraput	3.16	32.57	3.14	32.36
	1.4	Kendujhar	2.93	29.69	2.95	29.89
	1.5	Mayurbhanj	3.20	31.57	2.94	29.05
	1.6	Sambalpur	0.96	9.20	2.08	19.89
	1.7	Dhenkanal	2.04	17.94	2.02	17.76
	1.8	Gajapati	2.11	17.42	2.10	17.35
	1.9	Rayagada	1.92	16.84	1.90	16.68
	1.10	Malkangiri	1.70	13.39	1.69	15.84
2. Kerala	2.1	Pathanamthitta	17.52	297.00	17.70	300.00
	2.2	Alappuzha	1.51	15.54	1.53	15.70
3. West Bengal	3.1	24 Paraganas South	4.53	54.23	4.53	54.28
	3.2	Hooghly	1.82	16.60	1.82	16.63
	3.3	Birbhum	0.84	16.03	0.84	15.50
	3.4	Nadia	1.55	15.48	1.52	15.24
	3.5	Murshidabad	2.00	16.32	1.90	15.20
	3.6	Dinajpur Dakshin	1.51	15.15	1.50	15.01
	3.7	Bardhaman	1.62	14.62	1.62	13.88
	3.8	Medinipur East	1.11	13.10	1.11	13.11
	3.9	Medinipur West	1.11	11.50	1.11	11.56

(Contd.)

States	S. No.	Districts	2014-15		2015-16	
			A	p	A	P
4. Uttar Pradesh	4.1	Etah	3.36	44.41	3.39	44.89
	4.2	Budaun	1.31	17.30	1.32	17.49
	4.3	Fatehpur	1.21	15.99	1.22	16.16
	4.4	Sultanpur	1.05	13.82	1.06	13.97
	4.5	Kasganj	0.95	12.60	0.96	12.74
	4.6	Farrukhabad	0.85	11.27	0.86	11.40
	4.7	Shahjahanpur	0.78	10.36	0.79	10.48
	4.8	Hardoi	0.69	9.09	0.70	9.19
	4.9	Kanpur Nagar	0.52	6.87	0.53	6.94
	4.1	Kaushambi	0.52	6.87	0.53	6.94
	4.11	Unnao	0.50	6.66	0.51	6.73
	4.12	Mainpuri	0.50	6.62	0.51	6.69
5. Madhya Pradesh	5.1	Tikamgarh	0.53	8.42	0.54	8.13
	5.2	Chhindwara	0.17	4.20	0.18	4.58
	5.3	Katni	0.31	3.18	0.32	3.35
	5.4	Chhatarpur	0.19	2.87	0.19	2.71
	5.5	Shahdol	0.18	5.64	0.14	2.13
	5.6	Mandsaur	0.13	2.30	0.11	1.98
	5.7	Singrauli	NA	NA	0.11	1.83
6. Chhattisgarh	6.1	Surguja	0.57	5.67	0.57	5.67
	6.2	Kondagaon	0.58	5.11	0.60	5.29
	6.3	Balrampur	0.51	5.04	0.51	5.04
	6.4	Korba	0.18	3.37	0.18	3.47
	6.5	Raigarh	0.14	2.40	0.14	2.40
	6.6	Kanker	0.20	1.74	0.22	1.91
	6.7	Bilaspur	0.23	1.77	0.24	1.88
	6.8	Kabirdham	0.09	1.53	0.10	1.70
7. Karnataka	7.1	Belgaum	1.22	16.72	1.08	13.07
	7.2	Mandya	0.29	4.54	0.29	4.68
	7.3	Dakshin Kannad	0.29	3.00	0.32	3.79
	7.4	Chikmagalur	0.19	2.75	0.16	2.52
8. Assam	8.1	Nagaon	0.50	5.78	0.44	5.06
	8.2	Sonitpur	0.36	4.34	0.38	4.58
	8.3	Dhubri	1.19	3.09	0.94	2.45
	8.4	Barpeta	NA	NA	0.56	2.02
	8.5	Karbi Anglong	0.31	1.71	0.30	1.69
	8.6	Cachar	0.16	1.59	0.15	1.47
	8.7	Goalpara	0.50	2.94	0.23	1.35
	8.8	Kokrajhar	0.33	1.31	0.31	1.23
	8.9	Udalguri	0.21	1.76	0.13	1.10

Source: Horticulture Statistics Division, Deptt. of Agriculture & Coopn.

Table 4.43 Area and Production of Tapioca for Major Producing Districts

A = Area in '000 Ha
P = Production in '000 MT

States	S. No.	Districts	2014-15		2015-16	
			A	p	A	P
1. Tamil Nadu	1.1	Namakkal	17.38	632.98	22.65	837.60
	1.2	Villupuram	13.09	354.59	16.45	446.40
	1.3	Erode	4.94	195.31	7.18	283.75
	1.4	Dharmapuri	18.16	442.14	10.10	279.24
	1.5	Cuddalore	3.35	134.69	5.46	196.38
2. Kerala	2.1	Kollam	16.80	418.50	17.32	420.34
	2.2	Alappuzha	2.10	104.30	4.20	211.27
	2.3	Thiruvananthapuram	44.50	135.40	44.66	135.53
	2.4	Idukki	7.00	172.70	3.41	83.33
	2.5	Pathanamthitta	3.00	74.00	3.20	78.00
	2.6	Kannur	1.80	65.80	1.82	67.68
	2.7	Thrissur	1.70	53.00	1.50	58.00
	2.8	Kottayam	3.40	78.30	2.50	50.00
	2.9	Malappuram	2.40	37.60	2.60	41.40
3. Andhra Pradesh	3.1	East Godavari	16.55	251.74	10.50	157.44

Source: Horticulture Statistics Division, Deptt. of Agriculture & Coopn.

Table 4.44 Area and Production of Tomato for Major Producing Districts

A = Area in '000 Ha
P = Production in '000 MT

States	S. No.	Districts	2014-15		2015-16	
			A	p	A	P
1. Madhya Pradesh	1.1	Chhindwara	7.34	205.49	7.68	215.12
	1.2	Shivpuri	5.15	113.37	5.87	129.21
	1.3	Jhabua	1.81	108.72	2.14	128.40
	1.4	Shajapur	2.09	38.79	1.70	110.03
	1.5	Raisen	3.95	57.10	4.00	99.88
	1.6	Dhar	3.69	112.42	1.53	91.86
	1.7	Jabalpur	3.39	81.10	3.78	90.03
	1.8	Sagar	3.28	64.59	3.56	89.00
	1.9	Dewas	1.75	67.96	1.92	74.88
	1.10	Satna	2.56	48.62	2.94	55.92
	1.11	Hoshangabad	NA	NA	1.96	48.93
	1.12	Katni	3.02	54.66	3.07	47.00
	1.13	Vidisha	1.95	47.85	2.08	37.40

(Contd.)

States	S. No.	Districts	2014-15		2015-16	
			A	p	A	P
	1.14	Shahdol	1.43	37.21	1.43	37.18
	1.15	Gwalior	0.96	21.08	1.05	36.89
	1.16	Khargone	0.46	31.73	0.58	36.83
2. Andhra Pradesh	2.1	Chittoor	20.14	805.72	22.00	990.00
	2.2	Anantapur	9.52	380.88	13.43	537.04
3. Karnataka	3.1	Kolar	10.00	567.75	5.96	338.02
	3.2	Haveri	4.82	164.94	4.13	210.64
	3.3	Belgaum	6.22	189.95	5.85	186.78
	3.4	Chikballapur	2.70	67.85	2.80	179.85
	3.5	Mandya	3.92	169.49	3.98	177.62
	3.6	Tumkur	1.73	92.92	2.18	114.80
	3.7	Davangere	5.59	100.19	5.58	98.80
	3.8	Mysore	3.29	83.39	3.54	89.73
	3.9	Chamarajanagar	3.31	88.15	3.86	88.11
4. Telangana	4.1	Medak	7.21	144.20	14.93	408.68
	4.2	Adilabad	0.48	9.50	5.39	323.52
	4.3	Mahbubnagar	9.68	205.07	10.86	217.30
	4.4	Warangal	13.35	266.96	8.12	162.30
5. Gujarat	5.1	Banas Kantha	4.73	168.70	5.13	184.57
	5.2	Ahmadabad	5.20	145.86	5.07	143.19
	5.3	Mahesana	3.95	126.30	4.34	140.09
	5.4	Anand	3.57	111.92	3.60	113.60
	5.5	Kheda	3.43	111.13	3.16	97.81
	5.6	Chhotaudepur	NA	NA	2.85	83.45
	5.7	Valsad	1.96	48.58	2.16	53.88
	5.8	Vadodara	4.50	130.36	1.73	50.03
	5.9	Kachchh	1.78	47.35	1.83	49.24
	5.10	Sabar Kantha	1.20	36.60	1.47	45.60
6. Odisha	6.1	Kendujhar	9.12	140.81	7.40	114.26
	6.2	Khordha	6.49	95.34	6.17	90.64
	6.3	Kalahandi	5.55	76.55	5.23	72.11
	6.4	Mayurbhanj	5.26	74.27	4.93	69.54
	6.5	Kendrapara	4.51	63.26	4.48	62.74
	6.6	Balangir	4.19	57.98	4.12	56.62
	6.7	Ganjam	4.42	53.21	4.29	51.65
	6.8	Koraput	3.49	52.77	3.33	50.35
	6.9	Kandhamal	3.22	49.06	2.96	45.18
	6.10	Sundargarh	3.68	54.79	2.99	44.06
	6.11	Bhadrak	2.63	37.19	3.17	43.74
	6.12	Jagatsinghapur	3.28	44.69	3.17	43.18

States	S. No.	Districts	2014-15		2015-16	
			A	p	A	P
	6.13	Dhenkanal	2.96	42.22	2.90	41.35
	6.14	Anugul	2.81	41.49	2.77	40.86
	6.15	Cuttack	3.28	47.34	2.76	39.94
	6.16	Sonepur	2.83	38.74	2.81	38.47
7. West Bengal	7.1	Coochbehar	3.30	148.41	3.31	148.75
	7.2	24 Paraganas	4.20	112.82	4.18	134.53
	7.3	Nadia	4.73	107.99	4.80	109.14
	7.4	Murshidabad	4.78	75.31	4.87	101.56
	7.5	Alipurduar	3.38	100.31	3.39	100.52
	7.6	24 Paraganas	4.91	91.21	4.94	92.19
	7.7	Purulia	5.54	79.04	5.54	80.50
	7.8	Medinipur West	4.28	73.43	4.21	70.66
	7.9	Bankura	3.80	61.30	3.80	61.85
8. Bihar	8.1	Vaishali	3.90	92.00	3.94	92.92
	8.2	Muzaffarpur	3.70	82.80	3.74	83.63
	8.3	Nalanda	1.90	44.80	1.92	45.25
	8.4	Begusarai	1.90	43.50	1.92	43.94
	8.5	Patna	1.90	43.20	1.92	43.63
	8.6	Samastipur	1.90	41.80	1.92	42.22
	8.7	Purbi Champaran	1.80	40.80	1.82	41.21
	8.8	Bhagalpur	1.60	38.20	1.62	38.58
	8.9	Sitamarhi	1.50	33.90	1.52	34.24
	8.10	Saran	1.60	33.30	1.62	33.63
	8.11	Gopalganj	1.50	33.00	1.52	33.33
	8.12	Darbhanga	1.50	32.90	1.52	33.23
	8.13	Katihar	1.40	32.50	1.41	32.83
	8.14	Aurangabad	1.70	32.30	1.72	32.62
	8.15	Siwan	1.40	30.50	1.41	30.81
	8.16	Madhubani	1.30	29.80	1.31	30.10
	8.17	Saharsa	1.20	26.20	1.21	26.46
9. Maharashtra	9.1	Nashik	11.65	290.01	11.55	289.81
	9.2	Pune	13.40	334.76	10.30	257.50
	9.3	Nagpur	2.01	48.26	2.01	48.26
	9.4	Gadchiroli	6.43	80.37	4.20	45.90
10. Chhattisgarh	10.1	Durg	4.41	114.66	4.72	121.50
	10.2	Bilaspur	7.41	91.70	7.84	97.09
	10.3	Jashpur	5.14	78.75	5.20	79.62
	10.4	Raipur	3.55	62.62	3.70	65.23
	10.5	Bemetara	2.50	62.25	2.55	63.50
	10.6	Mungeli	1.55	50.26	1.80	58.32

(Contd.)

States	S. No.	Districts	2014-15		2015-16	
			A	p	A	P
	10.7	Raigarh	3.60	56.25	3.60	56.25
	10.8	Balod	1.73	44.63	1.91	49.09
	10.9	Janjgir-Champa	2.25	45.00	2.25	45.36
	10.10	Kondagaon	3.41	36.83	3.61	38.99

Source: Horticulture Statistics Division, Deptt. of Agriculture & Coopn.

Table 4.45 Area and Production of Watermelon for Major Producing Districts

A = Area in '000 Ha
P = Production in '000 MT

States	S. No.	Districts	2014-15		2015-16	
			A	p	A	P
1. Uttar Pradesh	1.1	Aligarh	1.18	52.97	1.19	53.55
	1.2	Firozabad	1.18	52.97	1.19	53.55
	1.3	Etah	1.11	50.00	1.12	50.55
	1.4	Farrukhabad	1.10	49.59	1.11	50.14
	1.5	Kasganj	1.09	49.01	1.10	49.54
	1.6	Unnao	0.96	43.34	0.97	43.81
	1.7	Kannauj	0.71	31.73	0.71	32.07
	1.8	Mainpuri	0.67	30.15	0.68	30.48
	1.9	Budaun	0.58	26.06	0.59	26.34
	1.10	Kanpur Nagar	0.49	21.87	0.49	22.11
2. Karnataka	2.1	Mysore	1.42	40.22	1.49	50.08
	2.2	Kolar	0.92	40.26	0.89	39.36
	2.3	Haveri	1.56	38.67	1.42	35.90
	2.4	Mandya	0.93	33.51	0.89	31.91
	2.5	Chamarajanagar	0.98	30.24	0.93	28.51
	2.6	Koppal	0.56	22.90	0.61	24.90
	2.7	Bagalkot	0.59	17.54	0.66	20.69
	2.8	Belgaum	0.52	16.10	0.55	19.19
	2.9	Chikballapur	0.26	8.72	0.23	13.00
3. West Bengal	3.1	Medinipur West	2.71	39.74	2.69	38.91
	3.2	Coochbehar	2.52	36.14	2.53	36.20
	3.3	24 Paraganas South	1.21	17.53	1.22	18.98
	3.4	Murshidabad	1.44	18.00	1.45	18.30
	3.5	Nadia	1.24	16.07	1.17	16.68
	3.6	Medinipur East	1.06	15.81	1.07	15.95
	3.7	Birbhum	1.00	15.08	1.01	15.13
	3.8	24 Paraganas North	0.65	9.85	0.66	13.82
4. Odisha	4.1	Dhenkanal	NA	NA	3.14	59.19
	4.2	Balangir	NA	NA	1.15	23.17

States	S. No.	Districts	2014-15		2015-16	
			A	p	A	P
	4.3	Deogarh	NA	NA	0.71	16.13
	4.4	Kendrapara	NA	NA	0.87	15.92
	4.5	Jajapur	NA	NA	0.61	13.17
	4.6	Sambalpur	0.22	4.33	0.63	12.16
	4.7	Sonepur	NA	NA	0.49	9.38
	4.8	Cuttack	NA	NA	0.42	7.92
	4.9	Kendujhar	NA	NA	0.30	6.05
5. Andhra Pradesh	5.1	Anantapur	2.85	85.50	3.12	93.60
	5.2	Chittoor	2.93	87.78	2.20	55.00
	5.3	Kadapa	0.37	18.65	0.36	18.00
	5.4	Prakasam	0.48	9.62	0.53	15.75
6. Tamil Nadu	6.1	Kanchipuram	1.18	44.63	1.52	57.21
	6.2	Thiruvallur	0.53	19.87	0.68	18.96
	6.3	Tiruvannamalai	0.36	13.57	0.48	18.62
	6.4	Dharmapuri	0.05	1.71	0.40	16.47
	6.5	Erode	0.14	5.13	0.35	15.40
7. Maharashtra	7.1	Raigad	2.63	83.74	2.63	83.74
	7.2	Nashik	0.45	9.90	0.45	9.90
	7.3	Aurangabad	NA	NA	0.19	8.55
	7.4	Sindhudurg	0.32	7.87	0.32	7.87
8. Madhya Pradesh	8.1	Khandwa	1.90	45.60	2.00	48.00
	8.2	Burhanpur	0.60	21.00	0.64	20.35
	8.3	Khargone	0.29	8.52	0.30	8.34
	8.4	Shahdol	0.24	15.14	0.24	7.53
	8.5	Barwani	0.07	2.52	0.22	6.03
9. Haryana	9.1	Sonipat	NA	NA	2.39	18.90
	9.2	Gurgaon	NA	NA	1.19	12.67
	9.3	Mewat	NA	NA	0.56	11.50

Source: Horticulture Statistics Division, Deptt. of Agriculture & Coopn.

Table 4.46 Area and Production of Onion for Major Producing Districts

A = Area in '000 Ha
P = Production in '000 MT

States	S. No.	Districts	2014-15		2015-16	
			A	p	A	P
1. Maharashtra	1.1	Nashik	138.78	2358.88	142.63	2424.63
	1.2	Ahmednagar	90.51	1178.94	124.93	1433.71
	1.3	Pune	48.82	684.88	56.88	841.75
	1.4	Aurangabad	8.01	136.17	23.94	556.32
	1.5	Jalgaon	13.33	197.60	31.11	529.99

(Contd.)

States	S. No.	Districts	2014-15		2015-16	
			A	p	A	P
2. Madhya Pradesh	2.1	Indore	14.35	428.53	14.74	440.29
	2.2	Sagar	7.10	213.11	11.46	332.28
	2.3	Shajapur	14.37	251.42	13.40	281.40
	2.4	Khandwa	10.15	253.63	7.18	172.32
	2.5	Ujjain	3.48	94.25	7.53	158.83
	2.6	Dewas	4.72	134.81	5.20	156.61
	2.7	Ratlam	5.39	82.20	7.64	126.51
	2.8	Shivpuri	4.90	107.76	5.70	125.29
	2.9	Agar Malwa	3.95	79.27	3.97	123.80
	2.10	Rajgarh	6.21	77.97	7.68	118.88
	2.11	Dhar	1.20	33.75	4.49	112.34
	2.12	Satna	2.81	59.58	3.60	90.77
	2.13	Khargone	1.18	29.88	3.42	82.97
	2.14	Chhindwara	2.63	65.70	3.23	80.85
3. Karnataka	3.1	Bijapur	16.67	333.36	24.89	497.82
	3.2	Bagalkot	15.42	299.79	26.64	422.32
	3.3	Chitradurga	28.91	570.05	19.19	374.59
	3.4	Dharwad	35.30	500.85	36.32	273.33
	3.5	Koppal	2.33	54.55	15.89	236.30
	3.6	Bellary	6.30	113.98	9.30	191.26
4. Rajasthan	4.1	Jodhpur	16.68	403.88	23.49	564.87
	4.2	Sikar	10.08	218.43	15.19	439.50
	4.3	Nagaur	9.27	191.89	15.82	246.27
5. Gujarat	5.1	Bhavnagar	27.10	700.54	35.50	926.55
	5.2	Amreli	1.00	25.85	4.20	107.94
6. Bihar	6.1	Nalanda	6.00	160.00	6.06	161.60
	6.2	Katihar	3.90	83.10	3.94	83.93
	6.3	Muzaffarpur	2.60	67.00	2.63	67.67
	6.4	Patna	2.60	63.00	2.63	63.63
	6.5	Pashchim Champaran	2.40	62.00	2.43	62.62
	6.6	Purbi Champaran	2.40	59.30	2.42	59.89
	6.7	Vaishali	1.80	44.00	1.82	44.44
	6.8	Purnia	1.80	42.50	1.82	42.93
	6.9	Begusarai	2.00	42.40	2.02	42.82
	6.10	Bhagalpur	1.60	40.30	1.62	40.72
	6.11	Sitamarhi	1.30	32.50	1.31	32.83
	6.12	Kishanganj	1.40	31.40	1.41	31.71
	6.13	Samastipur	1.40	30.40	1.41	30.70
	6.14	Araria	1.50	30.30	1.52	30.60

States	S. No.	Districts	2014-15		2015-16	
			A	p	A	P
	6.15	Sheikhpura	1.30	30.00	1.31	30.30
	6.16	Rohtas	1.20	27.30	1.21	27.57
	6.17	Bhojpur	1.20	26.00	1.21	26.26
	6.18	Aurangabad	1.10	25.40	1.13	25.67
7. Andhra Pradesh	7.1	Kurnool	27.55	454.54	32.18	625.78
8. Haryana	8.1	Mewat	7.52	160.38	7.40	162.63
	8.2	Ambala	2.77	61.65	4.05	93.58
	8.3	Yamunanagar	2.55	62.93	2.57	66.56
	8.4	Sonipat	1.76	44.85	2.07	49.72
	8.5	Panchkula	2.25	48.79	2.20	46.69
	8.6	Jhajjar	1.20	26.25	1.93	45.94
	8.7	Karnal	1.98	40.69	1.84	42.18
9. West Bengal	9.1	Murshidabad	3.25	47.32	11.58	208.37
	9.2	Hooghly	2.99	66.07	3.00	66.50
	9.3	Medinipur West	3.65	44.45	3.60	43.52
	9.4	24 Paraganas North	1.99	40.85	2.00	43.07
	9.5	Nadia	2.65	41.98	2.71	42.82
	8.6	Bardhaman	1.23	25.00	1.28	24.17
	8.7	Birbhum	1.98	21.75	2.10	22.11
10. Uttar Pradesh	10.1	Fatehpur	1.18	23.71	1.20	24.24
	10.2	Ghazipur	1.06	21.89	1.08	22.37
	10.3	Jaunpur	1.19	19.37	1.21	19.80
	10.4	Farrukhabad	1.14	19.29	1.17	19.72
	10.5	Kannauj	1.13	18.44	1.15	18.85
	10.6	Ballia	0.91	16.36	0.93	16.72
	10.7	Sonbhadra	0.82	16.22	0.83	16.58
	10.8	Mainpuri	0.89	14.25	0.91	14.57
	10.9	Gonda	0.86	12.85	0.88	13.13
	10.10	Hardoi	0.64	10.93	0.65	11.17
	10.11	Kanpur Nagar	0.66	10.62	0.67	10.85
	10.12	Bahraich	0.71	10.42	0.72	10.65
	10.13	Azamgarh	0.69	9.87	0.71	10.09
	10.14	Amethi	0.63	9.07	0.64	9.27
	10.15	Sultanpur	0.61	8.68	0.62	8.88
	10.16	Allahabad	0.46	8.57	0.47	8.75
	10.17	Unnao	0.48	7.82	0.49	7.99
	10.18	Pratapgarh	0.49	7.39	0.50	7.55
	10.19	Kaushambi	0.37	6.69	0.38	6.83
	10.20	Siddharth Nagar	0.39	6.49	0.40	6.63

(*Contd.*)

States	S. No.	Districts	2014-15		2015-16	
			A	p	A	P
	10.21	Mirzapur	0.35	6.43	0.35	6.57
	10.22	Varanasi	0.35	6.21	0.36	6.34
	10.23	Rae Bareli	0.31	6.12	0.32	6.25
	10.24	Bareilly	0.31	5.96	0.32	6.09
	10.25	Chandauli	0.30	5.85	0.30	5.98
	10.26	Shahjahanpur	0.31	5.76	0.32	5.89
11. Telangana	11.1	Mahbubnagar	5.28	94.99	7.67	138.12
	11.2	Medak	3.48	62.66	6.46	130.05
	11.3	Karimnagar	1.66	29.93	1.80	32.40

Source: Horticulture Statistics Division, Deptt. of Agriculture & Coopn.

Table 4.47 Export of Fresh Apple from India-Country Wise

(Qty in MT, Value in Rs. Lakhs)

Country	2014-15		2015-16		2016-17	
	Qty.	Rs. Lakhs	Qty.	Rs. Lakhs	Qty.	Rs. Lakhs
Bangladesh	13807.86	2968.66	11087.92	2473.22	12045.41	3182.29
Nepal	6172.21	2088.45	8066.6	2103.28	8713.6	2416.11
Unspecified	0	0	0	0	250	132.78
Saudi Arabia	40	31.44	0.29	0.2	51	38.96
Seychelles	0	0	0	0	5	5.09
Bhutan	0	0	0	0	20	0.7
Sri Lanka	0	0	0	0	0.06	0.32
Hong Kong	0	0	0	0	0.11	0.05
Kuwait	0	0	0	0	0.02	0.03
Singapore	22	32.28	0.08	0.09	0.02	0.02
Others	332.63	171.9	43.19	24.32	0.01	0.01
TOTAL	**20,374.70**	**5,292.73**	**19,198.08**	**4,601.11**	**21,085.23**	**5,776.36**

Source: APEDA Website

The vast production base offers India tremendous opportunities for export of Horticultural produces. Mangoes, Walnuts, Grapes, Bananas, Pomegranates account for larger portion of fruits exported from the country. Indian export basket is full with vegetables like Onion, Okra, Bitter Gourd, Green Chillies, Mushrooms and Potatoes. Though India's share in Global market is still one percent only, there is increasing acceptance of India's Horticultural produces due to concurrent developments in the area of states- of- the- art cold chain infrastructure and quality assurance measures. Onion and Garlic fresh as well as dehydrated are exported from India. The major destinations for Indian fruits and vegetables are UAE, Bangladesh, Malaysia, Netherland, Sri Lanka, Nepal, UK, Saudi Arabia, Pakistan and Qatar.

Table 4.48 Export of Fresh Bananas from India-Country Wise

(Qty in MT, Value in Rs. Lakhs)

Country	2014-15		2015-16		2016-17	
	Qty.	Rs. Lakhs	Qty.	Rs. Lakhs	Qty.	Rs. Lakhs
United Arab Emirates	19010.88	8755.03	24971.12	11728.78	24368.63	11897.81
Iran	3377.6	1246.5	20304.5	6762.92	15838.89	5505.06
Saudi Arabia	9314.43	4557.57	12676.77	6318.26	10280.67	5199.46
Oman	4926.01	2004.43	8486.03	3199.92	12937.97	4619.24
Kuwait	4684.43	2241.77	4974.11	2470.35	10463	4302.8
Nepal	8637.73	955.12	15573.4	1637.43	27406.09	2691.16
Qatar	3210.7	1671.64	2946.66	1780.95	3761	2249.54
Bahrain	1802.08	952.58	1940.23	1053.51	2610.09	1307.23
Maldives	664.28	207.02	1272.6	450.88	1518.51	469.15
Iraq	0.44	0.37	0	0	593.23	189.4
Others	7645.81	1602.74	527.68	210.04	1093.79	421.72
TOTAL	**63,274.39**	**24,194.77**	**93,673.10**	**35,613.04**	**1,10,871.87**	**38,852.57**

Source: APEDA Website

Table 4.49 Export of Oranges (Fresh/Dried) from India-Country Wise

(Qty in MT, Value in Rs. Lakhs)

Country	2014-15		2015-16		2016-17	
	Qty.	Rs. Lakhs	Qty.	Rs. Lakhs	Qty.	Rs. Lakhs
Bangladesh	14210.73	2156.01	27901.56	5924.94	35365.36	8777.32
Nepal	2845.36	661.37	6364.93	1076.84	9950.22	1921.79
Egypt Arab Republic	0	0	0	0	1320	445.95
United Arab Emirates	32.05	16.04	94.33	52.79	551.91	267.5
Oman	5.6	1.73	30.15	10.64	92.86	49.37
Russia	0	0	53.2	21.99	118.64	43.6
Qatar	4.82	3.98	2.95	2.21	51.33	41.47
Saudi Arabia	42.77	34.47	0	0	27	11.53
Kuwait	1.8	0.96	0.87	0.6	12.82	10.94
Bhutan	82.05	12.78	38	14.19	22.5	5.94
Others	6.27	6.66	38.55	25.61	26.76	9.3
TOTAL	**17,231.45**	**3,194.00**	**34,524.54**	**7,129.81**	**47,539.40**	**11,584.71**

Source: APEDA Website

Table 4.50 Export of Grapes from India-Country Wise

(Qty in MT, Value in Rs. Lakhs)

Country	2014-15		2015-16		2016-17	
	Qty.	Rs. Lakhs	Qty.	Rs. Lakhs	Qty.	Rs. Lakhs
Bangladesh	14210.73	2456.01	27901.56	5924.94	35365.36	8777.32
Nepal	2845.36	661.37	6364.93	1076.84	9950.22	1921.79
Egypt Arab Republic	0	0	0	0	1320	445.95
United Arab Emirates	32.05	16.04	94.33	52.79	551.91	267.5
Oman	5.6	1.73	30.15	10.64	92.86	49.37
Russia	0	0	53.2	21.99	118.64	43.6
Qatar	4.82	3.98	2.95	2.21	51.33	41.47
Saudi Arabia	42.77	34.47	0	0	27	11.53
Kuwait	1.8	0.96	0.87	0.6	12.82	10.94
Bhutan	82.05	12.78	38	14.19	22.5	5.94
Others	6.27	6.66	38.55	25.61	26.76	9.3
TOTAL	**17,231.45**	**3,194.00**	**34,524.54**	**7,129.81**	**47,539.40**	**11,584.71**

Source: APEDA Website

Table 4.51 Export of Guavas (Fresh/Dried) from India-Country Wise

(Qty in MT, Value in Rs. Lakhs)

Country	2014-15		2015-16		2016-17	
	Qty.	Rs. Lakhs	Qty.	Rs. Lakhs	Qty.	Rs. Lakhs
Sri Lanka	51	20.36	0	0	290.04	129.65
Nepal	87.4	35.03	528.2	154.1	403.85	112
United Arab Emirates	167.45	77.44	215.48	97.51	108.82	59.15
Saudi Arabia	290.22	92.73	899.08	316.13	136.4	52
Kuwait	9.66	5.93	10.22	7.91	98.22	39.88
Netherland	35	20.88	0.2	0.13	54.12	32.51
Qatar	9.96	6.22	27.97	19.79	38.83	28.68
Singapore	2.35	1.49	5.14	4.04	43.85	25.21
Malaysia	35.4	14.94	4.36	2.46	52.62	21.77
Bahrain	48.34	24.08	43.9	23.4	37.09	20.28
Others	171.09	72.77	170.01	86.92	161.55	97.84
TOTAL	**907.87**	**371.87**	**1,904.56**	**712.39**	**1,425.39**	**618.97**

Source: APEDA Website

Table 4.52 Export of Litchi from India-Country Wise

(Qty in MT, Value in Rs. Lakhs)

Country	2014-15		2015-16		2016-17	
	Qty.	Rs. Lakhs	Qty.	Rs. Lakhs	Qty.	Rs. Lakhs
United Arab Emirates	0.03	0.01	0.03	0.02	20.37	52.68
Nepal	44.6	17.73	9.4	3.84	53.96	30.77
Thailand	0	0	0	0	50	20.93
France	0	0	0	0	0.46	1.2
Kuwait	0.09	0.09	0.12	0.3	0.33	0.46
Canada	0	0	0	0	0.2	0.18
Qatar	0	0	0	0	0.05	0.06
Bangladesh	915	163.8	0	0	0	0
United Kingdom	0.08	0.2	0	0	0	0
Bahrain	0	0	0.2	0.16	0	0
Others	1.63	33.35	0.12	0.05	0	0
TOTAL	**961.43**	**215.18**	**9.87**	**4.37**	**125.37**	**106.28**

Source: APEDA Website

Table 4.53 Export of Fresh Mangoes from India-Country Wise

(Qty in MT, Value in Rs. Lakhs)

Country	2014-15		2015-16		2016-17	
	Qty.	Rs. Lakhs	Qty.	Rs. Lakhs	Qty.	Rs. Lakhs
United Arab Emirates	29231.9	21497.85	19973.6	19199.34	28751.23	24792.7
United Kingdom	329.8	606.38	1496.28	3205.72	3034.41	4963.36
Saudi Arabia	2171.49	1428.59	1399.08	1675.19	2394.1	2447.68
Qatar	998.1	810.81	1016.25	1023.29	2273.47	2163.99
Kuwait	787.28	1238.18	748.35	1298.33	1104.27	1918.11
United States	271.79	687.72	266.45	734.3	637.38	1601.04
Nepal	3574.93	695.42	8273.99	1393.53	9423.61	1597.47
Bahrain	658.71	505.36	747.79	633.01	1088.15	979.72
Singapore	562.95	588.31	579.96	626.27	829.07	869.78
Oman	605.2	469.27	426.84	412.5	901.92	824.96
Others	3806.16	1725.76	1400.41	1508.56	2739.65	2395.73
TOTAL	**42,998.31**	**30,253.65**	**36,329.00**	**31,710.04**	**53,177.26**	**44,554.54**

Source: APEDA Website

Table 4.54 Export of Papaya (Fresh/Dried) from India-Country

(Qty in MT, Value in Rs. Lakhs)

Country	2014-15		2015-16		2016-17	
	Qty.	Rs. Lakhs	Qty.	Rs. Lakhs	Qty.	Rs. Lakhs
United Arab Emirates	4378.81	1402.14	6009.22	2630.19	5145.19	2005.53
Saudi Arabia	2406.25	870.57	2421.26	1067.28	2215.02	1047.11
Netherland	691.79	473.85	895	685.63	608	463.07
Qatar	520.1	232.41	661.47	352.02	801.16	430.78
Kuwait	429.45	177.53	482.81	231.44	793.08	341.28
Bahrain	744.95	246.69	832.18	299.27	695.48	273.71
Oman	193.97	73.96	316.46	142.11	365.54	186.67
Nepal	1768.08	163.48	1311.39	125.35	1817.51	179.39
United States	1	2.49	14.8	5.26	11.63	46.01
Germany	127.05	85.38	104.63	69.66	48.55	34.66
Others	223.37	97.91	132.02	69.38	272.14	116.04
TOTAL	**11,484.82**	**3,826.41**	**13,181.24**	**5,677.59**	**12,773.30**	**5,124.25**

Source: APEDA Website

Table 4.55 Export of Pineapples (Fresh/Dried) from India-Country Wise

(Qty in MT, Value in Rs. Lakhs)

Country	2014-15		2015-16		2016-17	
	Qty.	Rs. Lakhs	Qty.	Rs. Lakhs	Qty.	Rs. Lakhs
Qatar	563.37	365.2	726.6	507.49	976.48	670.73
Saudi Arabia	679.17	406.83	271.61	161.46	545.57	382.43
United Arab Emirates	60.89	27.39	212.73	110.35	538.48	328.74
Maldives	270.06	162.74	366.99	368.75	453.44	316.61
Nepal	1440.71	209.3	1540.96	242.8	1828.18	311.25
Oman	192.97	131.01	294.52	203.48	313.32	224.27
Russia	248.87	205.16	187.54	149.96	187.29	156.12
Bahrain	191.05	116.41	199.76	106.84	223.57	120.77
Italy	0.2	1.09	0	0	1.63	51.15
France	0	0	0.28	0.17	1.05	25.68
Others	104.24	39.79	390.03	125.68	139.38	28.56
TOTAL	**3,751.53**	**1,664.92**	**4,191.02**	**1,976.98**	**5,208.39**	**2,616.31**

Source: APEDA Website

Table 4.56 Export of Fresh Sapota from India-Country Wise

Country	2014-15		2015-16		2016-17	
	Qty.	Rs. Lakhs	Qty.	Rs. Lakhs	Qty.	Rs. Lakhs
United Arab Emirates	535.43	351.02	442.95	477.53	756.68	430.5
Bahrain	357.33	170.07	204.38	189.67	260.9	143.76
Oman	45.18	20.82	61.05	40.63	107.91	64.93
Qatar	75.08	50.21	65.32	60.86	121.59	51.91
Saudi Arabia	50.46	39.06	50.44	43.69	129.11	47.03
Canada	117.89	62	35	41.16	93.24	43.29
United Kingdom	72.91	65.05	47.28	56.18	40.08	34.2
Singapore	16.74	9.48	18.71	17.8	25.45	17.18
Kuwait	22.64	11.94	23.88	18.37	25.1	13.67
United States	32.8	21.37	13.35	11.25	40.98	3.82
Others	45.65	17.07	10.28	10.02	5.91	4.37
TOTAL	**1,372.11**	**818.09**	**972.64**	**967.16**	**1,606.95**	**854.66**

Source: APEDA Website

Table 4.57 Export of Walnuts from India-Country Wise

(Qty in MT, Value in Rs. Lakhs)

Country	2014-15		2015-16		2016-17	
	Qty.	Rs. Lakhs	Qty.	Rs. Lakhs	Qty.	Rs. Lakhs
France	187	724.25	487	908.64	533.5	893.7
United Kingdom	392.5	1941.69	452.01	1693.91	261.88	811.91
United Arab Emirates	71.25	13.47	0.19	0.36	346.38	665.68
Netherland	254.66	1419.47	424.5	1310.5	156	568.96
Nepal	97.65	274.23	217.37	420.89	246.95	522
Germany	108.57	570.89	511.56	2040.31	141.56	467.32
Sweden	37	189.12	51.5	311.91	52	252.34
Belgium	40	201.93	105.5	398.18	71.5	237.09
Egypt Arab Republic	241	1631.15	252.6	1267.69	46	223.06
Spain	31.5	193.96	238	781.25	84	191.65
Others	1204.74	6485.07	551.49	2657.92	251.44	693.52
TOTAL	**2,665.87**	**13,645.23**	**3,291.72**	**11,791.56**	**2,191.21**	**5,527.23**

Source: APEDA Website

Table 4.58 Export of Cabbage& Lettuce (Fresh/Chilled) from India-Country Wise

(Qty in MT, Value in Rs. Lakhs)

Country	2014-15		2015-16		2016-17	
	Qty.	Rs. Lakhs	Qty.	Rs. Lakhs	Qty.	Rs. Lakhs
Malaysia	0	0	0	0	149	19.15
Seychelles	0	0	0	0	80	12.29
United Arab Emirates	36	6.6	161.87	31.98	32.2	5.36
Nepal	0.1	0.01	1	0.16	32.55	4.18
United Kingdom	0	0	0	0	2.05	1.65
Bangladesh	0	0	0	0	2.15	1.59
Kuwait	3.42	1.31	0	0	0	0
Canada	0.1	0.01	0	0	0	0
Singapore	0.47	0.21	0	0	0	0
Pakistan	152.86	28.99	0	0	0	0
Others	102.87	34.33	181.54	32.7	0	0
TOTAL	**295.82**	**71.46**	**344.41**	**64.84**	**297.95**	**44.22**

Source: APEDA Website

Table 4.59 Exports of Cauliflowers & Headed Broccoli (Fresh/Chilled) from India-Country Wise

(Qty in MT, Value in Rs. Lakhs)

Country	2014-15		2015-16		2016-17	
	Qty.	Rs. Lakhs	Qty.	Rs. Lakhs	Qty.	Rs. Lakhs
Nepal	18.47	2.46	0.6	0.14	1108.26	119.41
Russia	0	0	0	0	188.4	77.1
Saudi Arabia	8	3.41	0	0	38	14.43
Maldives	14.53	9.57	10.53	7.14	11.25	7.25
Singapore	8.72	3.67	5.87	3.21	6.63	5.04
Qatar	0	0	0.03	0.02	4.5	3.3
United Arab Emirates	6	2.7	8.06	3.02	8.9	3.17
Seychelles	0	0	0	0	15	2.25
Sudan	0	0	0	0	1.44	0.93
Canada	0	0	0	0	0.52	0.39
Others	6.17	1.17	4.02	8.53	0.2	0.13
TOTAL	**61.89**	**22.98**	**29.11**	**22.06**	**1,383.10**	**233.4**

Source: APEDA Website

Table 4.60 Export of Fresh Onions from India-Country Wise

(Qty in MT, Value in Rs. Lakhs)

Country	2014-15		2015-16		2016-17	
	Qty.	Rs. Lakhs	Qty.	Rs. Lakhs	Qty.	Rs. Lakhs
Bangladesh	456734.5	77964.61	251384.13	62384.22	847023.86	97611.19
Malaysia	215194.39	41621.67	244272.7	58641.91	374480.61	49523.6
United Arab Emirates	131630.19	24772.72	169675.48	32728.03	303706.31	40098.17
Sri Lanka	131646.45	25839.13	199150.44	44926.73	207526.57	26425.33
Nepal	70543.31	13940.27	70024.86	17337.88	128499.11	15821.58
Indonesia	45629.04	5788.98	11046	1754.65	82161.82	11451.29
Philippines	0	0	29617	7565.86	49395	10245.49
Qatar	25414.31	5303.42	33573.87	6910.57	67987.4	9145.81
Kuwait	24874.08	5026.99	36402.36	6651.77	65280.98	8910.67
Saudi Arabia	13692.64	3071.93	17668.34	3092.03	58874.29	7494.5
Others	122743.69	26724.42	138430.11	32747.4	230821.16	33922.46
TOTAL	**12,38,102.60**	**2,30,054.14**	**12,01,245.29**	**2,74,741.05**	**24,15,757.11**	**3,10,650.09**

Source: APEDA Website

Table 4.61 Export of Peas from India-Country Wise

(Qty in MT, Value in Rs. Lakhs)

Country	2014-15		2015-16		2016-17	
	Qty.	Rs. Lakhs	Qty.	Rs. Lakhs	Qty.	Rs. Lakhs
United Kingdom	96.38	162.8	125.26	222.2	163.69	251.59
Nepal	129.03	45.7	257.59	152.71	815.11	245.81
Russia	0	0	0	0	41.83	21.29
Australia	2.71	1.7	43.7	38.85	8.27	14.67
Ireland	2.8	7.56	10.42	11.86	9.54	13.16
United Arab Emirates	116.96	56.9	36.72	29.81	11.46	6.68
Bahrain	140.5	54.58	4	3.16	5.74	5.26
Saudi Arabia	155	117.35	48	42.09	13.25	4.58
Hong Kong	1.6	0.86	8.08	1.25	1.2	1.6
Belgium	0	0	0	0	1	0.9
Others	18720.9	4208.49	26.37	17.3	3.49	2.77
TOTAL	**19,365.88**	**4,655.94**	**560.14**	**519.23**	**1,074.58**	**568.31**

Source: APEDA Website

Table 4.62 Export of Tomatoes (Fresh/Chilled) from India-Country Wise

(Qty in MT, Value in Rs. Lakhs)

Country	2014-15		2015-16		2016-17	
	Qty.	Rs. Lakhs	Qty.	Rs. Lakhs	Qty.	Rs. Lakhs
Pakistan	174501.05	33836.82	118360.27	25232.82	190739.69	36845.47
United Arab Emirates	23173.56	7285.53	30571.76	10486.4	30789.61	8948.2
Bangladesh	13997.32	2114.34	2183.91	455.98	28927.18	6641.75
Nepal	4416.08	480.27	4983.88	558.66	14512.33	1611.13
Maldives	730.29	360.33	1023.72	641.28	1161.76	485.69
Oman	848.94	271.17	808.5	236.05	659.23	194.6
Qatar	47.87	10.46	13.37	5.04	158.82	44.67
Saudi Arabia	130.7	46.47	296.2	99.33	69.61	21.29
Kuwait	9.1	4.41	94.82	32.3	22.61	7.47
Bhutan	0	0	128.94	14.32	155.56	4.99
Others	144.42	51.54	39.22	10.73	2.09	0.78
TOTAL	**2,17,999.33**	**44,461.34**	**1,58,504.59**	**37,772.91**	**2,67,198.49**	**54,806.04**

Source: APEDA Website

Table 4.63 Export of Potatoes (Other than seeds) from India-Country Wise

(Qty in MT, Value in Rs. Lakhs)

Country	2014-15		2015-16		2016-17	
	Qty.	Rs. Lakhs	Qty.	Rs. Lakhs	Qty.	Rs. Lakhs
Nepal	162925.95	32828.07	178931.41	16723.21	298739.09	47225.17
Sri Lanka	34230.71	9112.96	27279.06	5550.96	32436.53	5714.35
Oman	7154.9	1670.09	9482.12	1848.32	16838.34	3276.78
Mauritius	7532	2050.38	10230	2185.48	7308	1770.61
Kuwait	7403.04	1605.03	6664	1114.07	7144.83	1478.88
Malaysia	1759.05	425.35	3401	722.02	6734.83	1360.09
Maldives	3838.5	1132.76	5060.65	1399.8	5309.36	1272.05
United Arab Emirates	9892.34	1875.54	6203.6	1122.66	4932.58	900.24
Seychelles	1160	320.85	1555	317.82	1436	323.91
Hong Kong	1337.46	327.75	1194	298.36	1192	303.99
Others	136697.67	32157.15	3134.65	733.03	2175.28	430.42
Total	**3,73,931.62**	**83,505.93**	**2,53,135.49**	**32,015.73**	**3,84,246.84**	**64,056.49**

Source: APEDA Website

Table 4.64 Export of Sweet Potatoes from India-Country Wise

(Qty in MT, Value in Rs. Lakhs)

Country	2014-15		2015-16		2016-17	
	Qty.	Rs. Lakhs	Qty.	Rs. Lakhs	Qty.	Rs. Lakhs
United Arab Emirates	344.16	102.2	293.5	96.09	246.46	72.68
Maldives	43.95	10.43	120.05	118.79	58.5	17.32
Nepal	8.4	1.07	83.76	8.81	126.94	15.84
Japan	0	0	1.54	0.85	1.62	1.03
Bahrain	5.55	1.56	4.07	1.15	0.12	0.23
Singapore	0	0	0	0	0.4	0.23
Qatar	1.51	0.76	0.46	0.2	0.27	0.12
Norway	0	0	0	0	0.1	0.1
Saudi Arabia	0	0	0	0	0.02	0.02
Italy	0	0	0.99	0.21	0.05	0.01
Others	23.46	9.78	0.06	0.75	0	0
TOTAL	**427.03**	**125.8**	**504.43**	**226.85**	**434.48**	**107.58**

Source: APEDA Website

Table 4.65 Export of Floriculture from India-Country Wise

(Qty in MT, Value in Rs. Lakhs)

Country	2014-15		2015-16		2016-17	
	Qty.	Rs. Lakhs	Qty.	Rs. Lakhs	Qty.	Rs. Lakhs
United States	5490	9813.61	5185.09	9679.11	3764.94	9917.01
United Kingdom	2557.24	5947.56	2197.52	5597	2473.07	6878.26
Germany	2240.04	5547.21	2336.2	5692.88	2443.71	6250.06
Netherland	2060.74	5125.1	1883.9	5567.55	1811.07	5803.41
United Arab Emirates	1582.65	2204.17	1499.63	2699.31	1441.43	3444.53
Canada	856.16	1538.45	945.88	1736.13	748.53	1793.18
Singapore	916.94	1067.92	1092.47	1337.17	1313.94	1610.41
Italy	561.65	1207.84	444.91	1135.73	555.1	1609.93
Japan	608.91	1467.35	421.97	1596.52	365.66	1481.75
Australia	474.73	1458.86	380.96	1330.67	285.39	1381.21
Others	5598.18	10699.16	6130.06	11569.93	6883.27	14704.21
TOTAL	**22,947.24**	**46,077.23**	**22,518.59**	**47,942.00**	**22,086.11**	**54,873.96**

Source: APEDA Website

Table 4.66 Import of Fresh Fruits and Vegetables in India-Country Wise

(Qty in MT, Value in Rs. Lakhs)

Country	2014-15		2015-16		2016-17	
	Qty.	Rs. Lakhs	Qty.	Rs. Lakhs	Qty.	Rs. Lakhs
China P Rp	71340.18	45132.39	64183.28	37400.05	199548.98	106975.95
Pakistan	135646.96	65166.37	116451.17	58673.39	165544.45	84893.78
Afghanistan	31387.54	82532.54	0	0	26476.1	80565.49
United States	66035.19	54079.53	104751.18	83520.84	76732.1	74770.67
Iraq	145756.66	36136.34	144674.19	37353.75	146921.92	39158.98
Chile	48431.05	33813.36	21674.18	16747.04	35769.99	27001.13
Sri Lanka	41037.97	57805.94	31740.7	52274.26	13683.6	24428.72
New Zealand	15073.63	14193.83	19015.61	18720.5	20566.6	19738.06
Egypt Arab Republic	44374.52	14395.29	64055.65	19471.94	49661.42	18408.01
Iran	32564.55	11590.64	37975.51	14463.19	42832.85	17680.67
Others	148875.05	126617.8	137622.19	113445.04	126844.67	96084.43
Total	**7,80,523.30**	**5,41,464.03**	**7,42,143.66**	**4,52,070.00**	**9,04,582.68**	**5,89,705.89**

Source: APEDA Website: Accessed on 15.06.2017

Table 4.67 Import of Processed Fruits & Vegetables in India-Country Wise

(Qty in MT, Value in Rs. Lakhs)

Country	2014-15		2015-16		2016-17	
	Qty.	Rs. Lakhs	Qty.	Rs. Lakhs	Qty.	Rs. Lakhs
Canada	2195137.87	659433.98	2510531.51	931525.09	2402515.04	778312.12
Australia	332562.75	115855.39	912311.92	415081.46	1174676.89	620870.94
Myanmar	625793.27	347283.99	507561.42	376054.76	466576.83	400905.15
United States	315612.62	109247.81	264733.95	108788.73	324183.96	132528.43
Russia	236893.08	63526.33	503490.34	142832.12	392273.5	121618.01
Lithuania	0	0	93260.7	20549.18	284964.39	65545.6
Tanzania Republic	61288.22	31201.62	49624.32	27658.17	103041.01	64031.52
France	576.68	2284.74	121242.61	29671.1	216829.33	56184.67
China P Rp	82563.24	53869.83	83870.95	52033.7	81424.61	55536.07
Ukraine	62067.98	15588.49	89248.36	21649.51	158322.07	40920.31
Others	135466.78	102555.95	235749.66	150587.89	357893.89	174133.96
TOTAL	**40,47,962.49**	**15,00,848.13**	**53,71,625.74**	**22,76,431.71**	**59,62,701.52**	**25,10,586.78**

Source: APEDA Website

Among imports, Cashewnut in shell are top of the list which is exported after shelling. India imports Almonds, Pistachios, Dates, Raisins, Dried Apricot, and Dried Fig in significant quantities for domestic consumption. Import of fresh fruits like Apples, Pears, and Grapes have started three years back due to initiative taken by growers of USA and other fruit producing countries. Fresh fruits are imported during the months when the fruits are not available from domestic crops. Import of fresh fruits has been growing steadily. However, the volumes are still very small keeping in view the large size of Indian markets.

India does not import prepared food in large quantities, the reason being low labour cost and processing cost in India. Some products like Cucumber and Gherkins are imported into India for re-export but not for domestic consumption.

Table 4.68 Import of Flowers in India-Country Wise

(Qty in MT, Value in Rs. Lakhs)

Country	2014-15		2015-16		2016-17	
	Qty.	**Rs. Lakhs**	**Qty.**	**Rs. Lakhs**	**Qty.**	**Rs. Lakhs**
Thailand	1904.92	14595.8	1717.3	17739.58	1773.79	17541.47
Egypt Arab Republic	10791	10096.66	10364	9391.22	9184	8580.49
Chile	141.31	10461.25	130.51	10667.69	138.68	8091.93
Netherland	1365.33	3901.56	1378.94	4942.25	1962.21	7194.12
United States	539.42	7581.96	850.66	7392.43	1279.28	6562.27
China	681.66	3876.65	666.29	6561.5	648.97	6322.95
Italy	914.17	3860.67	944.69	3098.79	1346.24	4248.62
Taiwan	25.44	3758.94	32.37	3863.61	21.41	4148.73
New Zealand	1137.81	1962.98	1425.6	2760.92	1855.27	3049.98
Korea Republic	77.45	1650.37	74.8	2322.03	43.39	2349.48
Others	1350.4	10727.05	1511.74	13003.53	1370.7	10624.18
TOTAL	**18,928.91**	**72,473.89**	**19,096.90**	**81,743.55**	**19,623.94**	**78,714.22**

Source: APEDA Website

India exports 57% of fresh fruits and 36% of processed fruits and vegetables. Among different fruits, Banana and Mango account for more than half of total production, whereas Banana alone contributes about 33%. Among different vegetables, the share of Potato is at the peak contributing 28% of overall production followed by Tomato-11%, Onion-10%, and Brinjal-8%. In case of plantation crops, Coconut contributes 92%. The share of Chilli among different spices and condiments is about 45% together with Garlic.

REFERENCES

APEDA Website

Fruits and Vegetables Availability Maps of India, Ministry of Food Processing Industries

Horticulture at a Glance 2017

Horticulture Statistics Division, Department of Agriculture, Cooperation & Farmers Welfare, Govt. of India

Indian Horticulture Database- 2014

Agricultural Statistics at a Glance, 2014, Directorate of Economics and Statistics, Ministry of Agriculture, Govt. of India. Website: http://www.dacnet.nic.in/eands.

FAOSTAT Website (http://faostat3.fao.org/home/E) accessed on 3 July 2015.

Singh, J. (2017). Fundamentals of Horticulture. Kalyani Publishers, Ludhiana

OUTCOMES ASSESSMENT

PART A

Answer the following questions (True or False).

1. Among fruits, banana and mango account for more than 1/3 of total production. (True/False)
2. India exports tomatoes to Pakistan. (True/False)
3. India does not import flowers from the Netherlands and New Zealand. (True/False)
4. India imports prepared food in large quantities. (True/False)
5. India is the largest producer, consumer and exporter of spices in the world. (True/False)

PART B

Answer the following questions.

1. India exports —% of fresh fruits.
2. What are important fruits exported from India?
3. Name two countries importing sweet potatoes from India in remarkable quantities.
4. Does UAE import fresh sapota from India?
5. Whether Nepal is an important importer of fresh and dried pineapples from India?

PART C

Write a brief note on each of the following.

1. Fruit growing regions in India.
2. Area and production of fruits and vegetables.
3. Export of Flowers from India.
4. Export of potatoes from India.
5. Leading fruit producing states of India.

Chapter 5

Nursery Techniques and Their Management

5.1 INTRODUCTION

The importance of the best quality planting material as an initial investment is a well-realized factor for persons engaged in horticulture field. So, nurseries have great demand for the production of plants, bulbs, rhizomes, suckers and grafts. But in general, good quality and assured planting material at reasonable price is not available many a times. Nursery is the basic need of horticulture. Plant propagation techniques and practices is the core of horticulture nurseries. The planting materials for horticultural plantations are raised from seeds and vegetative parts. The difficulty of procuring tree seeds and their rising cost makes it necessary to find means to increase seedling survival and growth.

Raising a nursery needs a great skill. Nursery business has become an agro-business for the unemployed youths. Generally, various commercial crop growers require a good quality saplings or grafts of genuine type. So, nursery raising is quite a remunerative enterprise.

Nursery is a special place where seeds are sown, plants are propagated, grown, nurtured to usable size for planting or for sale. Nurseries provide the necessary control of moisture, light, soil and predators and allow production of healthy and hardy seedlings.

5.2 REQUIREMENT OF NURSERY RAISING

- To get higher return.
- Many small seeded vegetables respond well to nursery raising followed by transplanting e.g., brinjal, chillies, tomato, cabbage, cauliflower, knol-khol, onion etc.
- Some species are not annual good seed bearer, but need to be planted annually. So, to meet the need of seedlings of such species, nursery is important.
- Slow growing species need nursery raising to avoid competition.
- Roadside and urban plantation always need a nursery for raising planting materials.
- The best method of introduction of exotic crops and their cultivars is only by nursery raising.
- Planting of nursery-grown plants is the surest method of artificially regenerating poor and barren sites.
- Casualty replacement is only possible by the plants grown in nursery.

5.3 ADVANTAGE OF NURSERY RAISING

- It becomes very easy and convenient on the part of the nursery man to look after baby seedlings which are tender enough against different biotic and abiotic stresses.
- As a small patch of land is selected for nursery raising, favourable growing condition can be provided easily.
- As the nursery area is small and compact, it is easy and convenient to manipulate growing condition.
- Nursery crops can be protected from heavy rain, too hot sun, low temperature etc.
- A large number of seedlings can be obtained per unit area.
- As a separate area is taken for nursery and one is waiting for about 30-45 days, by that time one can prepare the main field for transplanting of seedlings raised from that nursery.
- An early crop at least a month earlier can be produced to fetch good return.
- There are some crops of vegetables and flowers whose seeds are expensive; so, raising of seedlings through nursery is economical.
- In case of hybrid seeds which are costly in nature are always sown for raising seedlings.
- Nursery raising ensures proper utilization of labour, water, nutrients etc.
- By providing extra care we can raise off-season crops.
- A large number of uniform seedlings can be obtained easily from unit land area.
- Desirable plant types can be obtained from the nursery.
- Plants can be available cheaply and easily from the nursery.
- Due to better germination, seed requirement is reduced.
- Proper utilization of land is possible.
- Desirable healthy seedling plants can be obtained.
- Eliminate problems of difficult soils.
- Easy weed control.
- Reduced crop management cost.

5.4 CLASSIFICATION OF NURSERY

Nurseries are of various types. Selection and type of nursery depend upon market scenario and demand from surrounding areas. They can be classified as:

5.4.1 Based on Duration of Use

Temporary Nursery

It can be constructed either in open sky or under a tree. There is no need of protection wall. It is required for a short period. This type of nursery is developed only to fulfill the requirement of the season or a targeted project. The nurseries for production of seedlings of transplanted vegetables and flower crops are of temporary nature. Likewise, temporary arrangement for growing forest seedlings for planting in a particular area can also be done in temporary nursery. It is maintained for supplying stock for a short period after which it is abandoned.

Main features of this nursery are:

- Normally, it is constructed in the plantation area.
- It is suitable for hilly regions.
- Constructed for a short period of time and smaller is size.
- Manuring is not necessary.
- Mostly located near/inside the planting area and which is appropriate chiefly for casualty replacement.
- Elaborate soil preparation is not necessary.
- Cost of transportation of seedlings to the planting sites is less.
- Availability of different species of seedlings for mixed crops.
- Gap between lifting the stock from the nursery and actual planting is less.
- Special supervision is not required.

Advantages:

- Usually constructed in newly cleared sites fairly rich in humus and so, manuring is not required.
- Eucalyptus nurseries are an exception to the general rules.
- Minimum trouble with the weeds, destructive insects and diseases.
- Enables raising of species in their optimum altitudinal zone in hills.
- Cheap transport of planting stock without any serious damage or shock.

Disadvantages:

- Comparatively costly.
- Difficult to supervise.
- Proper supervision is not possible as it is made out of way places.
- Due to lack of irrigation facility, the growth of seedlings is, usually slow and heavy mortality.

Permanent Nursery

It needs a permanent protection wall and overhead covering to protect from sun, rain, temperature, cold, wind etc. This type of the nursery is placed permanently so as to produce plants continuously. These nurseries have all the permanent features. The permanent nursery has permanent mother plants. The work goes on continuously all the year round in this nursery. It is maintained for supplying nursery plants for a long time on a permanent basis. It is intended to meet the requirements of one or more ranges and it is relatively larger in extent.

Main features:

- Fit for large and intensive work and intensively managed.
- Established whereas all the facilities are available, i.e., easy supervision, communication facilities, labour, etc.

- Intensive manuring and soil working are done in perpetuity.
- Used for large scale afforestation works, or distribution to the villagers under community and private forestry programme.
- A large labour forces, tools and equipments are available.
- Original cost of formation is high but is cheaper in the long run.
- Regular skilled supervision is done.

Advantages:

- Varieties of planting stocks supply; such as root- shoot cuttings, grafted plants, layering, budding, poly pot seedlings, etc.
- Duration of service life is long and production cost is reasonable.
- Meet the requirement of more ranges.
- Supervision cost is low and can be easily supervised.
- Easy transport of nursery stocks due to nearness of roads.
- Plants are raised year after year for a long time on same site.

Disadvantages:

- Transportation of seedlings is difficult and costlier.
- Establishment cost is high.
- Manuring of beds annually and intensive soil working is essential.
- Requires large labour forces throughout the year which is difficult to be available in agricultural seasons

5.4.2 Based on Type of Production and Plants Propagated

Fruit Crops Nurseries

In this nursery seedlings and grafts of fruit crops are developed. Fruit nurseries are essential for production of grafts and rootstocks of pomological crops.

Forest Plant Nurseries

The seedlings of plants useful for aforestation like pine, oak, teak, eucalyptus, casuarinas are prepared and sold. Forest plants are essential for synthesis of gums, honey, timber and fuel. They are important in afforestation programmes.

Medicinal Plant Nurseries

It ensures production and sell of medicinal plants. Many wild plants have medicinal values and are exploited for this. Nurseries play an important role in preservation of such species; they also act as a resource for many pharmaceutical industries.

Ornamental Plant Nurseries

The seedlings of flowering plants like gerbera, carnation, petunia, salvia, rose, chrysanthemum, coleus, aster, and dianthus are developed in these nurseries. Ornamental and floricultural crops require special practices regarding their propagation and early growth.

Plantation Crop Nurseries
In this nursery different plantation crops are propagated.

Vegetable Crops Nurseries
In this type of nursery seedlings of cauliflower, cabbage, brinjal, tomato, onion etc. are raised up to proper age for transplanting. Seedlings are produced from seeds which are very expensive.

Hi-Tech Nurseries
Tissue culture and other modern techniques are used in such nurseries for raising of seedlings in controlled climate. Hi-Tech nursery fulfils the sudden increase in demand of certain commercial plants.

Miscellaneous Nurseries
In such type of nurseries plants with great economic value, rare and medicinal, herbal plants are propagated. In this nursery plants like geranium, rose, calendula, and marigold are propagated.

5.4.3 Based on Sale

Retail Nurseries
They sell all planting materials to general public. Retail nurseries raise a variety of plants for sale to the general public.

Wholesale Nurseries
They sell only to other nurseries and landscape gardeners. Wholesale nurseries usually grow a limited range of plants on large scale for large clients or retail nurseries.

Private Nurseries
They supply the needs of institutions and private estates. Private nurseries grow plants exclusively for a single client. These nurseries are owned or contracted by the client itself.

Mail Order Nurseries
Mail order nurseries are privately owned, retail or wholesale nurseries, which provide customers a facility of shipment right up to the door step.

5.4.4 Based on Ownership

Public Nursery
It is owned by state or central government. Nurseries are owned by some department of the government like Forest Department, Social Forestry Department, Agriculture Department, etc.

Private Nursery
It is owned by general public. Nursery business is owned by an individual.

Communal Nursery
Owned by a group of people of specific community.

Cooperative Nurseries

Nurseries are owned and managed by a group of individuals like Self Help Groups, etc.

Assisted Nurseries

Nurseries are developed under various schemes of Horticultural Board, Employment Guarantee Schemes, etc.

5.4.5 Based on Irrigation Facility

Dry Nursery

It is a nursery that is maintained without any irrigation or other artificial watering.

Wet Nursery

It is a nursery that is maintained by irrigation or other artificial watering during the dry periods.

5.4.6 Based on Size of Seedling

Seedling Nursery

A nursery which has only seedling beds, i.e., in which seedlings only are raised, no transplanting being done is called seedling nursery.

Transplant Nursery

A nursery which has only transplant beds, in which seedlings are transplanted for preparation for forest planting is called transplant nursery.

5.4.7 Based on Space

Field Nurseries

These nurseries mainly produce ornamental shrubs, fruit trees and perennial flowering plants.

Container Nurseries

Plants are grown in containers. Some containers are pot-in-pot where plant and containers are placed in permanent ground containers called socket pots. These are usually used to grow trees or large shrubs. Smaller plants and shrubs are grown in pots above the ground.

Greenhouse Nursery

It uses a combination of growing media to grow plants. In these greenhouses aeration and drainage are important considerations.

5.5 PLANNING OF A NURSERY

- One has to decide which type of nursery is to be started.
- Duration and type of plants propagated should be finalized.
- Accordingly, a neat plan of layout of the nursery is prepared.
- In the proposed plan details of beds, store house, path, office, compost pit, water source, packing and forwarding room etc. should be included.

5.6 ESTABLISHMENT AND MANAGEMENT OF NURSERY

Different components of a nursery are:

- Selection of site and its location.
- Seeds and sowing.
- After-care or management of nursery.

5.6.1 Selection of Site for Nursery and Its Location

Site is the basic requirement of a nursery. Site is a place upon which one can produce seedlings of plants.

- Site should be selected as per requirement.
- Site selection inevitably requires suitable adjustments.
- Proximity to (center of) future out-planting areas.
- Accessibility.
- Proximity to services.
- It should not be nearer to a full-grown tree or tall building.
- In permanent nursery there should be a watch post.
- Area should be adequately fenced against human beings, cattle and both pet and wild animals.
- Good water supply (accessibility, quantity and of reasonable quality).
- Reasonably flat topography or raised area providing natural drainage.
- No flooding risks.
- No erosion risks.
- Space for future expansion.
- It should be well connected with road for efficient transportation.
- Should be near habitat.
- Climate should be suitable.
- Neither shady nor exposed area.
- Site should receive sufficient sunlight.
- Soil condition should be suitable.
- Good transport facility.
- Selected site should be away from industries.
- Site selected should be free from water stagnation.

5.6.2 Seeds and Sowing

Seed is a matured ovule containing a seed coat and when put under favourable condition and substrates gives rise to a tiny plant. Seed is the basic input. A good seed in good soil reaps a bumper harvest. Seeds must be sound, healthy, high yielding and true to the type. Qualitative and quantitative food can essentially be produced from healthy plants which in turn are produced only when their seedlings/saplings are vigorous and healthy.

Sources of Seeds

Seeds should be purchased from reliable sources like National Seeds Corporation, State Agriculture University, State Seed Corporation, Central University, seed farms and nurseries of government etc. Seeds can also be purchased from private seed shops by thorough analysis of tag and label attached to the seed packets.

Types of Seed Treatment

Seeds sometimes carry pathogens and/or insects on its seed coat which later on cause damage to the seedlings. Hence seeds should be treated with appropriate chemicals against any surface borne diseases and insect pests. Seed treatment is done in three different ways, like seed disinfection, disinfestations and protection.

Disinfection of Seeds

- Disinfection is done to eliminate seed borne microorganisms.
- It can be done by hot water treatment, application of formaldehyde solution and by aerated steam treatment
- Both hot water and aerated steam treatment is known as thermotherapy.

i. Hot water treatment

- The dry seeds are soaked in boiling water at 49-57 °C for 15-30 minutes depending upon the tolerance level of seed.
- Hot water treatment effectively controls black leg and black rot disease of cabbage and *Alternaria* blight of cole crops and onion.
- To prevent any injury to the embryo the temperature of water and time of treatment should be monitored properly.
- Soon after treatment, seed should be cooled down in dry places and spread thinly under shade.
- Low vigour seeds should never be treated.

ii. Seed treatment using Formaldehyde solution

- Formalin solution @1.5–2% is prepared and seeds are treated with it for 15 to 30 minutes depending upon quality of seeds.
- After completion of soaking, seeds should be kept dry by spreading thinly under shade
- Seeds of low vigour and weak, should not be treated

iii. Aerated steam treatment of seeds

- This method is another alternative method of thermotherapy.
- In a special type of machine both steam and air are mixed together.
- The seeds are treated in this machine at a temperature of 46 to 57 °C for a time period of 10 to 30 minutes depending upon the organism to be killed and tolerance limit of the seed to this treatment.
- In this method injury caused to seeds is limited.

Disinfestations

- Chemicals those are used to eliminate the surface borne insects and pests are known as disinfectants.

- Calcium hypochlorite is an effective and most suitable disinfectant.
- A solution containing 2% calcium hypochlorite is used for seed disinfestations which is prepared by dissolving 10 g of calcium hypochlorite in 140 ml distilled water and kept for half an hour.
- The seeds are treated with calcium hypochlorite to eliminate surface borne pathogen.
- It is used for treating seeds for 5 to 30 minutes depending the tolerance limit of the seed.

Seed Protection

- The chemicals those protect the seeds from microbial infection present in the soil are known as seed protectants.
- Agrosan GN, Thiram, Captan, Bavistin, Vitavax etc. are examples of protectants.
- The protectants are used both for seed treatment as well as soil application in form of drenching.
- In general, the protectants protect the seeds from soil dwelling pathogenic fungi.
- The chemicals are used at the rate of 2.0 to 3.0 g per kg of seeds.

Methods of Seed Treatments

Usually seed treatment is done in three methods like dry dressing, dusting and slurry method.

Dry Dressing Method

- In dry method of treatment, seeds are taken in a rotating drum.
- Required quantity of the chemicals in powder form is added to the drum.
- The drum is rotated vigorously until the chemicals are uniformly adhered over the surface of seed.

Dusting Method

- In case of dusting method, the seeds are spread uniformly over the clean floor or tarpaulin.
- The chemicals in form of powder are dusted by the help of a duster over the seed.

Slurry Method

In this case a suspension of chemicals is prepared.

- The seeds are immersed in the suspension for 5 to 30 minutes.
- Sometimes seeds are treated using insecticides or combination of insecticide and fungicide together.
- During mixed chemical treatment care should be taken to see the compatibility of chemicals.

Soil Treatment

- The nursery soil may contain noxious weed seeds, nematodes, pathogen like fungi, bacteria etc.

- The noxious weed seeds may be in the nursery soil or may be carried by the farm yard manure
- They germinate in the nursery beds and compete with the seedlings for solar energy, nutrition, space etc.
- Root knot nematode cause knots in the roots of many vegetables.
- The micro-organisms cause various diseases in the crops.
- Fungi like *Pythium, Rhizoctonia, Fusarium* and *Phytophthora* cause damping off disease of the seedlings in the nursery condition.
- Bacteria cause black rot in cabbage, bacterial wilt in solanaceous vegetable crops.
- To get rid of these harmful elements, treatment of soil is necessary.
- The contamination of weed seeds can be solved by application of herbicides.
- Soil treatment can be done either through solar heat or fumigation using heat or chemicals.

Heat Treatment

i. Moist heat treatment

- Treating the nursery soil at high temperature to kill all possible microbes is termed as soil sterilization.
- As sterilization kills both harmful and beneficial microorganisms, it is better to pasteurize the soil at a lower temperature.
- A low temperature of 60 °C for 30 minutes is more desirable because it kills many pathogens and the beneficial organisms are escaped which is termed as soil pasteurization.
- Moist heat is used for the treatment of soil.
- Soil can be treated through injecting moist heat or through perforated pipes placed at 15 to 20 cm below the surface of covered soil.
- The soil to be treated should be only moist not wet.
- Moist heat at 82 °C temperature is injected for a period of 30 minutes.
- Electric heat pasteurizers can pasteurize the soil up to a depth of 40 cm × 40 cm × 40 cm.
- Large size hot air ovens may be employed for the purpose.

ii. Solarization

- During hot summer months of April-May, when the temperature is 40-45 °C, the soil is ploughed, heaped and moistened with water.
- A wide polythene sheet of 200-gauge thickness is spread over it for 4-5 weeks.
- The margin of polythene sheet is covered with mud to arrest the heated air inside the polythene sheet.
- After 5-6 weeks, the polythene sheet is removed and beds are prepared for sowing of seeds.

Treatment with Chemicals

Soil can be treated by chemicals through vapour or solution form. The process for soil treatment through vapour is termed as fumigation whereas in case of solution form of chemical it is termed as soil drenching.

i. Fumigation

Use of Formaldehyde for fumigation:

Formaldehyde is a very effective fungicide and can kill weed seeds. Formalin content of formaldehyde is 37 to 40%. Formalin solution is prepared by using 3.8 litres of formaldehyde dissolved in 190 litres of water and applied to the soil @ 21–42 litres/m^2 nursery area to make wet.

- Fumigation using formaldehyde should be done at least 2-3 weeks before sowing of seeds.
- Solution should saturate the soil up to 15-20 cm.
- The treated soil is covered immediately with polythene sheet of 200-gauge thickness with closing of sides properly by mud plaster for 24 hours.
- After 24 hours the polythene sheet is removed.
- The treated soil is left exposed for about 12 to 15 days for complete escape of odour of the chemical.
- Repeated turning of soil is required for the purpose.
- After complete drying of soil and escape of gas, beds are prepared for sowing of seeds.
- Other chemicals used for soil fumigation are vapam, chloropicrin and methyl bromide.

ii. Soil drenching

- Soil can also be treated by fungicides and insecticides.
- Fungicides like captan, thiram, ceresan, bavistin etc. are applied @ 5-6 g/m^2 area.
- It may be mixed in the soil or can be made solution and the soil is drenched.
- Soil should be kept saturated by the drenching solution up to 5 to 6 cm
- The insects damage the young seedlings after germination, so to control the persistent insects and pests on the surface of the seed insecticides are used to treat the soil.
- Insecticides like furadon, thimet, phorate, heptachlor are applied @ 4-5 g/m^2 area.
- The chemicals should be mixed well with the nursery soil up to a depth of 15 to 20 cm and then seeds are sown.

Nursery Bed

It is a special piece of land, normally of varying sizes (big or small) prepared for raising seedlings of certain crops which normally would not be sown directly into the field before they are transplanted. It acts as a temporary home for young plants where they are carefully reared until they are fit to be planted in the main field.

Preparation of Nursery Beds

- Nursery bed is made free from weeds, stumps, clods, debris, stones etc.
- Then it is cultivated to a fine tilth.
- Beds are prepared having a width of 0.75 to 1.0 m and length as per availability of land.

- But for convenience, the length of bed should be 4 to 5 m.
- In between two beds 60cm spacing is provided for irrigation/drainage channel.

Types of Nursery Beds

Nursery beds are prepared into three different types depending upon the season. They are:

Flat Beds

Flat bed type nursery beds are prepared during summer and spring season and where the soil is sandy and sandy loam in texture because there is no or less chance of accumulation of rain water. The entire field is cultivated thoroughly to a fine tilth. Farm yard manure or compost is applied @ 10-15 kg/m^2 area and mixed well in nursery soil. Beds are prepared with narrow bunds. In between two beds 30-45 cm width channel is prepared for irrigation and drainage. It is very simple and easy for preparation.

Raised Beds

These types of beds are made during rainy season because there is chance of stagnation of rain water inviting damping off disease. The beds are raised 10-15 cm height from the ground level. Between the beds 50–60 cm drainage channel is prepared which also helps during cultural practices.

Sunken Beds

This type of nursery beds is prepared during winter season. Beds are prepared 10-15cm down from the soil surface. It helps in passing the cold air through the surface of the soil without hitting young seedlings.

Method of Sowing

Generally, two methods are followed for sowing in nursery beds. They are:

Broadcasting

- Seeds are broadcasted over the entire nursery bed.
- Seeds those are bold in nature can be broadcasted.
- In case of seeds of very small in size and slippery in nature, they are mixed with equal amount of dry sand during sowing for proper and uniform distribution in the nursery beds.
- After sowing seeds are covered with loose soil or powdered farm yard manure or compost or sand or very light hoeing is done followed by careful leveling without disturbing the position of seeds.

Advantage:

- Easy and labour saving process.

Disadvantage:

- No symmetry is maintained.
- Seed rate is high.
- Expensive.

- Needs thinning to maintain optimum plant population.
- Due to crowded nature, difficulty in getting uniform seedlings.

Line Sowing

- Lines are drawn at 5cm apart in the properly leveled beds.
- Seeds are sown at a distance of 1 to 2 cm from each other along the lines uniformly.
- Seeds are placed at a depth of 0.3 to 0.5 cm.
- In case of very small and slippery seeds equal quantity of dry river sand is mixed with the seeds for uniform distribution.
- Seeds are covered with loose friable soil, sand or powdered farm yard manure or compost.
- The lines are very gently pressed to enable the seeds to come in direct contact of soil or soil mixture.

Advantages:

- All the seedlings get uniform light.
- Less competition for nutrients.
- Intercultural operations become easy.
- Checks damping off.
- Uprooting of seedlings become easier.

Disadvantage:

- Time consuming and expensive process.

5.6.3 After Care or Nursery Management Practices

Nursery plants require due care and attention after being emerged from the seeds or have been raised from other sources like rootstock or through tissue culture technique. Generally, they are grown in the open field under the protection of Mother Nature where, they should be able to face the local environment. It is the duty and main objective of a commercial nursery grower to supply the nursery plants with suitable conditions necessary for their development and growth. This is the major work of management in the nursery which includes all such operations right from the emergence of young plantlet till they are fully grown-up or are ready for uprooting and transplanting in the main fields. Nursery facilities such as irrigation, fence, shade, drainage, raised benches, potting media, storage and hardening beds are important factors in quality seedling production.

- Soon after sowing of the seeds, nursery bed is covered with mulching material and water is added to it by sprinkling through rose can. Mulching is a practice of covering soil surface with some extraneous materials like paddy straw, dry grasses, waste residues, polythene sheets, thin gunny bags, newspapers etc. to conserve soil moisture, regulate soil temperature, suppress weed growth, prevent attack by birds, reduce damage by splash etc.
- Mulching is done up to a thickness of 5 cm.

- Soon after sprouting of seeds the mulching material should be removed as the sprouts require plenty of sun shine.
- Watering through rose can should be given twice a day in the morning and evening till germination of seeds.
- Subsequent watering is given once a day.
- Maintain optimum moisture level for uniform germination of seeds.
- Use sprinkler system of irrigation using a rose can rather than flooding.
- Take necessary steps for drainage of excess water to avoid damping off disease.
- Apply pesticide and/or fungicide as a prophylactic spray as the insect-pests or microbes may harbour in the mulch materials.
- Regular weeding should be done to check the harbouring of insect-pest and disease-causing organism.
- Provide agro shade net, gunny bag, thatch or locally available grasses to protect seedlings from hot and dry summer.
- If polythene sheet is used as mulch, remove polythene sheet during day time everyday till germination.
- Use protective chemicals against insects, pests and diseases.
- Thin out the excess, weak, diseased and insect damaged seedlings to maintain uniformity, facilitating root aeration, better development of seedling, reducing competition between seedlings for light and nutrition.
- If required, pricking may be taken up when the seedlings are large enough to handle.
- Pricking is the process of transferring young seedlings from primary bed to secondary bed or seed pan or tray which enables fast and vigorous growth and minimizes the time taken for transplanting.
- At least 10 to 15 days before transplanting the seedlings are harden which can be done by withholding irrigation or removing the shade or thatch, fully exposing the seedlings to bright sunlight to be fit for the main field reducing percent of mortality.
- So due to adequate temperature, light and ample soil moisture seedlings perform better in field condition.

5.6.4 Management Practices in Large Size Commercial Nurseries

Potting the Seedling

Before planting of sapling in the pots, the pots should be filled up with proper potting mixture. Now-a-days different size of earthen pots or plastic containers are used for propagation. For filling of pots loamy soil, sand and compost can be used in 1:1:1 proportion. Sprouted cuttings, bulbs, corms or polythene-bag-grown plants can be transferred in earthen pots for further growth. All the necessary precautions are taken before filling the pots and planting of sapling in it.

Manuring and Irrigation

Generally sufficient quantity of nutrients is not available in the soil used for seedbed. Hence, well rotten F.Y.M/compost and leaf mould is added to soil. Rooted cuttings, layers or grafted plants till they are transferred to the permanent location, require fertilizers. Addition of fertilizers will give healthy and vigorous plants with good root & shoot system. It is recommended that each nursery bed of 10 m × 10 m area should be given 300 g of ammonium sulphate, 500 g of single super phosphate and 100 g of muriate of potash. Irrigation either in the nursery beds or watering the pots is an important operation. For potted plants hand watering is done and for beds low pressure irrigation by hose pipe is usually given. Heavy irrigation should be avoided.

Plant Protection Measures

Adoption of plant protection measures, well in advance and in a planned manner is necessary for the efficient raising of nursery plants. For better protection from pest and diseases regular observation is essential.

Disease control in seedbed: The major disease of nursery stage plant is damping off. For its control good sanitation conditions are necessary. Preventive measures like treatment with 50% ethyl alcohol, 0.2% calcium hypochloride and 0.01% mercury chloride is done. These treatments are given for 5 to 30 minutes. Other diseases like rust, powdery mildew, leaf spot, bacterial blight and yellow vein mosaic are also observed. For control of these diseases Bordeaux mixture, Carbendazim, Redomil can be used. *Tricoderma viridi*, a bio-fungicide can also be tried.

Weed Control

Weeds compete with plants for food, space and other essentials. So timely control of weeds is necessary. For weed control weeding, use of cover crops, mulching, use of chemicals (weedicides) are practised. Pre-emergence weedicides like Basaline or post-emergence weedicide like 2, 4-D and Roundup are useful.

Measures Against Heat and Cold

The younger seedling is susceptible to strong sun and low temperature. For protection from strong sun, shading with the help of timber framework of 1-metre height may be used. Net house and green house structures can also be used.

Packing of Nursery Plants

Packing is the method or way in which the young plants are tied or kept together till they are transplanted. So, they have to be packed in such a way that they do not lose their turgidity and are able to establish themselves on the new site. At the same time, good packing ensures their success on transplanting. For packing baskets, wooden boxes, plastic bags are used. In some parts of the country banana leaves are also used for packing the plants with their earth ball. This is useful for local transportation.

Sale Management

In general, the main demand for nursery plants is during rainy season. A proper strategy should be followed for sale of nursery plants. For that advertisement in local daily newspapers, posters, hand bills, catalogue and appointment of commission agents can be followed.

Management of Mother Plants

Role of mother plants is very primary and important. The fate of nursery depends on quality and truthfulness of mother plants. A good nursery entrepreneur does not depend on others for procurement of mother plants. Mother plants are required for both stock and scion. Mother plants should be selected on the basis of its genetic traits and other factors like availability and adaptation in the growing environment. Care of mother plants is necessary so as to get good quality propagules and scion. The mother plant should be properly labeled, certified, manured pruned, etc. In general, care of the mother plant should be properly monitored. New mother plants may also be purchased.

Propagation

It is an important technique of multiplying useful plants. We can propagate plants by sexual or by asexual means. Both the methods have certain advantages and disadvantages. So, we must select proper method according to our needs and situation. Seed propagation is the simple way of multiplying plants but for plants which do not set fertile seeds we must practice vegetative types of propagation. Vegetative methods of propagation are cutting, layering, grafting, budding and tissue culture. For propagation, medium is an essential factor. We should use proper medium for particular method of propagation and for particular plant.

5.7 OTHER REQUIREMENTS IN A NURSERY

Growing Media

Different rooting media are used for rooting of cuttings as well as sprouting of seeds. Characteristics of a good rooting media are:

- It should be firm enough to hold the seeds or cuttings in position.
- It should be light and porous for easy drainage of excess water.
- It should have good water holding capacity.
- It should not contaminate the seeds or seedlings.
- It should not change its form to any treatment.
- It should be neutral in reaction.
- The volume must be fairly constant when it is dry or wet
- The media should be free from weed seeds, pathogens, termites, nematodes etc.
- The media should be capable or suitable for getting sterilized without any ill effects

5.7.1 Different Commonly Used Growing Media

Sand (River sand)

It is the resultant of continuous weathering of parent rocks and minerals, sterile in nature, contain no organic matter and nutrient. The usual size of sand is from 0.05 to 2.0 mm. Sand is generally used in plant propagation media. The sand used in plastering is very much suitable for rooting of cuttings. The sand should be heated or fumigated before being used as media. Generally, sand does not contain any mineral nutrients and has no buffering capacity.

Soil

It is composed of sand, silt and clay in varying proportion. It is easily available, most common and cheapest growing media used in nursery. For raising of seedlings there is need of light soil for easy uprooting of seedlings but in case of transportation of seedling to a distant place as in case of fruits, plantation and ornamental plants heavy soil is required. The soil contains both organic and inorganic matters. The organic part is the residues of living and dead parts of plants, animals, and microbes. The liquid part of the soil is the soil solution containing water, dissolved minerals as well as O_2 and CO_2. The gaseous portion of the soil is important to keep the balance of air and water in proper and desired condition. The texture of the soil depends on the relative proportions of sand, silt and clay. Depending on their proportions, soils are classified as sandy, loamy sand, sandy loam, silt loam, clay loam and clayey soils. The soil structure refers to the arrangement of their particles in the soil mass. The nursery soil must have a good texture and structure.

FYM (Compost)

It is a well decomposed and rotten product of different farm waste materials of agri-horti sector. It contains organic matter. When added to the soil it changes the structure of the soil and plant grow healthy and vigorous. Compost contains the dung which might have dangerous weed seeds, if not properly decomposed.

Sphagnum Moss

It is a swampy land grass growing naturally in damp humid forest lands. The grass is dehydrated and shredded. It is sterile, light in weight and has a very high-water holding capacity It absorbs and retain water up to 10–20 times of its original weight. Sphagnum is acidic in nature and pH of the moss is 3.5–4.0. It is used as a rooting medium due to its high-water absorbing capacity. It also contains a fungistatic substance which is useful to inhibit damping off. Moss is soaked in solution containing fungicide and is impregnated with nutrient solution before being used for propagation. It is used for air layering in woody perennials like pomegranate and figs.

Perlite

It is a volcanic crude ore. After processing and heating to a high temperature of 760 °C, it is grey white in colour, silicaceous, sponge like and very light in weight. Water holding capacity is 3-4 times of its original weight. It is nearer to neutral in reaction i.e. 6–8. It is devoid of any mineral nutrition. It has no buffering reaction and it contains no mineral nutrients.

Vermiculite

It is the mica ore prepared after processing at 1090 °C and expands significantly when heated. It is very light in weight, spongy in nature, neutral in reaction, good buffering capacity, insoluble in water, good cation exchange capacity. It can hold nutrients in reserve and release slowly. Chemically it is hydrated Mg-Al-Fe-silicate. Vermiculite is available in 4 grades, out of which the Horticultural Grade No. 2 should be used for rooting and No. 4 for seed germination.

Peat

Decomposed aquatic marshy vegetation and sediments of water bodies constitute the peat. Depending upon the source of its origin it is divided into 3 types.

Peat Moss

Prepared from sphagnum and other mosses. Water holding capacity is 15%, acidity 3.2-4.5 and contain little amount of nitrogen.

Reed Sedge

Derived from grasses, swampy plants. Water holding capacity is 10 times more and pH 4-7.5.

Peat Humus

Water holding capacity very low. pH is 2-3.5.

Saw Dust

It is the waste from wood shavings and timber industry. Water holding capacity is good but it takes more time for decomposition and is attacked by fungus. It is a by-product or waste material from saw mills. The quantity and quality depend on the parent wood material. It is mixed while preparing media.

Leaf Mould

Prepared by decomposition of leaves of garden plants. Sometimes it is mixed with cow dung and urine. It is used in pot mixture. It may contain nematode, fungi etc. It is prepared by using fallen leaves of various tree species available locally, e.g., Oak, Silver oak, Maple, *Azadirachta*, *Ficus*, etc. It is prepared by stacking a few layers of leaves, then covering them with a thin layer of soil and cow dung slurry. Some live culture of decomposing organisms is added to hasten the process of decomposition. The medium is ready for use after about 12 to 18 months of decomposition.

Grain Husk

Several types of husks are available; paddy husk is one of the important wastages from rice mills. It is light in weight and cheaply available. It is suitable for mixing with other types of media.

Coco Peat

Coco peat, cow dung etc. are also used as media. A mixture of few media is always preferred and used in commercial nurseries. Many a times soil is one of the main parts for mixtures. Media must be selected on the basis of the availability, cost, ease in handling etc. The media should be procured and stored and kept ready for use in nursery.

5.8 A WELL-DEVELOPED COMMERCIAL NURSERY MAY CONTAIN THE FOLLOWING STRUCTURES FOR CARRYING OUT DIFFERENT OPERATIONS

5.8.1 Green House

Initial cost of the structure is high. Green house is most sophisticated. Inside the structure environment control is maintained either by manual or automatic. The structure is prepared by using wood or metal frames to which wood or metal sash bars are fitted. Cladding materials like plastic films, glass, polythene, fibre glass etc. are fixed as top. The structure controls temperature and ensures enough sunlight necessary for raising of seedlings, rooting of cuttings and proper graft union. Glass covered greenhouse proves better than plastic covered one. Glass panes are fixed to the frame through use of putty. Preferably translucent glass panes are used for availing uniform diffused light. There is enough arrangement for ventilation, temperature control and humidity inside the structure. The thermostatic device regulates a micro environment inside. In winter season to keep the structure warm there is provision of heating arrangement through blowers. Similarly, for escape of hot air there is arrangement of perforated polythene tubes of 30 to 60 cm long all along the length as well as hanging from upper cover. Size of the holes is 5 to 7.5cm. During summer forced air ventilation system holds good whereas in winter loss of heat is prevented by stretching a double layer polythene sheet. In some cases, there are movable screens between crop and roof top and walls. Large size evaporative cooling fan and pad system of cooling is installed to control temperature in summer. Crops like tomato, capsicum, cucumber, brinjal, mint, flowering plants like chrysanthemum, rose, carnation etc. are grown successfully. Night temperature of 13 °C to 15 °C is maintained inside the greenhouse which is favourable for most of the crops. Due to investment of large amount of money it is beyond the capacity of small and medium farmers to build this structure. So low cost structure like plastic green house, poly tunnel house can be prepared. The greenhouse can be constructed in very less space or up to a space of 100 acre or more.

5.8.2 Lath House

The structure is useful for protecting young tender plants from high temperature and high light intensity. It is constructed by stretching with agro shade net of different shading intensities. Inside the structure water consumption rate is less. The base of structure is almost similar to green house. Wood or iron poles are erected over which criss-cross arms are fixed. Over the structure agro shade nets are properly stretched.

5.8.3 Hot Beds

These are used during winter season for hardening the young plants before transplanting in the main field. The hot beds should face towards south or south-east direction for exposure of enough sunshine. Structurally it is a box of size 1.8 × 0.9 sq. m. The cedar wood/cement brick/stone can be used for preparing the box. Backside of the box is 30–45 cm above the ground and 15–30 cm in the front giving a slope appearance. During winter season beds are protected from cold wave by hanging old mats, carpet mats, or straw mats. The propagating materials in hot bed are heated by electric heating cables, steam pipes, hot air fumes or by fermenting the manures. But plastic covered soil, heating electric cables give most satisfactory result. In case of using manure method two parts of fresh horse manure and 1 part of straw with sufficient moisture is fermented for 10–15 days which liberate steam and heat the medium.

5.8.4 Cold Frame

Construction of cold frame is same as that of hot bed. It is useful during summer. Crops like celery, lettuce, radish, beet can be grown up to maturity but it should be protected from heavy wind. To cool the medium, it may be allowed to moist the bed. It utilizes heat received from sun through transparent covering.

5.8.5 Seed Bed/Planting Bed

It is used for raising of seedlings from seeds or rooting in cuttings. Length of bed may be 3–4 m and width may be 1–1.5 m. Cuttings/seedlings can be planted in 25 × 10 sq. cm polybags of 100-micron thickness arranged in the bed.

5.8.6 Net House

This structure is enclosed by agro-shade net or any other woven material. It allows sufficient sunlight, moisture, air as per the specification. It provides the required amount of shade.

5.8.7 Mother Plant/Scion Plant

It is the plant from which required planting/scion materials are taken for multiplication of cultivars through vegetative means. These plants are planted in separate blocks. The plants should be free from disease and pest, uniform in growth, of high production capacity and known pedigree, and commercially acceptable. Plant propagation techniques and practices is the core of horticulture nurseries. The planting materials for horticultural plantations are raised from seeds and vegetative parts. Role of mother plants is very primary and important. A good nursery entrepreneur does not depend on others for procurement of mother plants.

5.8.8 Packing Yard

All the sold plants or seedlings are packed here and the yard is located nearer to the store/office house.

5.8.9 Pot House

As the name indicates plastic pots, earthen pots or cement pots or poly bags of different size are stored in this house.

5.8.10 Compost Pit

Different waste materials from the nursery are converted to useful compost by fermentation method. The compost pit looks ugly that is why it is located in one corner of the nursery under the shade of the tree.

REFERENCES

Kumar N 2018. *Introduction to Horticulture*. Oxford and IBH Publishing Co. Pvt. Ltd., New Delhi.

Singh J 2018. *Basic Horticulture*. Kalyani Publishers, Ludhiana.

e-resources:

Abbas, J. (2016). Nursery Management and Certification System. Junaid Dairy Farm, PMAS Arid University, Rawalpindi, Pakistan.

https://www.researchgate.net management in horticultural crops for the officials/growers/farmers/students, CHES, IIHR, Chettalli, Karnataka)

Nursery management of horticultural plants: https://www.scribd.com

West Bengal Forest and Biodiversity Conservation Project, Central Nursery Manual Volume II, West Bengal Forest and Biodiversity Conservation Project, Directorate of Forests, Government of West Bengal.

OUTCOMES ASSESSMENT

PART A

Answer the following questions (True or False).

1. Sand is a common growing medium. (True/False)
2. Raising a nursery does not need any skill. (True/False)
3. Damping off is a common problem in nursery. (True/False)
4. Depending upon the source of origin peat is divided into 2 types. (True/False)
5. Mulching is done up to a thickness of 5 cm. (True/False)

PART B

Answer the following questions.

1. Why line sowing is preferred to broadcasting?
2. What is perlite?
3. Why should hot beds face towards south or south-east direction?
4. — and — are two growing media.
5. Define fumigation.

PART C

Write a brief note on each of the following.

1. Method of sowing.
2. Greenhouse.
3. Sphagnum moss.
4. Types of nursery beds.
5. Classification of nursery.

Chapter 6

Soil and Climate for Horticultural Crops

6.1 INTRODUCTION

The Climate and soil play important role in growth and development of the plant. Growers in general and fruit growers in particular should have adequate knowledge of the effect of various soil and climatic conditions on raising horticultural crops. Horticultural crops cannot be grown in all types of soil and climate. Hence zone-wise cultivation is made. Climate includes a number of parameters like temperature, rainfall, atmospheric humidity, wind, hail and light whereas soil covers factors like moisture content, texture, soil reaction, nutrient content, chemical composition and soil temperature.

6.2 SOIL

Soil is a thin outer covering of the earth surface. It is the natural resource directly developed by different natural forces acting on natural materials. Soil serves as a base and support for plant growth. It is a basic medium for plant growth and supplies nutrients for growing plants. Soil is the home of the plant root and the reservoir for essential nutrients and water for growth and development of crop plants.

6.2.1 Formation of Soil

Soils are the loose organic matter which are composed of minerals formed by weathering of rocks. These rocks pass through weathering processes and form soil. The weathering processes are carried out by physical, chemical or biological processes. These forces act alone or in combination with themselves. Water, wind and ice are the main physical weathering agents. Different chemical weathering agents are oxidation, reduction, solution, hydrolysis, hydration and carbonation. Man, animals and root of plants are the major biological weathering agents. Due to the action of all these biological forces soils are formed and are of different types. The soil horizon takes place due to action of various weathering forces with the passage of time. When the soil is formed at the site due to various weathering forces, the soil is known as sedentary soil. When the soils are formed at one place but transported through air, water, wind or ice, and deposited at other place other than its origin place is known as alluvial soil. Accordingly depending upon the origin, soils are of 7 types. They are:

Alluvial Soil

It is originated at one place and transported and deposited at other place in form of silt through flow of heavy water during rainy season. It is mostly found in river banks. There is deposition of a rich layer of alluvial soil. Alluvial soils are poor in organic matter but rich in potassium and lime. The soil reaction is neutral to slightly alkaline. In course of time where regular deposition of silt takes place due to precipitation of lime and poor internal drainage, a hard pan or layer of calcium carbonate is formed which is known as *kankar pan*. As the pan is hard it leads to development of saline-alkaline soil. In north Indian condition these types of soils are known as *usar* or *reh*. Horticultural crops grow best in these soils are guava, jackfruit, jamun, *khirni,* etc.

Tarai Soil

These soils have high water table and contain relatively high moisture content. Tarai soils are found in the foot hills of the Himalayan ranges mostly in the states of Uttarakhand, Bihar and West Bengal. Due to high soil moisture content crops like mango, litchi, orange, *Morus* spp., grasses, etc. grow luxuriantly.

Arid Soils

Arid soils are found in the water deficient zones of Rajasthan, Southern Haryana, Punjab and northern part of Gujarat. The soil is deficient of moisture. Mostly shifting sand dunes and undulating plains are major features of arid soil. There is presence of hard pan of calcium carbonate which prevents the cultivation of water loving/water requiring fruit plants. The soil contains soluble salts. pH of the soil ranges from 8.0 to 8.8 with very poor organic matter content. Due to shortage of moisture, generally bushy and thorny shrubs, ber, date, aonla, Indian cherry (lasoda), phalsa, khejri grow well in these soils.

Black Soil

As the name indicates the soil looks black in colour. The soils are found mostly in Maharashtra. The soil is loamy to clayey in texture, nearly neutral to alkaline in reaction with pH 7.2-9.0, poor in nitrogen and organic matter. When dry, it becomes hard and when wet it becomes highly adhesive and sticky. Crops suitable are citrus, banana, sapota and mango.

Red Soil

The colour of the soil looks red due to presence of iron oxides. These soils are found in almost all the states of the country except coastal strips and Deccan plateau. The soils are poor in nutrition, shallow in depth and contain low percentage of clay. Hence the water holding capacity of the soil is very poor. The soil may be of acidic or alkaline in nature. A wide range of crops like coconut, arecanut, black pepper grows suitably in these soils.

Laterite Soil

These soils are gritty in feel and porous in nature. Nutrient status is low as well as organic matter but rich in aluminium and iron oxide giving reddish colour to the soil.

Due to porous nature, water holding capacity is very poor. Soil reaction is very low. Crops like tea, coffee, rubber, cashew nut, etc. grow on these soils.

Marshy Soil

These soils are found in high rainfall areas of the country. The characteristics of the soil are acidic in reaction having pH 3.5, submerged in nature, black in colour and clayey in texture. These soils are rich in free aluminium and ferrous sulphate. The low acidity of the soil is due to formation of sulphuric acid and decomposition of organic matter under anaerobic condition. Crops of *Syzygium* and *Eugenia* spp. grow in these soils.

6.2.2 Properties of Soil

It can be studied into two groups like physical properties and chemical properties.

Physical Properties of Soil

Soil Structure

Soil structure should be uniform and favourable for water penetration, soil aeration and drainage.

Soil Aeration and Drainage

An aerated soil is helpful for growth of aerobic organism to promote metabolic activities of these organisms. Fruit and vegetables require well-drained soil. Poor performance is observed due to poor aeration and drainage. Therefore, well-drained soil is essential. Extreme wet and dry soil should be avoided.

Water Table

Availability of water at a certain depth is called as water table. High percentage of water can rise into water logging which affects growth of the plants. Therefore, depth of water should not be more than 2 m throughout the year.

Soil Depth and Organic Matters

Heavy soil causes water logging and poor aeration. Light soils are infertile due to leaching of nutrients. Therefore, the soils should be 2 to 2.5 m deep for growing of fruits and vegetables. Similarly, organic matters influence physical and chemical properties of soils. The organic matter content of soil helps in increasing the production of fruits and vegetables.

Soil Temperature

Soil temperature affects the root activity and is influenced by aeration and drainage. In cold soils chemical and biological activities are slow and availability of nutrients like N, P, S and Ca is limited. Nitrification would not start when the temperature is 4 °C. For successful growth of horticultural plants, the soil temperature should be within the range of 26 to 36 °C. Due to low temperature, absorption and transport of water and nutrient are adversely affected.

Chemical Properties of Soil

Soil Fertility

Soil fertility depends on nutrient contents such as N, P, K, Ca, Mg and S which are important elements required for growth and development of plants. Micro-nutrients like Fe, Mn, Zn, Bo, Cu, Mo, etc. are also required for crop plants in less quantity.

Soil Reaction

Soil analysis is important to find out the chemical composition of soil. The safe pH range is from 6 to 8. Some soils are problematic for plant growth like saline and alkaline soils. In alkaline soils, the concentration of sodium salts above 0.1% is harmful. Boron is deficient in alkaline soils and is unavailable in acidic soils. Iron is available in acidic soils whereas calcium and magnesium are deficient. In alkaline soils K, Mn, Fe, and Boron are deficient.

Soil Salinity

Information on salt tolerance is necessary to select salt tolerant cultivars and to adopt proper soil management practices.

1. *Salt Tolerant Crops*: Date palm, Phalsa, Sapota, Fig, Grape, Aonla, Wood apple, Ber, Chikory, Potato, Sweet potato, Watermelon, etc.
2. *Moderate Salt Tolerant Crops*: Pomegranate, Grape fruit, Lemon, Apple, Pear, Plum, Beans, Cucumber, Brinjal, Garlic, Radish, Pea, Tomato, Turnip, etc.
3. *Salt Sensitive Crops*: Orange, Peach, Avocado, Strawberry, Asparagus, Beet, Cabbage, Cauliflower, Palak, Leek, Lettuce.

In general, it may be stated that soils for fruit growing should be porous, deep and aerated. They should not be water-logged, marshy, saline, or acidic and there should be no hard pan at the bottom layers.

6.2.3 Soil and Water Management in relation to Horticultural crops

After laying out the orchard and planting of the fruit trees, the farmer is interested in optimum growth of trees and maximum production of fruit. Soil management, cultivation of inter-crop, irrigation and manuring are the factors to be considered for getting economic return from it, and maintain the health of trees.

Soil Management

Soil management practices such as cultivation, (interculture, weeding), mulching, sod culture, etc. are of various systems used in different parts of the country.

Cultivation (Clean Cultivation)

Cultivation of orchard soil is important to incorporate fertilizers and green manure, and to facilitate absorption of water in the soil and also increase the biological activities of soil due to better aeration. Deep tillage is not important in orchards because it may cause injury to the roots of the trees.

Mulching

It is the system in which materials like hay, straw, cut grasses or plastic sheet are spread over soil surface. Mulching prevents evaporation of water from soil and it also improves the structure and aeration by reducing rain drop impact (Fig 6.1).

Fig 6.1 Mulching

Sod Culture

This is a system in which fruit trees are grown in any tillage or mulching. The grass may remain without cutting but it is usually cut once in a year. This system is not followed in tropical and subtropical region whereas it is applied in temperate region (Fig 6. 2).

Fig 6.2 Sod culture

Weeding

There should be the removal of weeds to facilitate other operations like irrigation, manuring, etc.

Inter-Cropping

The crops which are raised in the orchard for increasing the income from the land are considered as intercrops, e.g., vegetables, pulses, short duration fruit crops like banana, papaya, pineapple, phalsa etc. and they can be grown in the orchard (Fig 6.3).

Fig 6.3 Inter cropping

1. These fruit trees, which are used as intercrops are also known as **filler crop.**
2. **Cover crop:** The crops are grown to cover the soil to protect it from soil erosion, e.g., grasses, pulses, moong, cowpea, peas.
3. **Green manuring crops:** The crops are grown in the orchards and after certain growth, they are buried in the soil for addition of organic manure, e.g., sunhemp, cowpea, dhanicha, etc.

Manuring

Fruit trees take large amount of nutrients from soil for their growth. So far maintaining fertility of soil in orchards, manures and fertilizers are added in the soil.

1. *Organic Manures*: Compost, FYM, oil cakes etc., improve the physical condition of soil and add some nutrients.
2. *Fertilizers*: Among chemical fertilizers urea, ammonium sulphate, super phosphate, D.A.P. are common.

Application of fertilizers depends on the soil and climatic conditions, and kind and age of crops.

6.2.4 Characteristics of Ideal Soil for Horticultural Crops

1. The soil should be well drained and fertile.
2. There should be lack of hard pan and the soil should be deep in nature.
3. The water table should be within 4 m deep from ground level.
4. It should retain nutrient and water.
5. The ideal soil depth for most of the horticultural crops should be 2.0 m.
6. The soil should be almost neutral in reaction and the pH should be in the range of 6.5-7.5.
7. Water should be below 2 m at all the times.
8. The soil should be free from high water table, high salt concentration and hard pan.
9. It should not be very heavy or very light in texture.
10. Soils with fluctuating water table should be avoided.

6.3 CLIMATE

Climate is the principal factor controlling plant growth. It refers to the average weather condition of the atmosphere for a long period over a larger area. The term weather is used to describe the current and temporary atmospheric conditions. *Climate is defined as the whole of average atmosphere phenomenon for a certain region calculated for a period of thirty years.*

For successful growing of horticultural plants, various components of climate like temperature, relative humidity, rainfall, solar radiation, wind velocity, sunshine hours, hail and frost should be carefully studied. Changes in climate takes place due to change in its composition.

6.3.1 Components of Climate

Temperature

Temperature is one of the most important components of climate. It plays a vital role in the production of horticultural crops. Each and every horticultural crop has its optimum temperature requirement which is suitable for its growth. It affects different activities of plant like growth and development, respiration, photosynthesis, transpiration, uptake of nutrients and water, reproduction (Such as pollen viability, blossom fertilization, fruit set etc.), carbohydrate and growth regulators balance, rate of maturation and senescence, quality, yield and shelf-life of the edible products. The above functions of the plant should be well when the temperature is at the optimum range. The activities of the plant are affected by very high or very low temperature. For every 1000 m increase in altitude there is a decrease of air temperature by 5–6 °C. The temperature range for most of the horticultural plants is:

Minimum 4.5° to 6.5 °C (40°: 43 °F)

Optimum 24° to 27 °C (75°: 85 °F)

Maximum 29.5° to 45.4 °C (85°: 114 °F)

During high temperature a plant does not perform proper functions of growth, whereas in low temperature physiological activities of the plant are stopped or liable to be injured to a more or less extent.

Relative Humidity

The atmospheric humidity also influences growth and development of plants. It plays a vital role in deciding the amount of moisture needed to produce crops. Low humidity has drying effect and enhances water requirement whereas high humidity favours fungal diseases, tastelessness and low keeping quality. High humidity combined with high temperature promotes rapid growth and higher yield but increase incidence of pests and diseases. High humid condition many times reduces the atmospheric temperature.

In hot, dry weather enormous amount of water is lost through transpiration. If the atmosphere is humid, even though hot, the crop needs less irrigation. The water is lost from the plant through transpiration by leaves. Transpiration depends on humidity, temperature, wind, light, etc. It is necessary to maintain the health of the plant by maintaining the balance between uptake and loss of water. Atmospheric humidity decides the water uptake of the plant. The plant absorbs water from the soil which depends upon water absorbing area of the tree, rate of evapo-transpiration and amount of water in the soil as per availability, basing on texture and structure of soil.

Rainfall

Rainfall is one of the most important factors for horticultural crops, and if a garden or orchard is to be established in a new area it is essential that the pattern of rainfall in the region should be studied before any decision is taken concerning the types of crop to be cultivated. A well-distributed and consistent rainfall is always desirable. Adequate soil moisture is necessary for good crop establishment, good yield and good quality. This moisture may be obtained from rainfall or irrigation. The amount and distribution of rainfall play a significant role in dryland vegetable cultivation. Excess rain causes flooding and invites disease and pest. Rain at the time of flowering is not suitable, because it washes away pollens in most of the tropical fruit crops.

Solar Radiation and Sunshine Hours

The sunlight is found to affect the quality of the horticultural crops, e.g., fruits exposed to light are found better in quality as compared to those receiving less sunlight. This is because of more amount of carbohydrates prepared in the leaves. In mandarin it has been observed that the fruits borne on the upper half of the trees and consequently receiving more light are richer in vitamin C content and also more sugars as compared to fruits on the lower half of the trees.

Light is an electro-magnetic radiation which is a form of kinetic energy. It comes from the sun to the earth as discrete particles called quanta or photons. Light is one of the most important factors affecting plant life. It is an integral part of the photosynthetic reaction which provides energy for the combination of carbon dioxide (CO_2) and water (H_2O) in the green cells having chlorophyll for the formation of carbohydrates with release of oxygen. The following equation is to explain the oxidation of water in photosynthesis.

$$CO_2 + 2H_2O \longrightarrow CH_3O + H_2O + O_2$$

$$6CO_2 + 13H \xrightarrow[\text{chlorophyll}]{\text{Light, radiation energy}} C_6H_{12}O_6 + 6H_2O + 6O_2$$

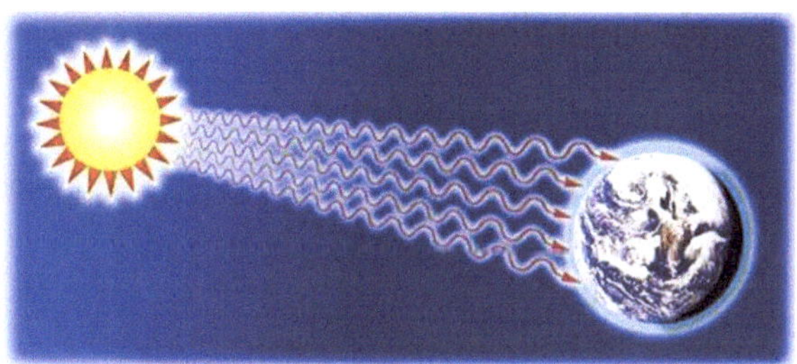

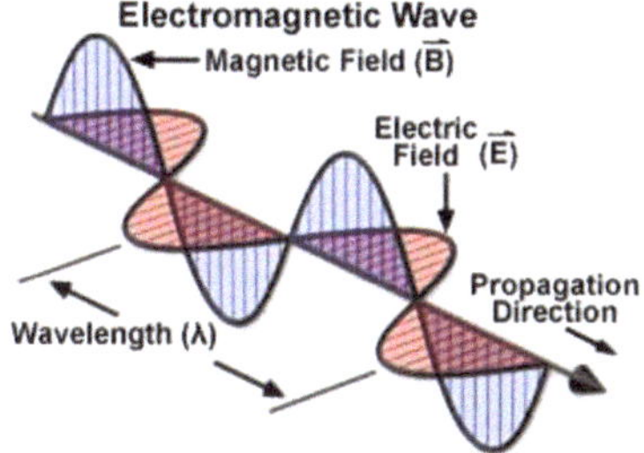

The performance of plants growth is influenced by three aspects of light (a) quality of light (b) intensity of light (c) duration of light.

(a) *Quality of Light*: It refers to the length of the waves. The visible part of spectrum of electromagnetic radiation ranges from wavelength 390 to 730 nm (nanometer). It is also called photosynthetically active radiation.

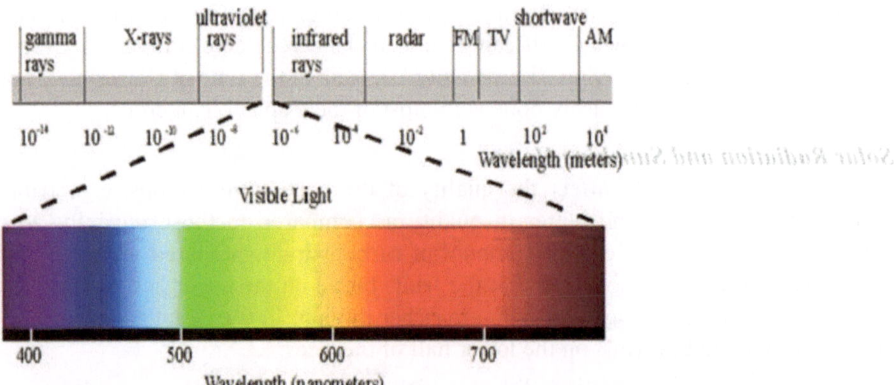

In general, red and blue light produce a greater dry weight. Green light inhibits plant growth. Red light promotes seed germination, growth and flower bud formation in long day short night plant. Photosynthesis is more in the red region.

(b) *Intensity of Light*: Light intensity refers to the number of photons falling on a given area or to the total amount of light which plants receive. The intensity of light varies with the day, season, distance of equator, dust particles and water vapour in atmosphere, slope of the land and elevation.

(c) *Duration of Light*: It refers to the period for which light is available. Duration of light required is also known as photoperiod. The response of plant to daily exposure to the light is known as **photoperiodism**. It affects the time of flowering, formation of tubers, fleshy roots, etc.

Wind

Wind is also one of the climatic factors influencing the plant growth. The effect of high wind on crops can be appreciable. Wind direction and velocities have significant influence on crop growth, both mechanically and physiologically. Different physiological effects of wind are increase in transpiration, plant desiccation, reduction in plant height because of reduction in cell elongation. Mechanically, it affects flower pollination, seed dispersal, tearing of leaves, lodging of plants, fruit drop, and uprooting of plants. In the case of deserts, valleys for avoidance of high-speed wind windbreaks and shelter belts are suggested. Wind breaks can help to reduce this problem. The wind break trees, like casuarinas, eucalyptus, *Inga dulsis*, etc. are grown around the orchard for protection.

Hail

Fruit belts of northern India are greatly affected by hail. It is very rare in Maharashtra. Hail causes shedding of young fruits and flowers while maturing fruits become almost unmarketable.

6.3.2 Classification of Climatic Zones

Basing on the variation of different components, climatic zones can be broadly divided into 3 types

Tropical Climate

Tropical climate experiences hot and humid summer and mild winter particularly found in the coastal belts of the country. There is no distinct summer and winter, not much fluctuation in day and night temperature. The area in this climate experiences high rainfall and high humidity. Average temperature ranges between 22-27 °C. Tropical climate is found within 300-900 m height from ground level. Different horticultural crops suitable for tropical climate are mango, banana, papaya, pineapple, cashew nut, jackfruit. Among vegetables onion, chilli, tomato, cassava, sweet potato, elephant foot yam, dolichos bean, curry leaves, ginger, pumpkin, snake gourd, cucumber are grown successfully. Flowers like tube rose, jasmine and gladiolus and plantation crops like coconut, areca nut, rubber, oil palm, tea, coffee and different tree spices can be grown successfully.

Subtropical Climate

Areas under subtropical climate experiences hot and dry summer and less severe winter. There is clear cut distinction between summer and winter season. There is high fluctuation in day and night temperature. The areas under subtropical climate receive less rainfall but the relative humidity reaches almost 100% during monsoon period. Temperature ranges between 25–30 °C. This climate is found at 900–1800 m height from the ground level. Crops like citrus, phalsa, fig, guava, pomegranate, pea, beans, cucurbits, tomato, brinjal, chilli, cauliflower, cabbage, knol khol, rose, chrysanthemum, tube rose, jasmine and different seed spices grow successfully in this climate.

Temperate climate

This type of climate is seen at 1800-3500 m height from the ground level. Areas under temperate climate experiences very low and chilling temperature below freezing point in winter and snowfall for 3–5 months in a year. Summer temperature ranges from 10-14 °C. Atmospheric humidity: 80–100%. Due to snow fall and chilling winter plants shed their leaves during winter as a mechanism for survival. Horticultural crops like apple, pear, peach, plum, walnut, apricot, potato, cole crops, root crops, peas, beans, celery, lettuce, globe artichoke, asparagus, mushroom, rose, gladiolus, orchids, lily, marigold, etc. can be grown successfully.

6.3.3 Horticulturally Potential Climatic Zones of the Country

Climate of a region is mainly influenced by the factors like latitude, altitude, topography, position related to continents and oceans, and large-scale atmospheric circulation patterns.

Almost all components of the climate influence horticultural crops. All are closely interrelated. The effect of each is modified by others. On horticultural point of view for cultivation of different horticultural crops, the climate of the country is divided into 6 potential fruit growing zones. They are:

Temperate Zone

Covering the union territory of Jammu and Kashmir, state of Himachal Pradesh, parts of Uttarakhand, Arunachal Pradesh and Nagaland, Nilgiris and Palani Hills of Tamil Nadu.

North Western Subtropical Zone

Covering the states of Rajasthan, Punjab, Haryana, parts of Uttar Pradesh, West Bengal, Madhya Pradesh and Chhattisgarh.

North Eastern Subtropical Zone

Covering the states of Bihar, Jharkhand, Assam, Meghalaya, Tripura, parts of Arunachal Pradesh and parts of West Bengal.

Central Tropical Zone

Covering the states of parts of Madhya Pradesh, Chhattisgarh, Maharashtra, Gujarat, Odisha, West Bengal, Andhra Pradesh, Telangana and Karnataka.

Southern Tropical Zone

Covers the states of parts of Karnataka, Andhra Pradesh, Tamil Nadu and Kerala.

Coastal Tropical Zone

Covers the coastal part of Maharashtra, Kerala, Andhra Pradesh, Odisha, Tamil Nadu, West Bengal, Tripura, Mizoram, parts of Gujarat along sea side and Andaman and Nicobar Island.

6.3.4 Effect of Climatic Factors on Crop Growth

Component of climate	Effect on crop production and quality
Temperature	Growth of banana below mean temperature (26.5 °C) causes reduction in rate of leaf production and delayed harvesting. Apple, pear, peach, almond, etc. are successfully grown in the regions of low temperature. In warm winter areas, chilling requiring fruit trees fail to complete their physiological rest period or meeting their chilling requirement leading to bud dormancy and leaves and blossoms do not appear on the trees in the following spring. Mango, sapota, papaya, banana, etc. are successfully grown in high temperature regions. Low temperature invites desiccation, chilling injury and freezing injury leading to protoplasm coagulation. High temperature reduces quality in citrus, radish, spinach, cauliflower etc. On the contrary, it increases quality in grapes, melons, tomato etc. Temperature below about 7 °C, many of the biennial crops like beetroot, broccoli, cabbage, carrot, cauliflower, celery, parsley, parsnip, spinach, Swiss chard and turnip may be stimulated into bolting. In high temperatures with dry or windy conditions, vegetable crops such as beans or tomatoes may shed some of their flowers, leading to poor fruit set. High temperatures may also detrimentally affect pollination of sweet corn, and give rise to poorly-filled ears. Monoecious cucurbits tend to produce mainly male flowers under high temperature conditions. Prevailing temperatures also play a role in the speed of germination and emergence of vegetable crops.

Component of climate	Effect on crop production and quality
Relative Humidity	High humidity is favourable for sapota, banana, mangosteen, jackfruit and breadfruit whereas low humidity is suitable for ber, grape, date palm, pomegranate, citrus, aonla and guava. The atmospheric humidity affects the juiciness of the fruits. High humid condition reduces colour and total soluble solid content and increases acidity in citrus, grapes, tomato etc. On the other hand, it is needed for better quality of banana, litchi and pineapple.
Rainfall	Vegetable crops such as cucurbits, prefer dry air and a high temperature, while leafy vegetables such as cabbage and lettuce prefer more humid conditions. High humidity is more conducive to invite heavy dew at night; which can be beneficial in reducing moisture stress, but which can favour the development of certain diseases, such as leaf rust and leaf spots on some crops. Rains causes cracking in grape, dates, litchi, limes, lemon, tomato and pomegranate. It reduces appearance and sweetness.
Solar radiation	Solar radiation is essential for anthocyanin pigments. Fruits exposed to sun light develop lighter weight, thinner peel, lower juice content and acidity and higher total soluble solid content in citrus, mango etc, than shaded fruits. Exposure of potato tubers to light causes solanine formation. High light intensity causes sunscald in tomato and citrus. It reduces pure white colour of cauliflower. Low light intensity causes thin and larger leaves in leafy vegetables. Solar radiation and sunshine hours have a tremendous effect on the productive capacity of vegetable crops like onion. Cloud or mist might also reduce the amount of light a crop receives, and thereby lower the potential yield of the crop. In apple the blue violet region is more important for the development of red pigments and colour. Low light intensity causes decrease in rate of photosynthesis with normal rates of respiration, decrease supplies of carbohydrates for growth and yield, leaf tips become discoloured, leaves and bud drop, leaves and flowers become light in colour. Due to high light intensity, the plant wilts and light coloured leaves may become gray in colour due to reduction in chlorophyll, the rate of photosynthesis is lowered down while respiration continues.
Wind	Areas exposed to wind causes evaporation of soil moisture and thereby necessitates more frequent irrigation. Hot wind at the time of blossoming may cause failure of pollination due to drying of stigmatic fluid and reduced activities of the pollinating insects. Complete physical destruction may result by uprooting even large trees. In many regions high winds can destroy the flowers, fruits etc. High wind is very much detrimental in case of leafy vegetables. Wind causes bruising, scratching and corky scar on citrus fruits and leafy vegetables.

REFERENCES

Singh J 2017. *Fundamentals of Horticulture*. Kalyani Publishers, Ludhiana

e-resources

https://www.agrilearner.com>soil-and-climate-of –horticulture-crops

ecoursesonline.iasri.res.in › mod › page › view

https://www.kzndard.gov.za › Documents › Horticulture › Veg_prod › cli...

Introduction www.agrilearner.com

OUTCOMES ASSESSMENT

PART A

Answer the following questions (True or False).

1. High temperature does not reduce quality in citrus. (True/False)
2. There is no distinct summer and winter in tropical climate. (True/False)
3. Sod Culture is followed in tropical and subtropical regions. (True/False)
4. Black Soils are found mostly in Maharashtra. (True/False)
5. Mulching only prevents evaporation of water from soil. (True/False)

PART B

Answer the following questions.

1. What is alluvial soil?
2. Central Tropical Zone covers the states of ………………..
3. What is soil?
4. Define climate.
5. Light is — radiation.

PART C

Write a brief note on each of the following.

1. Climatic zones of India.
2. Solar radiation.
3. Physical properties of soil.
4. Effect of climatic factors on crop growth.
5. Types of soil.

Chapter 7

Gardening

7.1 INTRODUCTION

Gardening is a practice of growing different kinds of horticultural crops in a particular land area as a part of horticulture either for consumption or sale, export, beautification etc. Basing on the crops grown, gardens are named accordingly, *e.g.,* for raising vegetables, it is vegetable gardens; for fruit crops, it is orchard; for plantation crops, it is estate plantation; for ornamental plants, it is gardening and landscape etc. The history of gardening is not new. In ancient time kings, emperors, rulers, sages were establishing different types of gardens for different purposes.

7.2 IMPORTANCE

1. Gardening supplies us healthy foods. They contain vitamins and minerals which are the source of a healthy body.
2. We cannot live long without food derived from plants.
3. Plants in the garden not only provide us food but also filter pollution, otherwise that would be killing us.
4. Plants are the source of life.
5. If there are no plants on our planet, we cannot breathe or eat.
6. Gardening for oneself is a great exercise by planting, weeding, watering etc.
7. Gardening burns calories by physical exercise. According to the Centre for Diseases Control and Prevention, just one-hour light gardening can burn 330 calories.
8. Garden can reduce the monthly budget for purchasing fruits, vegetables, spices, flowers etc.
9. Gardens provide habitat to bees and other pollinators.
10. People can get timely supply of produces from their own gardens.
11. Gardening at home builds up community, relationship and compassion by sharing its produces among friends, relations, neighbours etc.
12. Gardening helps to create a healthy environment.
13. We can have our surroundings greenery by gardening.
14. We can get pesticide free fresh foods.

15. Gardening can reduce the risk of stroke.
16. According to the National Heart, Lungs and Blood Institute, just a 30 minutes moderate level physical activity a few times in a week can prevent and control high blood pressure.
17. Gardening decreases the likelihood of osteoporosis.
18. Gardening is a stress reliever.
19. Being surrounded by flowers, it brings the positive effect on mood, happiness etc.
20. Gardening and flowers serve as a means of survival.
21. Gardens provide us aesthetic value.

7.3 FRUIT AND PLANTATION CROPS GARDEN

Planting and management of fruit trees in a regular manner in definite proportion in a piece of land for successive economic production is termed as orchard. Orcharding is a long-term venture and needs important considerations before establishment, because it requires heavy investments. Any lapses during planning and layout may lead to incur heavy loss.

7.3.1 Features of Orchard

Different features of orchard are climate of the area, soil, location, site and topography. Climate is the first and foremost natural determinant of successful orcharding. Different components of climate like temperature, rainfall, atmospheric humidity, wind, hailstorm, solar radiation influence crop growth.

Like climate, soil is the second important natural factor. Soil texture, structure, reaction, water holding capacity etc. affect crop growth. Soil should be fertile, deep and well drained. It should be free from water-logged condition. The pH of soil should be nearer to neutral (6.5–7.5). Some fruit plants like date, *ber*, *aonla*, grape, guava, and mulberry can be grown under high pH condition with proper care. Water holding capacity of soil should be satisfactory. There should not be any hard pan in the sub-soil within at least 4 to 6 feet deep. Soil should be rich in organic matter and sandy to sandy loam texture.

Similarly, orchard should be developed in already established fruit belt for getting technical support, sharing of experiences, availing government aids and subsidies, hiring or procurement of equipment etc. The site should be well-connected with roads because orchard produces are highly perishable in nature and need quick disposal. There should be a market close to the orchard area to reduce the transportation cost when one aims at selling his produce in the local market. The climate should be suitable for growing different fruit plants. Labour should be cheaply available to carry out different orchard operations. The area selected should be free from hotspots. Adequate irrigation facility should be there with good quality water. The site should be away from industrial areas. The site selected should not be in wind, cyclone, frost, hail and flood prone areas.

Topography expresses the contour of the land and its elevation. The side on the top of the ridges are unsuitable to establish orchard because of exposure to wind and erosion of soil. The bottom of the ridges is also unsuitable because of frost danger. The north facing slopes delays bud development and south facing slopes accelerate bud development. The north facing slope prevents the frost damage during spring whereas south facing slope are preferable because of early fruiting and maturity.

7.3.2 Systems of Planting Orchard Plants

Different systems of planting of fruit and plantation crops are *vertical row planting* which includes square system and rectangular system of planting and horizontal row planting which includes triangular system, hexagonal system, and quincunx system of planting. In contour system, plants are planted in hill slopes. In contour planting no definite row of planting is maintained because of narrow or stiff hill slopes. Besides the above, other intensive planting systems are high density planting and meadow orcharding where quite a larger number of plants are planted per unit area for efficient utilization of space, light, nutrient, irrigation water etc. for maximization of profit using certain special techniques without sacrificing the quality.

7.4 VEGETABLE GARDENING

Fruits and vegetables are food of the future. They play an important role in the balanced diet of human beings by providing not only the energy rich food but also promise vital protective nutrients. Vegetables are rich source of vitamins, minerals and dietary fibres. They have antioxidant properties. Role of vegetables in our daily diet is immense. Hence, it is necessary to grow different kinds of vegetables.

7.4.1 Types of Vegetable Gardens

Barley (1949) classified the vegetable garden into two broad categories:

1. Home scale production: It is meant for supplying vegetables for household consumption.
2. Commercial production: It meets the requirement of the local market, distant market and export market.

Later on, Thompson and Kelly (1979) classified the vegetable garden into 7 different categories like kitchen garden, truck garden, market garden, forcing garden, processing garden, seed production garden and floating garden.

Kitchen Garden

As the name indicates these vegetable gardens are located adjacent to the kitchen yard. The garden provides requisite quantity and kind of fresh vegetables for consumption of one's own family members throughout the year. So intensive care is taken to utilize every inch of land both in time and space dimension. The garden provides nutritious vegetables to a family throughout the year without searching for vegetables from outside source. So, the nutrition balance of the family members is maintained. That is why kitchen garden is otherwise known as nutrition garden. Generally, kitchen garden uses the waste water of the bathroom and kitchen with or without tap water. It is also called as backyard garden, home garden or amateur garden.

Advantages:

1. The garden occupies unutilized back-yard space of the residential building within the compound which would help in taking proper care, irrigation, harvesting and other operations.
2. The whole family can be engaged in it, where no great technical skill is required.
3. Working in a garden becomes a pleasure, an inspiration, a means of recreation and a possible family enterprise in which all members have due share to spend the leisure hours.
4. Fruits and vegetables obtained from market lack freshness and deteriorate in the food value besides their exorbitant price. Therefore, the best quality fresh produce can be had from one's own nutrition garden as the time interval between the harvest and the consumption becomes the least.
5. It is a place of recreation and healthy occupation of all ages of family members.
6. It cuts down the vegetable bill of the family.
7. Vegetables produced from this garden are fresh and nutritious.
8. The vegetables are free from residues of insecticides or fungicides.
9. It cuts down the water bill of the family as it uses the waste water of bathroom and kitchen.

Principles in Planning A Vegetable Garden:

1. It is convenient to layout rectangular plot than a square plot.
2. Garden should be well protected with suitable fence.
3. Perennial vegetables like curry leaf, drumstick and quick growing fruits like papaya, banana and lime should be planted along the border on north side.
4. Perennial vegetables like coccinia, chow-chow, etc., which require support should be planted at the rear end of the garden.
5. Long duration vegetables like tapioca, elephant foot yam, etc., may be planted together.
6. Suitable short duration companion crops such as radish, beetroot, carrot, etc. can be grown with the long duration crops. These crops can be grown on the bunds.
7. Crop rotation should be followed in such a way so that each plot will be planted with leguminous vegetable crop at least once in two years and also see that at least 4-6 kinds of vegetables are always available.
8. One plot should be kept reserved for raising nursery seedlings.
9. Knowledge of planting season is essential in planning the cropping pattern.
10. The entire plot should be divided into a number of small plots (sub-plots). The size and number of sub-plots can be decided based on area available, family size and crops chosen with convenience.
11. One or two compost pits may be dug in the shady corner of the garden.

12. Creeping vegetables like gourds and others may be trailed on the fence or erected pendals.
13. The area in between the perennial plants may be utilized for short duration shallow rooted annual vegetables or spices like garlic, coriander, etc.
14. If the land is limited preference may be given for growing those vegetables which are costly, highly perishable, and not easily available in the market and which can produce maximum edible vegetables per unit area.
15. The irrigation channel from the water source and path should be so planned and prepared that it covers the whole area of the garden for easy operation

Location:

1. It should be at the backyard of the house.
2. Preferably it should be in South -East direction of building.
3. The site should be nearer to water source
4. The selected site should be away from shade of the house or tall plant.
5. In urban areas where yard space is not available one can use the window seal and roof of house, terrace etc. if light is not a problem.
6. Size of the kitchen garden depends upon size of the family member, availability of area etc. As a thumb rule 250 m^2 area is required for a family member of 5-6.
7. Every inch of land should be efficiently used for growing crops; utilizing irrigation channel and bund.

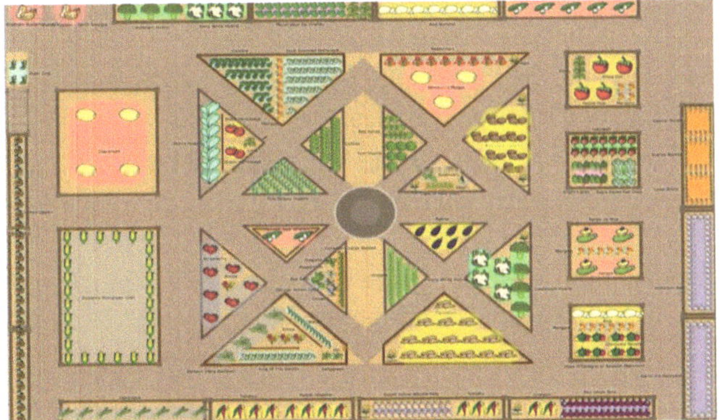

(*Source:* The Old Farmer's Almanac, Garden Planner)

Fig. 7.1 Model of Kitchen Garden

Design:

Design of a kitchen garden largely depends on availability of land. While designing the following principles should be followed (Fig. 7.1).

1. The land area should be rectangular in size rather than square.
2. The size of the garden may depend on the area available in the compound, the time available for its care and daily requirement of fresh fruits and vegetables of a family.

3. It should be in backyard of residential building.
4. The sub-plot should be small to accommodate good number of vegetables.
5. Path should be narrow.
6. Perennial quick growing trees like banana, papaya, lime, drumstick etc. should be planted on the north side as they will not only shade the vegetables but the roots may compete for moisture and nutrition if planted in between other plants.
7. Compost pits should be dug at one corner to convert the kitchen waste into organic manures.
8. Vine vegetables should be trained over the fences or compound walls.
9. The layout should be so designed to appeal to the eyes.
10. Crops which are yielding a greater number of fruits per plant should be grown.
11. Preference should be given to early maturing and afresh consumed vegetables.
12. Succession sowing should be done to ensure steady supply of vegetables throughout the year.
13. Some quick growing vegetables can be taken up in the inter spaces of slow growing late maturing crops.
14. Selection of crops largely depends on size of the garden, crops suitable to the region and choice of the family members.
15. Crops should be rotated throughout the year.
16. The ridges and water channel can be used for growing root crops like radish, beet, carrot, knol-khol etc. and climbing vegetables like basella, cowpea, French bean etc.
17. Depending on the space available and the family size the planning may be done either for a big or a medium or a small size garden.
18. Entire plot should be divided into annual, biennial and perennial blocks.
19. To produce 300 g of vegetables per day, all the year round, about 50 $m^2$of area is required.
20. Growing coriander, ginger, chilli, turmeric in the interspaces of perennial crops like banana, pineapple, drumstick, curry leaf, lime etc.

Truck Garden

The word truck is derived from French word "*troquer*" means to barter. It does not mean that the produce is being transported by trucks. Quite a large amount of produce is produced in these vegetable gardens applying intensive method of cultivation. The selection of the site and its location depend upon prevailing climate. The site is located away from consumption market, means cities, so the produce is transported to distant market through road and railways. One or two vegetables are grown in extensive manner in larger quantities for distant market. Suitability of vegetable depends upon climate of that region. Produce is traded by middle men. Maximum profit goes to the pocket of the middle men. Still the production is cheaper as the land is less costly due to distance from the city and labour is available cheaply. Mechanical farming is practised. Vegetables which can resist the transportation shock like onion, garlic, potato, pumpkin, ash gourd, dioscorea, elephant foot yam, colocasia etc. are grown extensively.

Market Garden

Now-a-days people are interested for city dwelling. The city dwellers have no time to spend in their own vegetable garden. So, there is development of vegetable farming in the vicinity of cities within 10–20 km distance to supply vegetables to the nearby market. To supply sufficient quantity of vegetables to the city dweller there is unavailability of big size plot and the land area is very costly. Hence by adopting intensive method of cultivation in small land holdings good quantity of vegetables is produced. The grower produces a variety of vegetables suitable to the prevailing climate as per the season in the year. The grower takes all attempts for continuous supply of vegetables to the city market. Hence to catch high profits, the grower produces early maturing vegetables early in the season as well as off season vegetables. No middle man is involved in trading. So, a higher profit goes to the pocket of the grower. There is assured supply of vegetables to the local market. These types of gardens came into existence due to urbanization because the urban people do not have time and land to grow vegetable for daily use. The growers of the garden utilize a small piece of land and produce most efficiently and intensively different early maturing vegetables, those reach at market early to fetch maximum price per unit land area. Marketing of vegetables is performed by the grower himself or herself.

Forcing Garden

It refers to the cultivation of vegetables out of their normal growing season. So, the vegetables come to the market early in the season. It is a common practice in western countries where the environment is not favourable for growing vegetables in their normal season. In India there is no need of forcing vegetable gardens because of prevalence of diverse climatic conditions. But due to growing demands in different seasons for supply some special commercial vegetables like capsicum, cucumber, gherkins they are cultivated in forcing gardens. Hence, to meet the growing demand and consistent supply of vegetables, vegetable forcing gardens are developed. Vegetables are grown in cold bed, hot bed, cellars, shade, shade net house, ploy house, lath house, plastic house and green houses. In some protected structures there is control over temperature, relative humidity and light intensity. The vegetable garden supplies fresh produces because of its location is nearby the centre of consumption. As cultivation is practiced out of normal season expertise skill is necessary. Generally high demand, high priced vegetables, e.g., tomato, capsicum, cucumber, mushroom etc. are grown in forcing garden by manipulating the temperature, humidity and light intensity in climate controlled protected structures.

Processing Garden

To reduce the market-glut the processing industries takes care of the market surplus vegetables but the processing industry cannot run relying on market-glut. So, there is development of processing vegetable garden for steady supply of fresh and appropriate quality vegetable to the processing unit. The produces can be processed and sold in the market during the time when price in the market is high. The processing vegetable gardens are mostly found in around the processing industry area to supply processing quality vegetables regularly. The garden supplies bulk fresh produces throughout the season. Specific crops and their cultivars are grown in the garden. The industry may purchase vegetables directly from the grower or through contract or middle man or by establishing own processing gardens. Very common vegetable suitable for processing are Arkel cultivar of pea for dehydration; Early Badger of pea for canning; Roma, Punjab Chhuara, S-152 of tomato for ketch up and sauce; Kufri Chipsona-1, Kufri Chipsona-2, Kufri Chipsona-3, Kufri Chipsona-4, Kufri Frysona of potato for chips making; broccoli, lima beans, spinach for freezing; onion and garlic for drying and dehydration.

Seed Production Garden

Good quality seeds with high yield potential and tolerant to abiotic and biotic stresses are demand of the day. It aims at producing and supplying quality seeds. It is a highly specialized farming system covering larger areas and taking due care at different stages of crop growth. Care is taken in selection of land, choosing quality of seeds, maintaining isolation distance, rouging, bee keeping, harvesting, collection of seed matter etc. The cost of production of seed is high as compared to vegetables for consumption. It is practiced on commercial scale. Proper isolation distance is maintained to avoid cross-pollination. Strict rouging is practiced. To increase seed production by artificial pollination, honey bees are reared in this vegetable garden. So, chemical insecticide and pesticides are applied judiciously to save the honey bees. As land area is occupied for quite a long period, it increases the cost of watch and ward.

Floating Garden

It is the most fascinating vegetable gardening found in Dal Lake of Kashmir. The floating aquatic grass typha found in the lake serves as the floating base or substrate. Every operation is performed by using boats. Vegetables produced in these gardens are supplied to Srinagar (Fig. 7.2).

Other vegetable gardens are roof top garden, vertical garden, hydroponic garden and aeroponic gardens, terrace garden and organic vegetable garden.

Fig. 7.2 Floating Garden (***Source:*** The Borgen Project)

Organic Vegetable Gardening

In 1980, organic farming was defined by the USDA as a system that excludes the use of synthetic fertilizers, pesticides and growth regulators.

Organic Farming relies on the following:

- Quality food
- Environment
- Soil fertility
- Soil flora and fauna
- Organic Agriculture is a unique production management system which promotes and enhances agro-ecosystem health, including biodiversity, biological cycles and soil biological activity and this is accomplished by using on-farm agronomic, biological and mechanical methods in exclusion of all synthetic off-farm inputs.
- Approaches and production inputs of organic farming
- Strict avoidance of synthetic fertilizers and synthetic pesticides
- Use of crop rotations, crop residues, cover crops, mulches, animal manures, composts and green manures
- Organic fertilizers and soil amendments
- Bio-stimulants, humates and seaweeds, compost teas and herbal teas
- Marine, animal, and plant by-products
- Bio-rational, microbial, and botanical pesticides, and other natural pest control products
- The Organic Foods Production Act, a section of the 1990 Farm Bill, enabled the USDA to develop a national program of universal standards, certification accreditation, and food labeling.
- In April 2001, the USDA released the Final Rule of the National Organic Program. This federal law stipulates, in considerable detail, exactly what a grower can and cannot do to produce and market a product as organic.

Hydroponics and Aeroponics

The word "hydroponic" has been derived from the Greek word "hydro" means water and "ponos" means labour or work. Hydroponics refers to technology of growing plants in nutrient solution with or without the use of substrate (*e.g.*, gravel, sand, vermiculite, rock wool, peat moss or sawdust) to provide mechanical support to the plants (Fig. 7.3).

The word "aeroponic" is derived from the Greek word "aero" means air and "ponos" means labour or work. The roots hang in the air and are misted with nutrient solution (Fig.7.4).

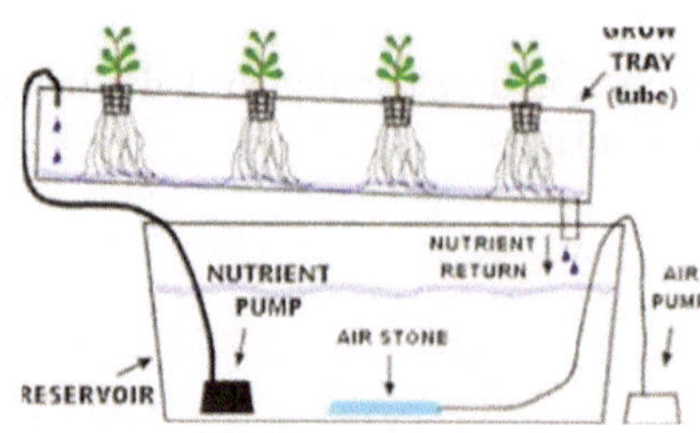

Fig. 7.3 Basic Hydroponic System

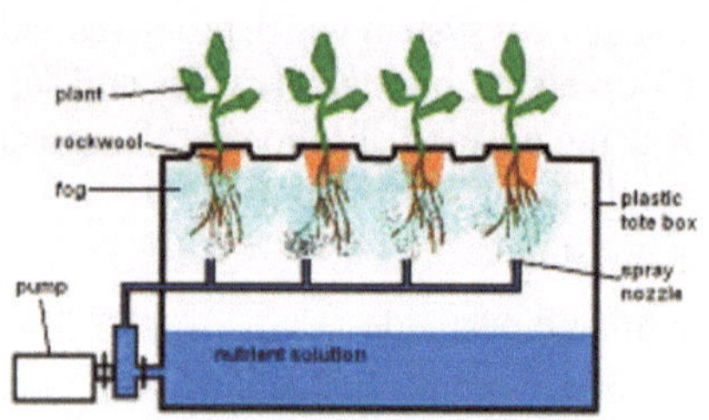

Fig. 7.4 Basic Aeroponic System

Vertical Garden

It is constructed in terrace or utility space of the house. Vegetables are grown both in time and space dimension where sky is the limitation. In general, it is constructed in urban areas, cities, metros where land is very costly and no enough yards are available around the building. The design of the garden is based on size suitable for terrace, preference of vegetables of family members, pots suitable, height of the structure for easy handling, light available and utilisation of maximum areas. Different structures used are base frame, main central support, support for pots or grow bags or trays.

Roof Top Gardening

Due to scarcity of land area the roof top of the building can be utilised for vegetable gardening, although it is a common practice of growing ornamental plants on the roof top. For monkey menace, shade net structure may be erected all-around the gardening area. Vegetable plants can be planted in bigger size pots or concrete structure filled with soil. There is provision of drainage of excess water.

Terrace Gardening

It is similar to vertical garden where trays, pots or grow bags are used for growing of vegetable crops.

7.5 ORNAMENTAL GARDENS

An ornamental garden is a place where plants are arranged in an aesthetic manner. Plants can be grouped together in various ways to give an aesthetic effect. In modern cities with growing slums and factories, gardens are essential to improve the environment and to provide healthy air for the inhabitants. They are really the lungs of the city. Gardens serve to beautify the country. Gardens are said to be the yardstick of the culture. This is true to some extent as they reflect the aesthetic taste of the people and are the chief pieces of art that confront a visitor and help him asses the cultural standards of the region. The purpose of gardening is both adding beauty and utility.

7.5.1 Features of Ornamental Garden

Some of the important features found in most of gardens are:

Fence

Fence is the outer most boundaries to prevent trespass and to ensure privacy in home gardens. Fences can be created either by using closely planted thorny plants, hedges and shrubberies or structures where wood, bamboo, barbed wires, netting and chain links are used. Various climbers can be trained over the fences to enhance attractiveness.

Hedges

Hedges are useful to divide the garden into sections and to line the drives so as to direct the visitors to a central object. They are sown and grown in the same manner as the fence and plants are pruned to a height of 3-4 feet.

Edges

These are row of plants which do not exceed one-foot height. They are grown along with paths and around the flower beds. Non-living materials like bricks, tiles are also used for this purpose. Live hedges are more in harmony with the garden than features. The foliage of hedges is not trimmed.

Drives and Paths

Drives and paths facilitate easy and purposeful movement within a garden, providing access to all features within a garden or can be included for their decorative value. It is usually designed to provide a visual line between separate areas. Cobbles, granite, flagstones, brick, concrete or wooden materials may be used. They should be laid with easy gradients and perfect paving and levelling.

Lawn

Lawns focus the background colour in the garden picture against which the colour of shrubberies and flower beds is brought into relief. Whether the garden is big or small it must have a lawn. In fact, a lawn and a mass of flowers can constitute a garden. The lawn should be sown only to a single species of grass so as to give a uniform colour. The most common lawn grass is ***Cynodon dactylon.*** It prefers slightly acidic soils of pH 5.5–6.0. It does not grow well under shade of trees. ***Dichondra ripens*** a new type of ground cover can stand sun and grows well under shade. ***Festuce*** is the quick growing and finest of lawn grasses.

Shrubberies and Shrub Borders

When the plants are grown in a row but not trimmed, the feature is called border. Borders are planted to different species of plants, while hedges are generally planted to a single species. When borders are of herbaceous plants, they are called herbaceous borders; or, when they comprise of shrubs, they are called as shrub borders. The shrub borders may be grown along wall or in front of fence of tree. The border of the shrubbery consists of more than single row of plants.

Flower Beds

These are also known as **annual flower beds** as they are planted with annuals or herbaceous perennials which are treated as annuals. They should be planted to a single species and cultivar so that each bed is of single colour. A flower bed should be behind a lawn or in the middle or at least should have a strip of lawn in front of it.

Carpet Beds

Plants of different colour foliage which can be clipped close to the ground are chosen for planting in an intricate design on the ground. Such a feature is known as carpet bed. The design may be conventional, geometrical ones or map or clock or a sundial.

Topiaries

Certain plants which can stand severe and constant pruning and which possesses small foliage and relatively short internodes can be trimmed in to globes, ovals or in to fancy shapes of animals etc. These are generally found in formal gardens.

Arches and Pergolas

Arches can be semi-circular or rectangular shaped and are used to link one part of the garden with another. Arches are constructed near the gate or over paths. Its proper place is nearer a path and its purpose is to support climbing plants. Pergola is a narrow vista consisting of a series of arches connected with climbers preferably leading to some other interesting feature of a garden. Pergolas are constructed over pathways. It brings height to the flat of a level compound.

Fern House (Fernery)

Plants of the humid tropical, subtropical and temperate regions cannot be grown in the open in the plains exposed to high sun. Such shade loving plants are grown in a structure called the fern house. In the centre of fernery, a cement tub is constructed and is filled with water so as to increase the humidity inside and also facilitating watering of potted plants. The beauty of fernery depends on the proper arrangement of plants grown in pots and kept on galleries. Small pots with orchids are hung from the roof.

Orchids

Orchids are humid tropical and subtropical plants loving shade. These are becoming great favourites because the flowers have gorgeous colour. They often assume shapes of birds, moths and butterflies, and last longer.

Pot Galleries

Circular galleries are constructed of masonry and on the steps of which potted plants are arranged. The height of each step and pot on the lower step should be the same.

The plants grown should be taller than the height of steps so that the pots and the masonry structures are both hidden behind plants and present the appearance of a mound of plants.

Lily Pool
Aquatic plants are grown in lily pools which may be dug in the ground and abetted with stones so as to look natural or may be constructed in cement of regular shape. In cement pools there should be an inlet at the bottom and an outlet, a few inches below the top of pool, so that a constant level of water is maintained.

Rockery
Plants growing on rocky situation are grown in the garden in rockery. The rockery is constructed by keeping up manured soil to a desired height and embedding rocks into it. The plants are set in the crevices between rocks. It can be raised under the tree or separately. Generally, both foliage and flowering succulents as well as xerophytes are grown.

Single Specimens
In an extensive lawn the monotony can be broken by single beautiful tree of exiting quality without blocking the view of the other features beyond.

Baradari
It is arbour like structure made up of stone and masonry with pucca roof and raised platform for sitting. They are provided with 12 or more doors and they are used to watch the dances.

7.5.2 Principles of Ornamental Gardens

In planning a garden, several factors like the size of the house and the space available for garden, availability of water, cost of the laying the garden and its maintenance have to be taken into consideration. A garden is planned primarily to suit the tastes of the people of the house hold and locality. There is no rigid system in garden planning and each system is open to modification to suit the environment and other factor. An ornamental garden is planned basing on some principles.

Different garden principles are:

1. *Axis*: It divides the garden into two equal halves.
2. *Focal Point*: It is the point of attraction. Only one object should be placed either at the centre or at extreme end.
3. *Mass Effect*: A large number of plants of same cultivar at one place to break the monotonous arrangement.
5. *Unity*: It should have an artistic look and brings overall impression. It can be done by combining different objects, trailing of creepers over windows, walls, porches etc.
6. *Space*: Arrangement of objects should be done by borrowing different objects, or by constructing paths in such a way that the garden looks larger than its actual size.
7. *Line*: Lines are powerful designing elements. Lines must be prominent enough with gentle curve.

8. *Scale*: It defines relationship between masses. Scale changes the feeling.
9. *Forms and Texture*: Forms of objects should be columnar, oval, vase, pyramidal, weeping or rounded.
10. *Time and Light*: It relates to position of sun with quality and quantity of light daily and seasonally.
11. *Tone and Colour*: It should give greatest appeal and evoke greatest response. Colour should be soothing in nature.
12. *Mobility*: The objects should speak anything.

7.5.3. Styles of Gardening

A gardener may think that a landscape garden can be laid out only on a gently undulated land, but it is not so. The goal in landscape gardening is to improve landscape with an idea of developing view or design. The other two familiar terms, which associate landscape gardening, are formal and informal gardens.

Formal Gardens

It is laid out in a symmetrical or geometrical pattern. Everything is planted in straight lines. If there is a plant on the left-hand side of a straight road similar plant must be placed on the right-hand side also. Flower beds, borders and shrubbery are arranged in geometrically designed shapes. Trimmed formal hedges, topiaries are the typical features of a formal garden. Land is prepared according to design (Fig. 7.5).

Fig. 9.1 Square System

Informal Gardens

The whole design looks informal. Features are arranged in a natural way without any hard and fast rules but here also the work has to proceed according to a well-set plan. The idea behind this design is to imitate nature. Design of garden is prepared according to land. So, it looks natural (Fig. 7.6).

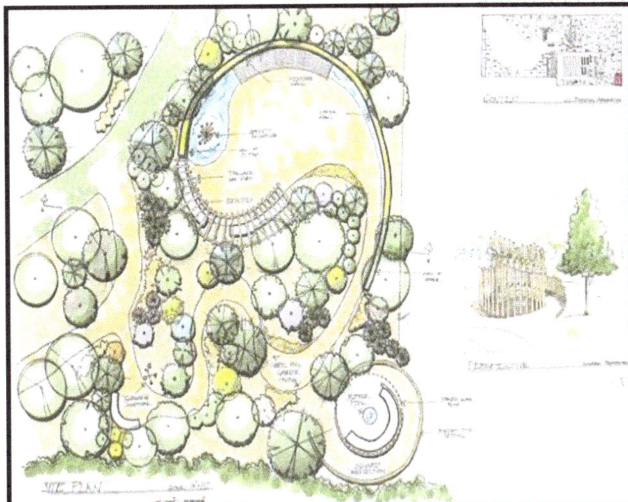

Fig. 7.6 Informal Garden

Wild Garden

William Robinson in the last decade of 19th century made the idea of wild garden. His main idea was i) to naturalize plants in shrubberies, ii) grass remains unmoved as in nature and iii) few bulbous plants should be grown scattered.

A garden enthusiast has to study the different styles available in the world to gain some knowledge. In India even though we were interested in the gardening since ancient times there was no style to denote as Indian style of gardening.

Even the famous garden style of India i.e. Moghul garden is a replica of ancient Persian garden.

7.5.4 Major Ornamental Garden Styles of The World

1. English garden
2. Moghul garden
3. Persian garden
4. Italian garden
5. French garden
6. Japanese garden

The Moghul, Persian, Italian and French styles fall in the category of Formal gardens whereas the English and Japanese garden are classified in the informal style of garden.

7.5.5 Features of Mughal Gardens

The main features of Mughal gardens which are largely borrowed from the Persian style are site and design, walls and gates, nahars (running water), baradari, tomb or mosque, trees and flowers.

7.5.6 Features of English Gardens

The main features of English gardens are flower beds, topiary, curved path, streams, artificial waterfall, clipped hedges and flowering annuals, lawn, mixed border especially of herbaceous annuals as well as herbaceous perennials, shrubbery and rock gardens. Most of the flowering annuals that we see today in the Indian gardens, with few exceptions of Amaranths, Balsam, Gomphrena, Marigold etc., were brought here by the British.

7.5.7 Features of Japanese Gardens

Ponds, streams, waterfalls, fountains, wells, Islands, bridges, water basins, stone lanterns, stones, pagodas, fences and gates are the main features of Japanese gardens.

Types of Japanese Gardens

Hill garden, flat garden, tea garden (the outer tea garden, the inner garden), passage garden and sand garden.

7.6 LANDSCAPE GARDENING

It may be described as the application of garden forms, methods and materials with a view to improve the landscape. The art of designing is known as "Landscape Architecture" although the older term "Landscape Gardening" is also popular. It is an imitation of nature.

7.6.1 Important Considerations of Landscape Gardening

1. A garden has to be one's own creation and not an imitation, giving due consideration to the local environment.
2. Overcrowding of the plants should be avoided.
3. Take advantage of natural topography while designing garden.
4. Perfect harmony of different components is the essence in landscape gardening.
5. Before planning a design one must be sure for what purpose the garden is – utility or beauty or both.

7.6.2 Principles of Landscape Gardening

A good designer should design the landscape in the available space. The natural topography should be retained. Fencing, as far as practicable, should not obstruct any natural view. For example, if there is natural forest scenery or a hillock just outside the boundary, it should be incorporated in the garden design in a thoughtful manner so that it appears to be a part of the garden.

7.6.3 Pattern of Landscape Gardens

1. *Circular Pattern*: Series of circles can be utilized to create circular. It is used in formal and informal gardens.
2. *Diagonal Pattern*: Draws a grid line at 45° to the boundary. It is also used in formal and informal gardens.
3. *Rectangular Pattern*: It is utilized in formal garden in a symmetrical manner.

7.6.4 Landscape Gardening Symbols

Garden symbols are nothing but the pictures, which are used for representing the various garden components (features). They are mainly used in preparing the garden plan or design. The symbols cover primarily the garden surfaces such as steps, different types of shrubs, trees, climbers, perennials and other plants. On plan, the symbols are drawn to scale wherever possible and make the transition from broad outline plan to detailed structural and planting plans. When drawing symbols, generally it should look as much as possible like the shape of the plant or structure it is representing.

7.6.5 Other Different Types of Landscape Gardens

Shade garden, sunken garden, terrace garden, rock garden, gravel gardens, roof gardening, paved garden, marsh or bog garden, landscaping of places of public importance (highways, stations and railway line, bus terminus and airport, banks of rivers and canals, city, town and countryside, city parks, country home, crematories and burning ghats, home ground, historical importance, temples, gurdwaras, mosques, churches, public buildings, educational institutions and factories).

REFERENCES

Arora JS 1998. *Introductory Ornamental Horticulture.* Kalyani Publishers, Ludhiana.

Bhattacharjee SK 2004. *Landscape Gardening and Design with plants.* Aavishkar Publication, Jaipur.

Bose TK and Mukherjee D 1977. *Gardening in India.* Oxford & IBH Publishing Co., New Delhi.

Indian Horticulture (English) 2019 March- April issue. DIPA, ICAR, New Delhi.

Randhawa MS 1961. *Beautiful Trees and Gardens.* ICAR, New Delhi.

Randhawa GS and Mukhopadhyaya A 1986. *Floriculture in India,* Allied Publisher Pvt. Ltd, Kolkata.

Singh J 2017. *Fundamentals of Horticulture.* Kalyani Publication, Ludhiana.

e-resources:

- *http://earthg.com/*
- https://www.almanac.com/content/kitchen-garden-layouts-potager
- https://borgenproject.org/how-to-make-a-floating-garden/
- *www.unex.berkeley.edu/cat/course412.html*

OUTCOMES ASSESSMENT

PART A

Answer the following questions (True or False).

1. The English and Japanese gardens are formal ones. (True/False)
2. Hydroponics and aeroponics are two different concepts. (True/False)
3. A garden is an imitation. (True/False)
4. Axis divides the garden into two equal halves. (True/False)
5. The Moghul gardens are formal ones. (True/False)

PART B

Answer the following questions.

1. What is the difference between hedge and edge?
2. What is an orchard?
3. Differentiate between formal and informal gardens.
4. Kitchen garden is also known as — and —.
5. Define fernery.

PART C

Write a brief note on each of the following.

1. Principles of ornamental gardens.
2. Market garden.
3. Floating garden.
4. Features of orchard.
5. Landscape gardening.

Chapter 8

Orchard Management

8.1 INTRODUCTION

Orchard is a place or a piece of land where different kinds of fruit plants are planted in orderly manner and managed efficiently for production of successive yield to get assured economic return. An orchard being capital-intensive and long-term venture careful planning, layout, selection of site, preliminary operations etc. are to be done. It is easy to establish an orchard considering its features critically but difficult to manage afterwards. It needs high managerial skills and technically skilled manpower to carry out different orchard operations. Any lapses in management can lead to incur heavy loss.

8.2 PLANNING

Careful planning is necessary for efficient economic management of orchard. The principles of planning are:

- Optimum spacing should be provided to accommodate maximum number of fruit trees per unit area.
- At least 10% of the total area should be left for buildings, roads, paths, storehouse, tube well, water channel etc.
- The farm building or office should be located at the centre.
- The watch post should be at entry gate.
- Roads of 10–15 feet width should be provided and each should cross at 90° angle.
- Provisions should be there for store house, packing house etc.
- Irrigation system should be properly planned.
- Deep wells should be dug at convenient places at a rate of one for every 2 to 4 hectares of land.
- In the absence of bore well a deep farm pond with adequate water storage capacity should be constructed to reserve water from canal or rain for assured irrigation facility.
- For efficient utilization of every drop of water drip irrigation facility should be provided.
- Permanent fencing with brick wall, iron pole and barbed wire should be there to protect from stray animals, wild animals and thieves.
- In absence of permanent fencing temporary fencing of three closer rows with live thorny plants having dense foliage, easy to raise from seeds, quick growing and drought resistant capacity can be preferred.

- From few years back construction of solar fencing is getting momentum.
- Planting of 2 alternate rows of wind-break plants like jamun, desi cultivar mango, neem, sisoo, karanja, deodar, eucalyptus at 5m interval in north-west side to protect the orchard from hot and cold wave.
- Sufficient space should be maintained between row of wind break and first row of fruit plant by digging a trench of 90 cm deep at a distance of 3 m from the wind breaks.
- To avoid competition with fruit plants, roots of wind breaks should be pruned around the trench area and trench filling with pruned materials in every three to four years.
- Each kind of fruit plants should be assigned to separate blocks.
- There should be 7.5 cm gradient for each 30 m distance of irrigation channel.
- Fruits ripening at the same time should be grouped together.
- In case of dioecious fruit plants, every 3rd tree in every row should be a pollinator.
- Short growing trees should be allotted to the front and the tall trees at the back for easy watch and ward and beauty.
- Plant evergreen trees in the front and deciduous plants at back rows.
- Fruits and berries which are attracting birds and animals should be close to the watchman shed.

8.3 LAYOUT

The proper layout of orchard aims at accommodating maximum number of trees per unit land area, maintaining adequate space for proper development of the trees and convenient orchard management practices.

The methods of planting are broadly divided into 2 types:

i. Vertical row planting: Here trees are planted in rows exactly at perpendicular to each other in their adjacent rows, e.g., square system, cluster system and rectangular system of planting.
ii. Alternate row planting: In this method the plants are set at right angle in alternate rows only not to adjacent rows, e.g., triangular system, hexagonal system, contour system and quincunx system.

8.4 PLANTING SEASON

Planting time depends upon the locality, type of fruit crops to be grown, season of growing, soil type etc. Under Indian condition we follow 2 seasons of planting, i.e., monsoon planting during June to August and spring planting between February to March. Crops suitable to monsoon planting are citrus, mango, sapota, guava and all others can be planted during spring. Deciduous trees may be planted during dormant period.

8.5 DIGGING OF PITS

Planting is done in pits, opened earlier by using planting board at appropriate distance as per the recommended spacing. Pits are dug in the field marked earlier at least one month before as per the system of planting using planting board. Planting board is a wooden plank of 1.5 m long, 10 cm wide and 2.5 cm thick. Pits of size 30 to 60 cm^3 are opened as per the seedling to be transplanted. Pits are left open for about 15-30 days for proper solarization. During digging the top soil and subsoil are kept aside. After solarization the top soil is filled first and the subsoil mixed with equal amount of FYM and a handful of methyl parathion or quinalphos is filled then. The pit is filled up to 5 cm above ground level. During filling a peg is fixed at the centre of pit to mark the position of plant. Light watering is done for settling down of soil and FYM mixture. If any depression is observed after irrigation it should be brought to ground level by adding extra soil and manures.

8.6 PLANTING METHODS

The planting materials like rooted cuttings, budded seedlings or grafted materials are planted at the centre of pit by removing the guiding peg. Soil at the centre of pit is scooped out and the planting material is planted after tearing the poly bag exposing the ball of earth or bare roots. Evergreen plants are planted with a ball of earth enclosing the roots and deciduous plants as bare rooted. During planting application of 1 kg neem cake or 50 g bone meal per pit may be applied. In sodic soil 5–8 kg of gypsum and 20 kg of sand is applied during pit filling. During planting care should be taken not to bury the graft union or bud union inside soil. It should be slightly above the ground. They should stand erect in the field as they were in nursery. Slightly pressing the soil around the base of plant helps in contact of roots with soil. Soon after planting light watering is done. If planting is done during rainy season soil around the plant is raised in a sloppy manner. But during spring planting a shallow basin is prepared. Staking should be done wherever necessary. To avoid scorching sun temporary shade may be regulated by covering the plants with paper caps.

8.7 PLANTING DISTANCE

Plants are planted at appropriate distance to encourage proper growth. Considering the canopy coverage, soil type, climate, type of plant, techniques of cultivation etc. plants are planted at a specific distance. It varies from plant to plant. Closer spacing encourages attack of disease and pest and creates problem in intercultural operations. It badly affects the growth, flowering and fruiting. Recommended spacing of different fruit plants are as follows:

Fruit Crop	Spacing	Fruit Crop	Spacing
Apple	5–6m × 5-6m	Jamun	8m × 8m
Aonla	8m × 8m	Kair	3m × 3m
Avocado	7m × 7m	Karonda	2m × 2m
Bael	8m × 8m	Khirni	6m × 6m
Banana	1.8m × 1.8m	Litchi	8m × 8m
Ber	6m × 6m	Loquat	6m × 6m
Cashewnut	8m × 8m	Mango	8m × 8m

Fruit Crop	Spacing	Fruit Crop	Spacing
Citrus	5-6m × 5-6m	Mango (HDP)	2.5m × 2.5m
Coconut	7.5m × 7.5m	Mulberry	6m × 6m
Custard apple	6m × 6m	Papaya	2m × 2m
Date palm	6m × 6m	Phalsa	2.5m × 2.5m
Fig	5m × 5m	Pineapple	25cm × 25cm × 90cm
Guava	6m × 6m (HDP-1.2m × 1.2m)	Pomegranate	5m × 5m
Grape	3m × 3m	Sapota	8m × 8m
Jackfruit	8m × 8m	Tamarind	8m × 8m

8.8 TRAINING AND PRUNING

When a plant is tied, staked, supported, propped or allowed to trail or spread over a structure, the specific operation is termed as training. It is done to give the plant a strong frame work and shape. Basic training types are central leader system, open centre system and modified leader system. Specific types of training are bush system, pyramid system, espalier system, cordon system, tatura trellie, head system, kniffin system, telephone system and bower system. Pruning refers to judicious removal of plant parts like roots, leaves, stems, branches, flower buds, flowers, even immature fruits for good yield and quality fruits. Different types of pruning are heading back, thinning, ringing or girdling, notching, nicking, root pruning, smudging, bending, coppicing, pollarding, lopping, pinching, disbudding, containment pruning and alternate differential pruning.

8.9 THINNING

Fruiting is an exhaustive process when there is heavy fruit set. Due to heavy fruit load and all the fruits are allowed to mature, the tree becomes devitalized and produces inferior fruits. In order to ensure the amount of new shoot growth and flower bud formation, thinning is practised in certain fruit plants. It is a practice of removal of part of flower buds, flowers or certain fruits to encourage fruit size, quality like colour, TSS, nutritive value and to increase annual and marketable yield. It may be done manually or chemically. Manual thinning though expensive is done in date palm in which some bunches or strands from each bunch are removed. Chemical thinning is practised in some cultivars of grapes by application of plant growth regulators particularly NAA. The amount of thinning of fruits depends upon crop set, response of the cultivar, nature of thinning, age of the plant and value of increased size of fruit in the market.

8.10 CROPPING

For successful cropping effective pollination plays a significant role. The plant may be self-pollinated or cross-pollinated. Various factors that affect significant pollination are dioecious nature of plant, peculiar structure of flowers, dichogamy, functionally unisexual flower, self- incompatibility, requirement of artificial pollination, impotence of pollen grains, abortive flowers etc. Cropping also depend upon fruit set. The plant may be fruitful one or unfruitful or barren. Unfruitfulness is associated with some external factors like nutrient supply, pollinizers, training and pruning, locality, season, temperature, light, disease and pest, unwanted spraying of chemicals and fungicides during blooming.

8.11 SEEDLESSNESS

There is increasing demand among consumers for novel and seedless fruits in certain crops. Seedlessness occurs due to parthenocarpy. But it can be induced by use of growth substances like GA_3 at higher concentrations and changing the ploidy level by crossing. Most promising effect is seen in guava, watermelon and loquats.

8.12 CONTROLLING FRUIT DROP

Fruit trees usually bear a large number of flowers in a panicle or solitary. When the fruit set is much more than the tree can normally carry to maturity, there is drop of fruits at different stages of development. The fruit drop starts from shortly after opening of flower, at pea stage or at marble stage. Sometimes there is pre-harvest fruit drop also which may be due to formation of abscission layer, high temperature, low humidity, very low temperature, heavy wind, hailstorm, fluctuation in soil moisture content, deep digging or deep ploughing, lack of available nitrogen, pathogen activity, characteristics of the cultivar. Unwanted fruit drop can be reduced to some extent by timely irrigation, manuring, plant protection measures, provision of pollinizers, planting wind breaks, exogenous application of auxins like NAA, 2,4-D and 2,4,5-T at low concentrations.

8.13 ALTERNATE BEARING

Alternate bearing or biennial bearing or irregular bearing of fruit trees is an age-old problem. It is more common in perennial trees. This problem is very common in mango, olive, coffee, tamarind, apple, and certain cultivars of plums, pears, and coconut. Possible causes of this malady are genetic characteristics, bearing habit, old and senile stage of plant, C: N ratio or endogenous hormonal level.

8.14 ORCHARD SOIL MANAGEMENT

Soil is one of the most important factors associated with the success or failure of fruit production. Orchard soil management aims at conserving soil moisture, soil nutrition, controlling soil erosion, little disturbance to soil and reduction in cost of cultivation. Most of the soil management practices in the orchard enhance nutrient and water supply needed for growth and production of fruit trees. Orchard soil can be managed by improving soil nutrient status, conserving soil moisture, altering of soil properties like soil pH and soil structure, maintaining soil organic matter contents and useful soil organism.

Clean Culture

It envelops regular and deep ploughing to remove weeds in the orchard which is a regular and common practice in India. Clean cultivation though keeps the land weed free for more days in a year it brings lot of problems like depletion in humus content, injury to feeding roots, depletion in nitrogen content due to leaching, creation of hard pan and soil erosion. So, avoidance of deep and frequent cultivation should be practised.

Sod Culture and Cover Crops

Complete sod cover on the orchard floor causes a reduction in vigour and growth of the fruit plantations because of reduced nitrogen uptake. In order to maintain soil fertility status, mowing down of sod should be done regularly in addition to adequate nitrogen fertilization. Sod culture usually increases leaf phosphorous concentration in apples. The extent of soil moisture depletion varies with the vigour, rooting depth, and frequency of mowing of the orchard sod.

Shallow rooted grasses such as Kentucky bluegrass or annual blue grass deplete less moisture from the orchard soil profile than deep rooted sods such as clover and ryegrass. Competition of sod with fruit trees also modify the pattern of water depletion with more water being used from deeper in the profile. In summer, regular mowing of grasses is suggested to conserve moisture in soil by reducing water use by grasses. On the other hand, sod improves other physical properties as reflected by decrease in bulk density and increase in soil porosity. Due to decrease in bulk density there is increase in total pore space available for root growth and increase in water holding capacity. Such changes are likely to increase earthworm populations, reduce surface moisture runoff and erosion, and conserve moisture.

Soil under permanent grass may begin to lose water 15-30 days earlier in the spring as compared to the cultivated soil. Grasses in orchard require a large proportion of water for maintaining growth, and therefore in non-irrigated orchards this competition of water with fruit trees can become critical. Potassium returns are high from sod clippings.

Mulching

Organic mulches add significant amount of nitrogen to the soil and consequently improve soil fertility. Leaf phosphorus concentration is often reduced due to reduced under-tree vegetation in mulched trees. Large amount of potassium is added to the soil by many mulching materials. It is well recognized that conservation of soil moisture is the most significant advantage of mulching. This useful effect is more pronounced during the drying periods of May-June.

Cultivation

Cultivation accelerate the mineralization of organic matter and increase NO_3-N availability, however total nitrogen tends to decline under cultivation. Cultivation improves leaf size and colour, due to mineralizing of nitrogen from the soil. Removal of weed and cover crop competition by cultivation increases soil moisture content relative to permanent orchard vegetation. However, moisture content of cultivated soil may not differ significantly from mulched soil, and may be less than herbicide-treated soil. Also, the moisture content of cultivated soils may show considerable seasonal variation depending upon the time of cultivation and cover crop or weed re-growth.

Soil Moisture and Water Uptake

The efficiency of orchard soil management method is generally judged by its ability to conserve soil moisture and make it available for use by the trees for growth and development. In water-logged areas, sometimes soil is dried out towards maturity of fruits for the purpose of hastening maturity and in heavy compact soil a light tilling is done to provide better aeration.

Herbicides

In general, herbicide-treated orchard soils have a lower soil moisture deficit in comparison to grassed or cover cropped soil.

Cropping System

It includes intercropping, mixed cropping and multitier cropping. Intercropping is generally followed in young orchards when the canopy coverage of main plant is very low. To get some extra income from the land not covered by the main plant through its canopy by planting vegetables, tapioca, turmeric, ginger and banana in the interspaces. It not only provides extra income but also covers the soil checking weed growth. By manuring and irrigation to the intercrop the main crop also takes the advantages of it. Choice of intercrop depends upon the duration of intercrop, distribution of roots, water requirement etc.

Mixed cropping is a practice of growing certain perennial crops in the alley spaces of main orchard plants. It takes care of utilization of available land and increase the net income per unit area.

In multitier cropping, crops of difference in their height are planted in orchard for efficient utilization of land both in time and space dimension. As fruit trees remain alive as long as fifty years, the land remains occupied for that period. The system effectively utilises the land, water and solar radiation.

Organic Farming

Excessive use of chemical fertilizers and plant protection measure through chemicals boost up the production no doubt. But considerable damage is caused to human being, soil health and the environment. It is a topic of world concern. There is necessity of transferring conventional farming to organic farming which is ecologically sound, viable and sustainable. This production system largely excludes the use of synthetically compounded inorganic chemicals. It entirely relies on crop rotation, crop residue, animal manures, legumes, green manures, off-farm organic waste, bio-fertilizers, bio-pesticides, mechanical cultivation, biological pest control methods etc.

8.15 WATER MANAGEMENT IN ORCHARD

- Water is the driver of life.
- It is an important essential natural land resource affecting growth and production of fruit crops.
- Fruit trees require irrigation water for maintaining adequate growth, fruit quality and yield particularly in dry months.
- In spite of high rainfall, fruit plants suffer from drought at any stage due to erratic rainfall pattern and for the want of irrigation water.
- In orchard, water management can be done by *in-situ* moisture conservation, water harvesting and efficient utilization of conserved/harvested water through improved cultural practices.
- Mulching is the best practice of *in-situ* conservation of moisture.
- Rainfall and low water springs and rivulets are the natural water resources that can be conserved.
- Water harvesting can be done through *in-situ* rainwater harvesting and water harvesting in farm ponds.
- In fruit plantations, rain water can be harvested in the tree basin prepared prior to rain through infiltration of water and subsequent shallow hoeing.
- For arid and semi-arid regions, water can be conserved by 'crescent bund with open catchments, pits, trench systems, V- ditch etc.
- The run-off water from orchards and water flowing from other sources like spring and small streams can be effectively stored in suitable small or big reservoirs.
- Small ponds may also be used as water reservoirs in which water is stored during lean period and used when required by crops.
- The provision of a small pond in one corner of the orchard to collect runoff water during high intensity rains and its utilization as life-saving irrigation or during critical periods of the crop is an age-old practice.
- The water conserved/harvested from natural resources must be used very efficiently for fruit crops through pressurized irrigation system *e.g.,* drip irrigation.
- Water can be most effectively used by adopting high density planting, early cropping, efficient harvesting and canopy management particularly conical tree shapes, telephone system, tatura trellis, V-system, spindles, kniffen system and using efficient root stock.

8.16 CROP REGULATION

- Crop regulation is the basis for the regular and quality crop which can be achieved through manual thinning, chemical thinning, selective harvesting, training, summer and winter pruning, prevention of pre-harvest fruit drop, etc.
- During heavy crop load the fruits cannot be adequately sized to meet market requirement for higher price. So, there is selective removal of some fruits at pea stage.
- Though thinning may reduce the total yield, net return increases as larger fruits fetch much higher price in the market.

- Use of chemicals including plant growth regulators (PGRs) as chemical thinning agents is gradually replacing hand and mechanical thinning in stone fruits with certain considerations.
- Usually 30–40 leaves per fruit are sufficient to produce fruit of good quality.
- Irrigation management during the critical stages of fruit development is essential.
- Evaporation of water from the shoots cools the buds and therefore delays their break. So, misting of water is done during day time at frequent interval from bud swell stage until bloom.
- Application of growth regulators like GA_3 at 200 ppm and ethrel at 200 ppm delay flowering in spring by 3–7 days depending upon species and thus markedly increase fruit set.
- Autumn application of GA_3 and ethrel cause delaying of bloom in spring in stone fruit, which can be a useful practice to increase fruit-set in frost affected areas.
- Application of GA_3 and ethrel help in better synchronization flowering period and cross-pollination.
- In blossom thinning, chemicals are applied at bloom, whereas in fruit-lets thinning, chemicals are sprayed a few days after petal fall, i.e., after fruit set.
- Fruit crops grown in different climatic zones — tropical, subtropical and temperate— are required to be trained and pruned accordingly, depending on the type of plant and specific objectives of fruit growers.
- In mango approximately only 0.1% of perfect flowers develop fruits to mature because of high rate of fruit drop.
- Extent of fruit drop in mango can be controlled by regular irrigation during fruit development period only after fruit set.
- Biennial bearing in mango can be controlled by training the plant to open centre system or through application of paclobutrazol and KNO_3
- Smudging enhances fruit set in mango.
- *Bahar* treatment in guava regulates flowering and fruiting.

8.17 REJUVENATION OF OLD ORCHARDS

After few years of productive returns the orchard becomes unproductive or unfruitful due to the following reasons:

1. The plants become old and unproductive.
2. Diseased limbs.
3. Injured or damaged limbs by rodents.
4. Damaged by winter injury.
5. Damaged by implements.

To overcome these, rejuvenation is required. It is practised in already established old plants. It is done by top grafting. Top grafting is achieved by top working and frame working. It is done for following reasons:

1. To change the top portion.
2. To test the performance of a seedling, hybrid.
3. To renovate the injured parts and to renovate unproductive orchards.

Top working is a practice of converting an undesirable plant to desirable one by budding or grafting the top portion when seedling fruit trees are at active growth stage. In the frame working, almost all the primary scaffold branches are retained on the tree and small laterals arising from secondary scaffold are grafted by side grafting method. It is done at higher locations.

REFERENCES

Kumar N 2010. *Introduction to Horticulture.* Oxford and IBH Publishing Co. Pvt. Ltd., New Delhi.

Singh J 2017. *Fundamentals of Horticulture.* Kalyani Publication, Ludhiana.

http://ecoursesonline.iasri.res.in/course/view.php?id=151

OUTCOMES ASSESSMENT

PART A

Answer the following questions (True or False).

1. Plants are planted at appropriate distance to encourage proper growth. (True/False)
2. Application of GA_3 helps in better synchronization flowering period. (True/False)
3. Top working is a practice of converting an undesirable plant to desirable one. (True/False)
4. Smudging enhances fruit set in apple. (True/False)
5. *Bahar* treatment in guava regulates flowering and fruiting. (True/False)

PART B

Answer the following questions.

1. Cropping system includes intercropping, — and —.
2. What is crop regulation?
3. What do you understand by blossom thinning?
4. What is the function of organic mulch?
5. Alternate bearing is common in —, — and —.

PART C

Write a brief note on each of the following.

1. Cropping system.
2. Thinning.
3. Sod culture and cover crops.
4. Crop regulation.
5. Rejuvenation of old orchards.

Chapter 9

Planting System and Planting Density

9.1 INTRODUCTION

Establishment of an orchard incurs long term investment for which it needs a very critical planning. As it is a long-term investment, plants planted in the orchard should suit to the prevailing climatic condition of that area. The land remains occupied as long as 40 to 50 years. Any mistake committed during initial year is hard to rectify later on. Proper layout of the orchard ensures optimum utilisation of land and resources for successive economic production. It becomes easy for different inter-cultural operations, harvesting and transportation of the produces. Uniform distribution of plants in the selected land in various systems of planting is done after proper layout.

9.2 LAYOUT

The proper layout of orchard aims at accommodating maximum number of trees per unit land area, maintaining adequate space for proper development of the trees and convenient orchard management practices.

The method of planting is broadly divided into 2 types:

i. ***Vertical Row Planting*:** Here trees are planted in rows exactly at perpendicular to each other in their adjacent rows, e.g., square system, cluster system and rectangular system of planting.

ii. ***Alternate Row Planting*:** In this method the plants are set at right angle in alternate rows only not to adjacent rows, e.g., triangular system, hexagonal system, contour system and quincunx system.

9.2.1. Vertical Row Planting System

Square System

- The system of planting is most easy and popular (Fig. 9.1).
- The trees are planted at each corner of the square accommodating 4 number of plants in each square. So, the row to row and plant to plant distance remains same.
- Adequate space is available for different intercultural operation. The extra space at the early stage of plant growth can be utilized for growing some quick growing vegetable crops for getting extra income or remuneration.

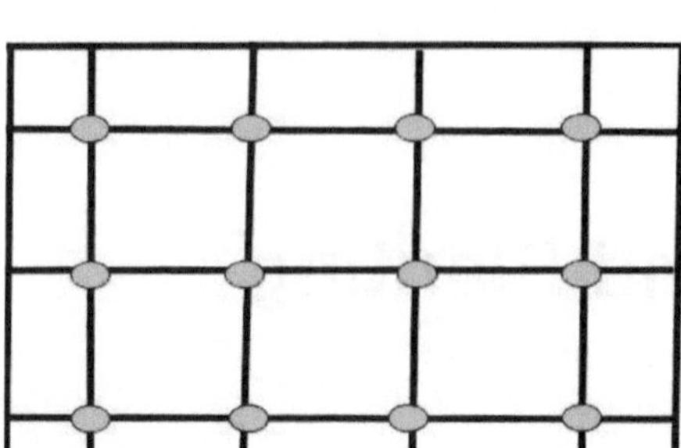

Fig. 9.1 Square System

Rectangular System

- The planting system is similar to square system but it differs on the basis of row to row spacing and plant to plant spacing. It gives a rectangular space (Fig. 9.2).
- The plants are planted at four corners of a rectangle
- Inter-cultural operation can be done through both ways.
- The plants get proper sunlight and space for their development. As interspaces are wider mechanical operations can be carried out easily.

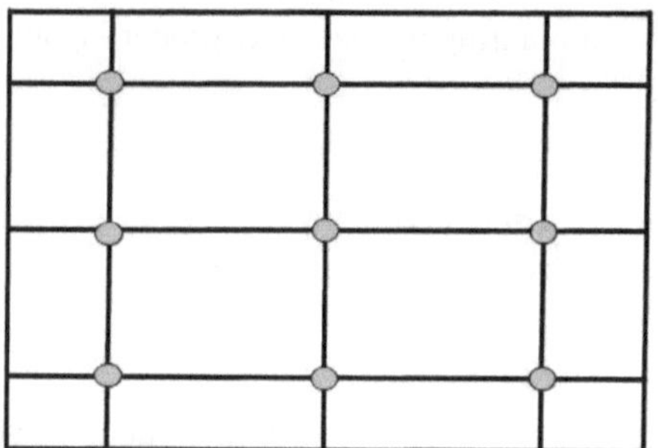

Fig. 9.2 Rectangular System

Cluster System

- Trees are planted at each corner of the square forming a cluster.
- Each cluster is set a part at double the distance of trees planted in the cluster.
- Here there is no equal distribution of space per tree.
- Due to wider distribution of trees different orchard operations become easy.
- This system accommodates nearly twice the population of plants than square system.

9.2.2 Alternate Row Planting

Triangular System

- Fruit trees are planted as in square system (Fig. 9.3).
- Three trees represent at three angles of an isolateral triangle.
- Trees in the even number of rows are midway between those in odd numbers.
- The distance between any two adjacent trees in a row is equal to the perpendicular distance between any two adjacent rows.

- Vertical distance between immediate two trees in the adjacent rows is equal to the product of 1.118 × distance between two trees in a row.
- The system accommodates 11% less number of plants as compared to square system but each and every tree get more space.

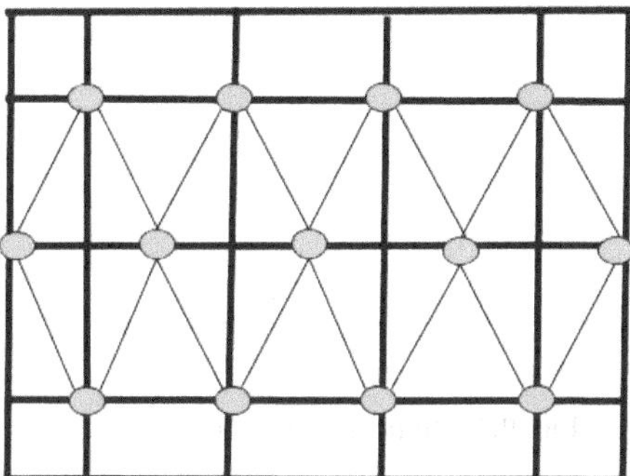

Fig. 9.3 Triangular System

Hexagonal System

- Plants are planted at each corner of an equilateral triangle (Fig. 9.4).
- So, six plants at six corners make a shape of hexagon.
- Sometimes a 7th plant can be planted at the center of hexagon.
- So, the system is also known as septule as it accommodates the 7th plant.
- Each tree can get equal space but layout of the system is difficult.
- The perpendicular distance between any two adjacent rows is equal to the product of 0.866 × the distance between any two trees.
- It allows three directional inter-cultural operations which is not so easy.
- The system is very intensive one and carried out in soil having high fertility status.
- In the vicinity of cities where land cost is high this system is worth adapting.
- This system of planting accommodates 15% more plants than the square system.

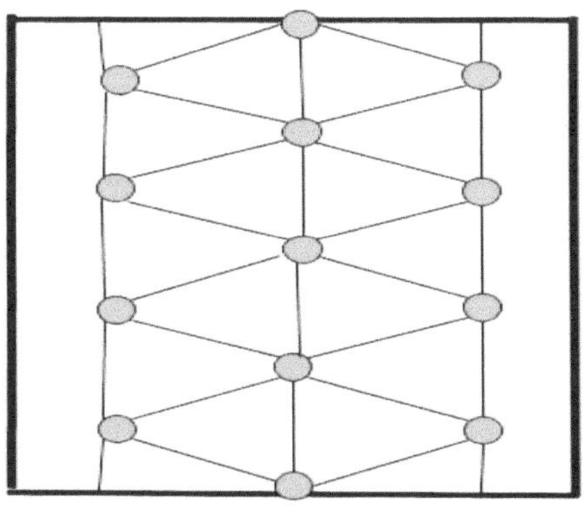

Fig. 9.4 Hexagonal System

Quincunx or Diagonal System

- The system is similar to square system except that a fifth plant is planted at the center of diagonal cross (Fig. 9.5).
- The fifth plant is known as filler plant.
- The filler plant should be short-lived, early in fruiting, early in growth and generating extra income before the main plant come to flowering and fruiting.
- The filler plant efficiently utilizes the extra space.
- This system holds good where the distance between two adjacent plants is more than 10 m.
- The filler plant is removed when the main plants come to usual flowering and fruiting stage.
- This system accommodates almost double plants over square system.

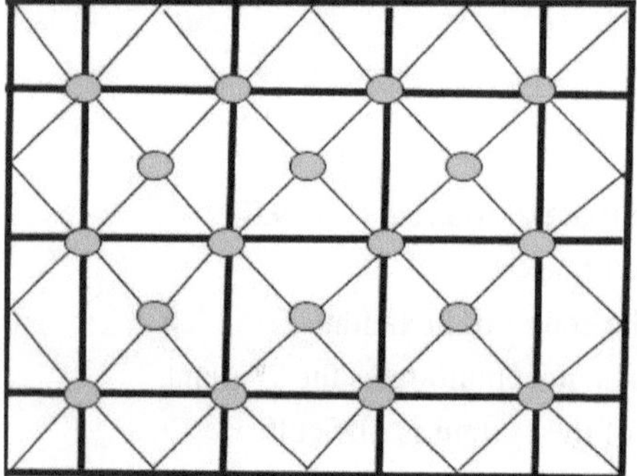

Fig. 9.5 Quincunx System

Contour System

- It is practised in hillocks or hill slopes where land available is undulating and there is greater danger of soil erosion (Fig. 9.6).
- As irrigation in the hill slopes is difficult, rain water harvesting structures should be constructed or drip irrigation can be provided.
- Drip system proves better in this condition because it operates at low pressure.
- Planning for irrigation should be in such a way that the water applied should slowly penetrate into soil without causing soil erosion.
- Plants are planted at the space available along the contour across the slope.
- For layout of the system along the contour the altitude of the area is marked; points coming under same altitude are grouped together and plants suitable for that altitude are planted.
- This system utilizes the rainwater conserved.
- This system checks the soil erosion and utilizes the rain water.
- The system looks irregular in shape as planting distance is not uniform.
- In terrace system, planting is done in flat strips of land formed across a sloping side of a hill.
- Width of contour varies according to the nature of slope.

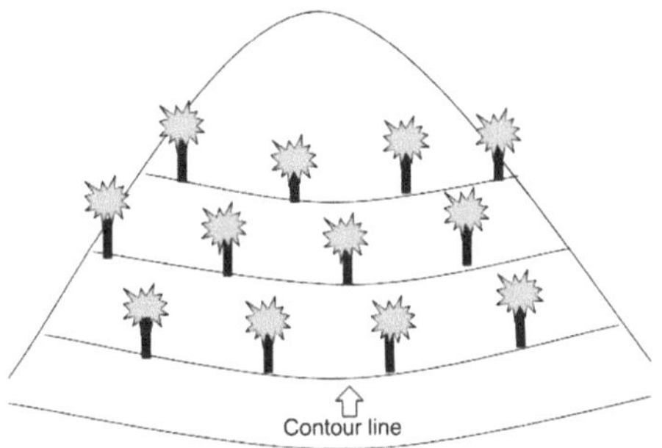

Fig. 9.6 Contour System

Fig. 9.6 Contour System

9.3 INTENSIVE PLANTING SYSTEM

This system deals with how to include a greater number of plants per unit area for efficient utilization of space, light, nutrient, irrigation water etc. for maximization of profit. Different intensive planting systems are:

9.3.1 High Density Planting (HDP)

It refers to planting of fruit trees at a closer spacing than the recommended one using certain special techniques for getting maximum productivity per unit land area without sacrificing the quality. This system is more useful in perennial fruit plants. It helps in efficient utilization of land, other resources, better canopy management, farm mechanization, convenience in spray of chemicals, harvesting etc. It is one of the most promising methods of increasing productivity. By adopting HDP yield can be boosted two to three-folds than the normal. The special techniques to achieve this are:

- By using dwarf root stocks.
- By using dwarf inter stocks.
- By inserting dwarf scion cultivars.
- By using growth retardant.
- By regulating shape and size by canopy management and by adopting suitable management practices.

The system was first developed in Europe during 1960's in the crop apple. Now it is being practised in all apple growing areas of the world, i.e., Europe, America, New Zealand, Australia. In India HDP is found successful in crops like apple, peach, pear, plum, banana, papaya, pineapple, mango, citrus, guava, and sweet cherry. In case of coconut and arecanut plantation HDP is advocated by using multi-crop species cropping system.

Types of HDP

- Low HDP: Accommodate <250 number of fruit trees per hectare.
- Moderate HDP: Accommodate 250-500 number of fruit trees per hectare.

- High HDP: Accommodate 500-1,250 number of fruit trees per hectare.
- Ultra HDP: >1,250 fruit trees per hectare.
- Super HDP: Accommodate about 20,000 fruit trees per hectare.

Advantages of HDP

- Early cropping with high yield for longer time
- Reduces labour cost
- Improves fruit quality

Characteristics of HDP

- The fruit tree should have maximum number of fruiting branches and minimum number of structural branches
- The tree should be trained in central leader system and fruiting branches grow horizontally
- Pruning should be done in such a way that each branch should provide minimum amount of shade on other branches
- The height should be maintained at one and half times of the diameter of plant at base

9.3.2 Meadow Orcharding

Original home of meadow orcharding is Israel. The system ensures planting of fruit trees at ultra-low spacing of 45-75 cm² within row and 210-270 cm² between the rows accommodating about 30,000 to 1,00,000 plants per hectare. Management is very essential, so plants are pruned at regular interval at a height close to ground level. The growth retardants play an important role in maintaining such type of orchard system.

9.4 CALCULATION OF NUMBER OF PLANTS IN DIFFERENT SYSTEMS OF PLANTING

9.4.1 Square System

$$\text{Number of plants per hectare} = \frac{\text{Area in square metre}}{\text{Planting distance}}$$

$$= \frac{10{,}000\,\text{m}^2}{\text{row to row spacing(m)} \times \text{plant to plant spacing(m)}}$$

If the spacing is 10 m × 10 m then,

$$\text{Number of the plants accommodated in one hectare of land} = \frac{10{,}000\,\text{m}^2}{10\text{m} \times 10\text{m}} = 100$$

9.4.2 Rectangular System

$$\text{Number of plants per hectare} = \frac{\text{Area in square metre}}{\text{Planting distance}}$$

If the plant to plant spacing is 8 m and row to row spacing is 10 m then,

$$\text{Number of the plants accommodated in one hectare of land} = \frac{10{,}000\,\text{m}^2}{10\text{m} \times 8\text{m}} = 125$$

9.4.3 Triangular System

$$\text{Number of plants per hectare} = \frac{\text{Area in square metre}}{\text{Row to row distance (m)} \times \text{plant to plant distance (m)}} \times 0.866$$

9.4.4 Hexagonal System

$$\text{Number of plants per hectare} = \frac{\text{Area in square metre}}{\text{Planting distance}}$$

Suppose Area is one hectare = 10,000 m^2 or 100 m × 100 m.

Planting distance: If plant to plant distance is 10 m, the row to row distance appears to 8.66 metre as specified under the head calculation of planting distance in hexagonal system ahead.

Hence, $\text{Number of plants} = \frac{100 \times 100}{10 \times 8.66} = 115 \text{approx.}$

9.4.5 Quincunx system

As the plants are planted additionally in the centre of the square, hence, first the number of plants is calculated for square system of planting and then number of additional plants planted in the centre of each square is calculated. Both the numbers are summed up to get total number of plants planted in Quincunx system of planting.

$$\text{Number of plants in square system of planting} = \frac{\text{Area in square metre}}{\text{Planting distance}}$$

$$= \frac{100 \times 100}{10 \times 10} = 100$$

Additional plants planted in centres of the squares =

(Number of rows length wise – 1) × (Number of rows width wise – 1)

In 100 × 100 m^2 field, if planting distance is 10 × 10 m, number of rows length-wise and also width-wise will be 10.

Hence, Number of the plants = (10 – 1) × (10 – 1) = 9 × 9 = 81

Total number of plants = plants planted in square system of planting + Additionally planted plants in the centre of square = 100 + 81 = 181

Thus, in quincunx system of planting almost two times plants (181) are accommodated in comparison to square of planting (100).

REFERENCES

Kumar N 2010. *Introduction to Horticulture*. Oxford and IBH Publishing Co. Pvt. Ltd., New Delhi.

Singh J 2017. *Fundamentals of Horticulture*. Kalyani Publication, Ludhiana.

https://www.agrihortieducation.com/2016/09/systems-of-planting.html

http://ecoursesonline.iasri.res.in/course/view.php?id=151

OUTCOMES ASSESSMENT

PART A

Answer the following questions (True or False).

1. Establishment of an orchard incurs short-term investment. (True/False)
2. The proper layout of orchard aims at accommodating maximum number of trees. (True/False)
3. Square System of planting is most difficult. (True/False)
4. Original home of meadow orcharding is USA. (True/False)
5. High density planting refers to planting of fruit trees at a closer spacing than the recommended one. (True/False)

PART B

Answer the following questions.

1. What is intensive planting system?
2. Define vertical row planting.
3. Triangular system, hexagonal system, contour system and quincunx system belong to — system.
4. What do you understand by contour system of planting?
5. Quincunx and Diagonal Systems are the same/different.

PART C

Write a brief note on each of the following.

1. Square system.
2. Hexagonal system.
3. Meadow orcharding.
4. High density planting.
5. Rectangular system.

Chapter 10

Training and Pruning of Fruit Crops

10.1 INTRODUCTION

Most botanists identify trees by their leaves, flowers and fruit. However, the plant architecture specialist is interested in trunks, branches and twigs. All types of fruit tree do not require pruning, e.g., mango, sapota, while some fruit trees can grow well naturally, e.g., pineapple, papaya as they do not require pruning. Most deciduous trees like apple, pear, almond etc. and grapes, ber, fig, citrus, pomegranate, guava etc. require pruning to train them for desired shape.

10.2 TRAINING

Training is an operation by which the plant is made to develop an orderly structure or frame work. This is done by staking, tying, supporting, propping, trailing or spreading on certain structure with or without pruning of plant. Training is usually done when plants are young. Bending, twisting, decapitations are done by using springs of hard wire, where it is required. It is a physical technique which controls the shape, size and direction of plant growth known as training *or* in other words, training in effect is orientation of plant in space through techniques like tying, fastening, staking, supporting over a trellis or pergola in a certain fashion or pruning of some parts.

10.2.1 Objectives

1. To provide the basic tree form and to aid the development of a strong tree framework by encouraging strong wide crotch angles.
2. To ease cultural practices including inter-cultivation, plant protection and harvesting.
3. To maximize utilization of light as it plays key role in flower induction as well as in fruit development through carbohydrate synthesis.
4. Avoidance of build-up of micro-climate congenial for pests and diseases.
5. Controlling excessive vegetative growth which reduces bloom and fruit set.
6. Preventing upright growth and developing horizontal laterals.
7. Developing rigid, strong and self-supporting laterals.
8. Maintaining fruiting branches in desired direction.
9. Increasing the orchard labour-use-efficiency.
10. To expose maximum leaf surface area to sun.
11. To permit proper levels of light and air distribution to the tree interior.

12. To direct the growth of the tree so that various cultural operations such as spraying and harvesting can be done easily.
13. To achieve orchard uniformity.
14. To encourage a structurally sound tree by removing branches with weak narrow crotch angles or those with diseased, damaged, or unproductive wood.

10.2.2 Methods of Training

Method of training of a plant is determined by the nature of plant, climate, purpose of growing, planting method, mechanization, etc. and therefore, intelligent choice is necessary.

10.2.3 Factors Determining the Shape of Plants

1. Location of the growing points on the main stem from which branches arise.
2. Subsequent orientation of the branches.
3. Branches may be kept around the stem to produce a natural shaped plant or they may be oriented in a single plane to provide a flat shape known as 'Espalier'. There are many variations in the two general shapes that they differ in height of the stem before the first branch arises from the main stem.

10.2.4 Training of Woody Perennials

The woody perennials, which are widely spaced and remain on a place for a long duration, are trained to develop strong framework for sustainable production of quality produce and for ornamental beauty in different shapes (topiary). In these plants following types of training are followed.

Open Centre System (Vase Shaped)

In this system the main stem is allowed to grow to a certain height and the leader is cut to encourage lateral scaffold from near the ground giving a vase shaped plant. The main trunk is allowed to grow up to 75 cm by cutting within a year of planting. All side branches are headed back and resemble a conical shape (Figure 10.1). This system lacks strong crotches and provides weak frame work and as such is less satisfactory.

Fig. 10.1 Open centre system of training

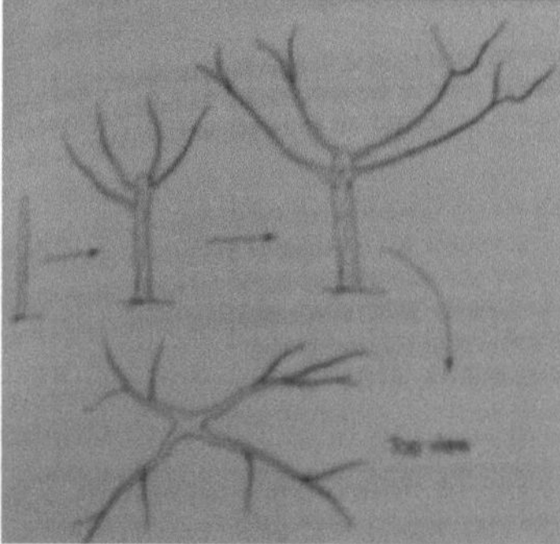

Fig. 10.1 Open centre system of training

However, it may be favoured for more sunlight for better colouration of the inside fruits. But under sub-tropical conditions where sunshine is plenty and there are strong winds during summer, this system is not suitable in view of the weak framework and other obvious defects. It is for low density planting system (250–500 trees/hectacre). This is common in peaches, apricots and ber.

Central Leader System (Closed Centre)

In this system the central axis of plant is allowed to grow unhindered permitting branches all around. This system is also known as closed centre system. Here proper selection and proper training of the scaffold branches is done. The first scaffold limb should be at least 45 cm from the ground. Additional scaffold limbs should be separated along the trunk by a minimum of 20 cm and be well distributed around the trunk (Figure 10.2). When the leader becomes too tall to harvest, it should be headed back into two-year old wood. Branches that have shade in excess should be removed. Maintain the cone shape of the tree. Remove shoots that are not productive. It is a medium density planting system accommodating 625 to 1000 plants per hectare and is commonly practiced in apple, pear, mango and sapota.

Fig. 10.2 Central leader system (closed **centre)**

Modified Central Leader System

This system is in between open centre and central leader system wherein central axis is allowed to grow unrestricted up to 4–5 years and then the central stem is headed back and laterals are permitted. It is first trained like central leader by allowing stem to grow for the first two years and then headed back at 75 cm height. Lateral branches are allowed to grow and cut back as in open centre system. With this system, the height of tree is lowered as the length of the main trunk is reduced. The scaffold branches are encouraged to become larger and grow to greater length. The trees possess strong crotches and a durable frame work (Figure 10.3). The system suits most of the commercial fruit trees because the height of the tree is comparatively less, which facilitates operation like spraying, pruning, harvesting. It is medium density planting system accommodating 625 to 1000 plants per hectare. It is common in apple, pear, cherry, plum, guava.

Fig.10.3 Modified central leader system

Cordon System

This is a system wherein espalier is allowed with the help of training on wires. This system is followed in vines incapable of standing on their stem. This is the easiest system to use for dwarf type apple tree. The cordon tree is a single stem with pruned, short side shoots (fruiting spurs).The sapling is planted at an angle of 45°.The cordon is built up from three 25 mm wires that are 60 cm apart from each other, making three levels from the soil surface, which are respectively 60 cm, 120 cm and 180 cm height from the ground. Once the wires are fixed, trees are tied to the wire. For each tree, a stake is fixed approximately 2.4 m long (Figure 10.4). All side shoots longer than 10 cm should be pruned after the third bud (summer pruning). Many of the production systems need a permanent support structure. This can be trained in single cordon or double cordon and commonly followed in crops like grape and passion fruit.

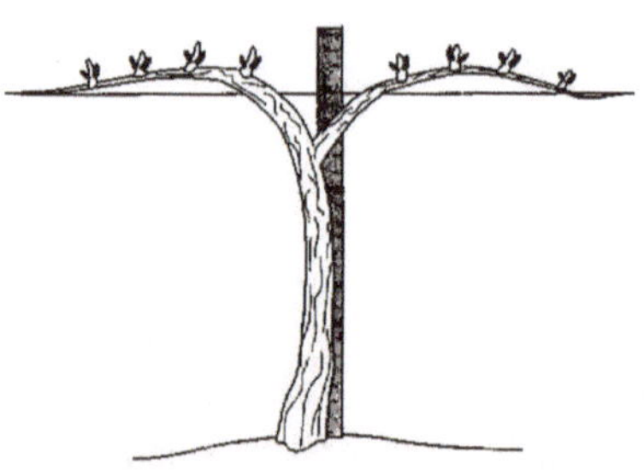

Fig. 10.4 Cordon system of training

Fig. 10.4 Cordon system of training

Tatura Trellis (Australia) (V & Y Trellis)

It is a high-density planting system which can accommodate (1000-4500 trees/ hectare). It results in early and high yields. Tree spacing is reduced to control vigour (Peaches). In this system of training the canopy volume of the tree is filled up quickly. There is orderly array of branches and uniform distribution of leaves which facilitates easy light interception (Figure 10.5). This system is easy to carry out and also facilitates the use of equipment and mechanical aids for carrying out intercultural operations like spraying and harvesting. In high density planting, this system of training of plants is very popular being very yield efficient. Trees are planted at a spacing of 5 × 1 m or 6 × 1 m. At the time of planting, one-year old plant is headed back to 20 cm above the ground level. In next growing season two limbs or branches are selected in opposite directions and these branches are trained across the inter-row space at an angle of 60° from the horizontal, forming V-shaped canopy. The canopy is supported by a permanent trellis constructed of high tensile galvanized steel fence posts. The secondary branches are developed along each primary branch forming fruiting canopy. Since fruit hangs underneath canopy, it results less sunburn and bruising of the fruits. However, it incurs higher establishment costs (trellis system).

Fig. 10.5 Tatura trellis system of training

Horizontal Espalier

The horizontal espalier training system consists of a set of horizontal wire trellis attached to walls or fences. When starting from a seedling with a single slender stem, it is cut just below the first wire. This practice causes new shoots to develop from lateral bud. Two lateral shoots of equal vigour are selected and trained on stake tied diagonally to the wire (Figure 10.6). Once established, side shoots will be produced, first on the lower limbs. These shoots are pruned in summer to form fruiting spurs in the next season. As fruiting spurs increase in number over the year, they should be thinned out to avoid overcrowding and reduced plant vigour.

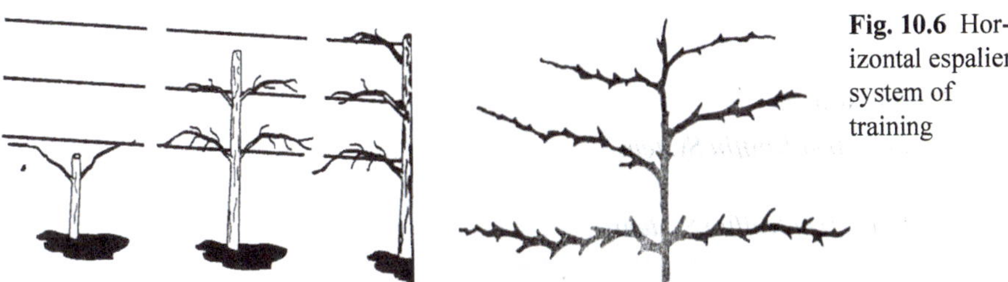

Fig. 10.6 Horizontal espalier system of training

Fig. 10.6 Horizontal espalier system of training

Palmette System

The palmette training system is the variation of the espalier system whereby plants are trained at about a 40° angle instead of having horizontal branches. The candelabra palmette training system uses a lattice framework consisting of horizontal and vertical arms to create balanced and attractive trees (Figure 10.7). This system improved orchard productivity inducing earlier bearing and enabling higher planting densities. This method has been widely adopted in modern commercial planting in Italy, France and other European countries. Rows are spaced 5-6 m apart with a distance of 3-4 m between trees. Height of the tree will remain around 5 m. Trees are trained on 4 wire system, the wire being spaced at one-meter intervals. The trees when fully grown consist of a central leader and four pair of framework branches that are supported by the four wires. It is a semi-high-density planting system accommodating 1000-2400 trees/hectare. It produces high quality and high quantity fruits. It is well suited to all the conditions (species, environment, rootstock/cultivar etc.). There is lesser need for summer pruning.

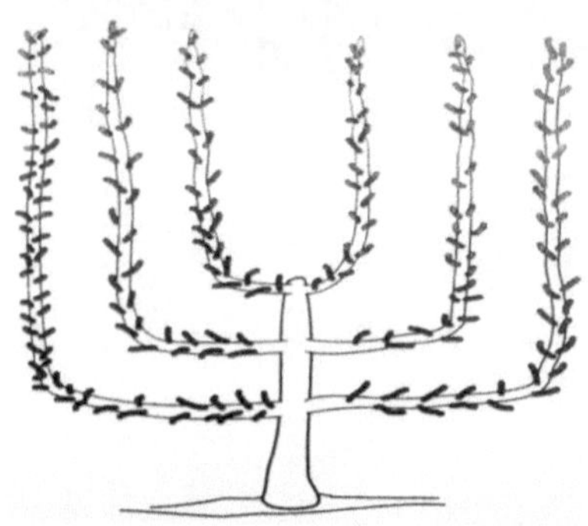

Fig. 10.7 Candelabra palmette system

Other Methods of Training

1. *Caldwall System of Training*: The branches of the plant are bent and pegged to the ground. This makes the plant to come to bearing early and yield better. No pruning is involved in this method of training, e.g., erect growing cultivars of guava.
2. *Fan System*: This method of training is suitable for wall side planting in home gardens. A fan shaped frame is developed by allowing the branches to grow in only one plane parallel to the wall, e.g., ornamental banana.
3. *Kniffin System*:
   (a) *Two Arm Kniffin System*: Two arms are allowed on the main stem at a height of 5-6 ft and they are trained on to wire on either side, e.g., grape.
   (b) *Four Arm Kniffin System*: Here there will be four arms two each on either side of the trunk. The first pair of arms arise at a height of 3-4 ft and the second at 5-6 ft, e.g., grape.
4. *Over Head Trellis or Telephone System*: The wires will be fixed horizontally on some posts just as telephone wires. The plant will be trained on to the wires with a suitable frame work, e.g., grape, passion fruit

5. *Pendal or Bower or Arbour System*:

   A pendal will be erected with the help of pillars and wires and the plant will be trained on to the pendal with suitable frame work. In this system vines are generally planted at a distance of 3 m × 3 m, thus accommodating 444 plants per acre. In this system, the vines are trained as single stem up to a height of about 6 feet. Subsequently trained in such a way to have at least 2 primary arms up to a maximum of four, so as to have uniform network of secondaries and tertiaries along all the sides of the bower, e.g., grape. (Figure 10.8).

Fig. 10.8 Bower system/Pergola/Pendal system in grape

6. *Y-Trellis System*:

   The Y- trellis system is designed as per the shape of Y, an English alphabet. The first and the last Y of a row should be supported with galvanized (G.I.) wired rope fixed in a concrete base. Six wires (three on each side of Y) are passed through to give support and train the growing vines. (Figure 10.9)

   Advantages of Y-trellis over Bower system:

   (a) Early fruit maturity

   (b) Better fruit quality

   (c) Low incidence of diseases

   (d) Ease in cultural operations

   (e) Less health hazards

Fig. 10.9 Y-trellis system in grape

7. *Single Stake or Umbrella System*: The main trunk will be supported by a stake. The trunk is beheaded at a height of 5-6ft. The branches which arise on the trunk will be hanging freely, e.g., grape.

10.2.5 Training in Different Shapes

Generally ornamental bushes are trained in different shapes for the purpose of enhancing beauty of places. These shapes could be vase, cone, cylindrical and rectangular box, flat and trapezoid. Presently for the convenience of mechanization these shapes are being utilized in fruit trees. Such shapes are given to adjust the geometry of plantation like hedge row system, box, and unclipped natural in fruits like guava, mango, sapota and citrus.

10.2.6 Details of Training

Height of The Head

This is the height from ground to first branching or scaffolding. Depending on the height the trees could be divided in three groups.

(a) *Low Head*: Here the height of first branching is maintained at 0.7-0.9 m from ground. This is common in windy areas. Such plants are easy to maintain.

(b) *Medium Head*: Here the height of first branching is maintained at 0.9 m to 1.2 m from ground. This is the most common height which combines both effects and ability to stand against wind and for easy management.

(c) *High Head*: Here the height of first branching is maintained at more than 1.2 m. It is common in tropics in wind-free areas. Operations under the canopy are easy to perform.

Number of Scaffold Branches

It refers to allowing of number of scaffolds on the primary axis of the tree which vary from 2 to 15 but extremes are undesirable. In fruit trees 5 to 8 scaffolds are preferred to make the tree mechanically strong and open enough to facilitate cultural operations.

Distribution of Scaffolds

Scaffolds should be distributed in all the directions spaced at 45-60 cm allowing strong crotches through wide angles of emergence. A well-trained tree is an asset to the farmer and therefore, efforts should be made for training trees appropriately in formative years for sustainable production. In fact, the process should have begun from nursery itself.

10.3 PRUNING

10.3.1 Definition

Pruning refers to cutting away a portion of the tree to influence its growth, flowering and fruiting with a view to improve fruit quality and maintain a balance between vegetative and reproductive growth. It refers to removal of plant part like bud, shoot, root etc. to strike a balance between vegetative growth and production. This may also be done to adjust fruit load on the tree.

Pruning may also be defined as the art and science of cutting away of portion of plant to improve its shape, to influence its growth, flowering and fruitfulness and to improve the quality of the product. It is done to divert a part of plant energy from one part to another part of plant.

Pruning is the part and parcel of the cultural management of orchard, farm and plantation. Pruning is essential for crops like grapes and temperate fruits – apple, pear, peach etc. In addition, features like climatic conditions (seasons, temperature and rainfall) and nutritive condition of the soil etc., also contributes for the increased productivity of the plants. For example, in case of banana and pineapple, the pruning refers to the de-suckering of the clump, it is to regulate the number of bearing plants per acre or unit area.

10.3.2 Principles of Pruning

The principle underlying pruning is the encouragement of the plant sap to flow towards certain desired parts of the plant such as the stem, leaf or roots to promote their growth and vigour by removing certain other parts which are not wanted. Removal of certain parts of the plant results in the lessening of struggle for existence among the remaining parts of the plant. Thus, pruning is invigorating processes which produce a definite effect in the formation of shoot, flower, fruits and roots etc.

10.3.3 Objectives of Pruning

1. To maintain the growth and vigour of the trees and to have a balance between the vegetative vigour and fruitfulness, so as to be conductive for production of optimum crop of best quality.
2. To regulate the size and quality of the fruits by way of proper distribution of the fruiting area.
3. To regulate the succession of crop and to have the crop where it can be managed easily and cheaply.
4. To remove the dead, diseased and over-aged wood.
5. For effective spraying of pesticides to the crop.
6. To minimize biannual bearing and consequent risk of die back.
7. Establishment of transplant where leaves/shoots are pruned to strike a balance between roots and shoot so that plants lose less water against restricted root system lost during lifting of plants.
8. Elimination of non-productive vegetative growth like water sprouts, suckers, dead and diseased wood.
9. To convert a small tree into a shrub or to make certain climbers bushy in form.
10. In case of forest trees production of knot-free timber.

10.3.4 Timing of Pruning

1. Pruning should be done when the plants are at least active in growth or resting.
2. Deciduous plants are mostly dormant from the time they shed their leaves till they break out new growth. So pruning is done one month after the leaf fall, especially in winter.
3. The best time to prune evergreen shrubs that have their vigorous growth and flowering during hot months and during the cool dry season is when growth is at minimum. December or early January is a good time for pruning of such trees.
4. Never carry out pruning operations if there is a drought or when strong scorching winds are prevalent.

10.3.5 Types of Pruning

1. *Frame Pruning*: This process starts right from the nursery itself and continues up to fruiting stage. This is done continuously irrespective of the season. The basic purpose of this pruning is to develop strong frame work and desired shape.
2. *Maintenance Pruning*: To maintain the level of production for uniform performance. It is done annually.
3. *Renewal Pruning*: Renewal pruning is employed to rejuvenate old plants by removing old, unproductive branches, which allows for fresh and vigorous replacement growth (Figure 10.12).

Pruning consists of two basic methods, *viz.,* thinning out and heading back.

Thinning Out

It refers to removal of excess vegetative growth to open the plant canopy and reduce the number of fruiting branches for larger fruits. The principal objective of thinning is to open up the canopy for light to penetrate to lower branches for better fruit set and increased productivity. In thinning, remove certain branches, such as those that are inward growing. The limbs are evenly spaced on the stem. Thinning is a common orchard management practice for keeping fruit trees in the best shape for high productivity (Figure 10.10).

Heading Back

The method of heading back involves the removal of the terminal parts of branches. Even though it appears as though the plant is being trimmed, it is not done haphazardly. Its primary effect is to promote secondary branching. It is an easy method to employ but care must be taken to trim the plant such that the topmost bud (at the end of the branch below the cut) point in the right direction (to produce outward growth). After this procedure, the plant is reduced in height and size and may have a new shape (Figure 10.11).

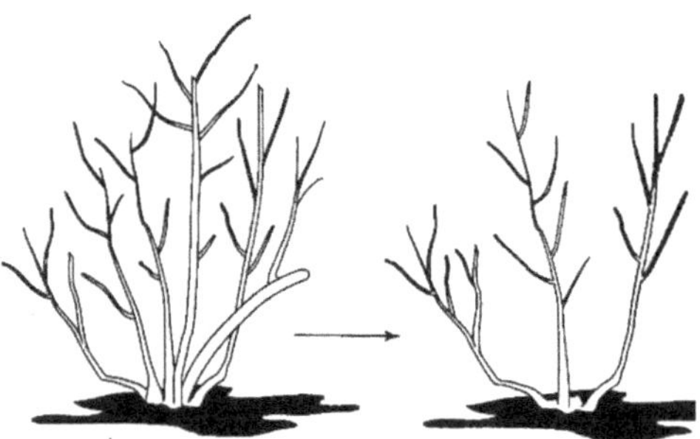

Fig.10.10 Thinning out

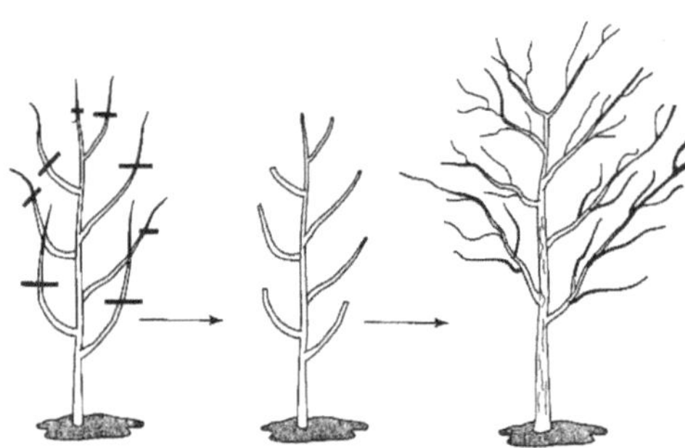

Fig. 10.11 Heading back

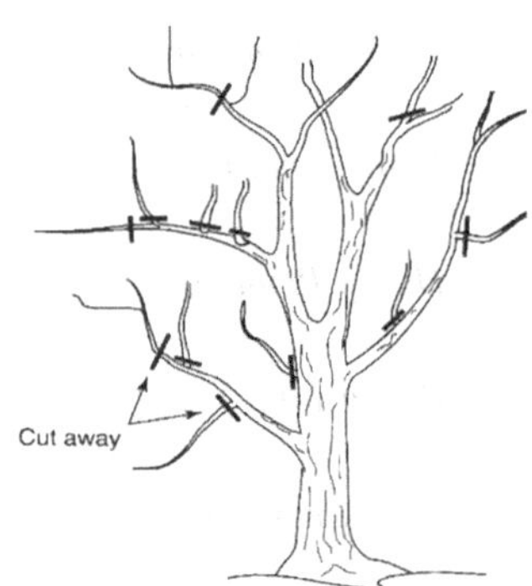

Fig. 10.12 Renewal pruning

10.3.6 Pruning Trees

Pruning of Bearing Trees

The primary consideration in pruning of bearing trees is to maintain proper balance between vegetative growth and production. It may be remarked that excessive growth takes place at the expense of fruit production.

Pruning of Old Trees

The old trees bearing heavy crops of small sized fruits and making a short extension growth require to be pruned more severely than young trees. By heavy pruning, new growth is encouraged and vigour of the fruit spur is maintained. The dead and diseased wood should be removed.

Pruning of Evergreen Trees

Evergreen trees such as citrus and mango require little pruning and only the interfering or drooping branches should be removed. Besides this, the dead and diseased wood and suckers/water sprouts should be removed. In most of the evergreen trees where the fruits are borne on the current season growth, they are automatically pruned during harvesting when the fruits are plucked with fruit stalk or twigs bearing them.

10.3.7 Techniques of Pruning

1. *Thumb Nail or Finger Pruning*: This is really heading back shoots by plucking off two or three levels from the tips. Disbudding, as employed with carnation, chrysanthemum and so on is a form of finger pruning.
2. *Heading Back*: This means that the upward growth of a plant is cut back to induce a spread of branches below the cut. Here the terminal portions of twigs, cane or shoot are removed, but the basal portion is not to be pruned. In general, heading back stimulates development of more growing points than a corresponding thinning out.
3. *Thinning Out*: The entire twig, cane or shoot is removed to reduce their number as per the age of the plant.
4. *Removal of Unwanted Parts*: This means cutting out dead woods, blind shoots and interlocking growth.
5. *Light Pruning*: It includes not only the removal of twig growth but also the cutting back of wood of pencil thickness.
6. *Heavy Pruning*: This consists of cutting back thick wood which forms the frame of the plant. The main objective of this pruning is to lower the height of the plant.
7. *Pollarding*: Pollarding means cutting off the top of a tree so that it may produce a dense growth of new shoots.
8. *Root Pruning*: It is considered to as a last measure in case of fruit trees which show a tendency to make tree growth and to produce very little or no fruit.

10.4 TOP WORKING

It is a technique or method of rejuvenation where in the objective is to upgrade seedling plantations of inferior cultivars with superior commercial cultivars or hybrids suitable for domestic or export market or the desired cultivar of the grower. The technique involves grafting with procured scions of desired cultivar on shoots emerged on pruned branches by adopting softwood grafting during monsoon season (season of top working slightly varies from species to species and it also depends on availability of good shoot and scions). The scion shoots and the emerged shoots should be of same thickness. Top working technique can be successfully followed in crops like mango, sapota, *aonla*, cashew, guava, tamarind, jackfruit, etc.

Advantages of top working:

1. Increase the tree productivity/orchard productivity.
2. Conversion of old and senile orchards into productive orchards.
3. Conversion of seedling or inferior cultivar plantation/orchard into new orchard with desirable cultivar or cultivars through top working.
4. Possibility of grafting several cultivars on the same plant.
5. Increasing the fruit set of orchards by grafting few shoots with pollinizer cultivars.
6. Additional income by selling the pruned wood during non-bearing season or period.

Disadvantages of top working:

1. Chances of death of plant if not done properly or on severe pruning.
2. Need good management during post-pruning period.
3. Loss of crop for 2-3 years
4. Chances of pest and disease occurrence (stem borer, anthracnose etc.)
5. Need skilled labour for thinning of shoots, removal of side shoots etc.

REFERENCES

Kumar N 1990. *Introduction to Horticulture*. Rajyalaxmi Publications, Nagarcoil, Tamil Nadu.

Singh J 2018. *Basic Horticulture*. Kalyani publishers, Ludhiana.

ecoursesonline.iasri.res.in

OUTCOMES ASSESSMENT

PART A

Answer the following questions (True or False).

1. Removal of excess vegetative growth is heading back. (True/False)
2. Open centre system is vase shaped. (True/False)
3. Training and pruning are synonymous. (True/False)
4. The bearing trees are pruned to maintain proper balance between vegetative growth and production. (True/False)
5. Renewal pruning is employed to rejuvenate any plants. (True/False)

PART B

Answer the following questions.

1. Define pruning.
2. There are 2/3/5 types of pruning.
3. What is maintenance pruning?
4. What is training?
5. Open centre system training is for low/high density planting system.

PART C

Write a brief note on each of the following.

1. Timing of pruning.
2. Modified central leader system
3. Top working.
4. Techniques of pruning.
5. Objectives of training.

Chapter 11

Growth Regulators in Horticulture

11.1 INTRODUCTION

Plant growth regulators have remained an important component in horticulture from time immemorial because they were effective means of quantitative as well as qualitative improvement in growth and development of crops. *Plant Growth Regulators (PGR) are diverse group of organic compounds other than nutrients, produced artificially or by plants which in low concentrations promote, inhibit or modify the physiological process in plant.* The term includes both naturally occurring phytohormones as well as synthetic compounds. These are produced naturally in higher plants, controlling growth or other physiological functions at a site remote from its place of production and active in minute amounts. Plant Growth Regulators are used in different forms like liquid, powder, paste etc. "Hormone" is Greek word derived from "hormao", which means to stimulate. Thimone (1948) suggested the use of term 'phytohormones' as the organic substance, which are produced naturally in plants, synthesised in one Part And usually translocated to other part wherein every small quantity affects the growth and other physiological function of the plants. To distinguish them from the animal hormones they are termed as phytohormones. Auxin was the first hormone to be discovered in plants and at one time considered to be only naturally occurring plant growth hormone. The role of plant growth regulators in various physiological processes such as seed germination, flowering, fruiting, seed development, fruit ripening and yield etc. in different crop plants is well established (Kiran *et al.*, 2011). There are five classical phytohormones: auxins, gibberellins, cytokinins, ethylene and abscisic acid. Besides classical phytohormones a variety of other compounds that play roles in plant growth and development have also been identified, including jasmonic acid, salicylic acid, brassinosteroids and polyamines (Basra, 2000). The application of plant growth regulators in agriculture has started in 1930 in the United States. Ethylene, a naturally occurring substance, is one of the first plant growth regulators being discovered and used successfully for enhancing flower production in pineapple (Fishel, 2006). Its toxic effects to human beings are low.

11.2 CLASSES OF PLANT GROWTH REGULATORS

Auxin	IAA, NAA, IBA, 2-4D, 4-CPA
Gibberellins	GA_3
Cytokinins	Kinetin, Zeatin
Ethylene	Ethrel

Abscisic acid	Dormins, Phaseic Acid
Phenolic substances	Coumarin
Flowering hormones	Florigin, Anthesin, Vernalin
Natural substances	Vitamins, Phytochrome
Synthetic substances	Synthetic Auxins, Synthetic Cytokinin
Growth inhibitors	AMO-1618, Phosphon-D, B-999

11.3 PLANT GROWTH REGULATORS AND THEIR ASSOCIATED FUNCTIONS

Plant growth regulators	Associated functions
Auxins	Apical dominance, root induction, control fruits drops, regulation of flowering, parthenocarpy, phototropism, geotropism, herbicides, inhibit abscission, sex determination, xylem differentiation, nucleic acid activity. Can induce fruit setting and growth in some plants, involved in assimilate movement toward auxin possibly by an effect on phloem transport, delays fruit ripening, promotes (via ethylene production) femaleness in dioecious flowers and stimulates the production of ethylene at high concentrations
Gibberellin	Stimulates cell division and elongation, stimulates germination of seeds, stimulates bolting/flowering in response to long days, prevention of genetic dwarfism, increase flower and fruit size, dormancy, induces maleness in dioecious flowers, extending shelf life, parthenocarpic (seedless) fruit development.
Cytokinin	Promotes cell division, cell enlargement and cell differentiation, stimulate bud initiation and root growth, translocation of nutrients, prolong storage life of flowers and vegetables, prevent chlorophyll degradation, morphogenesis, lateral bud development, delay of senescence
Ethylene	Induces uniform ripening in vegetables, stimulates leaf and fruit abscission, stimulates the release of dormancy, stimulates shoot and root growth and differentiation (triple response), may have a role in adventitious root formation, stimulates leaf and fruit abscission, stimulates flower opening.
Abscisic acid	Acts as plant stress hormone, dormancy induction of buds and seeds, induces seeds to synthesize storage proteins, dormancy, seed development and germination, stomata closing.

Source: Prajapati *et al.* (2015).

11.4 GROWTH INHIBITORS

These are substances which suppress the growth of plant, *viz.*, acceleration of degreening, induction of abscission, suppression of vegetative growth, induction of flowering, enhancement of flowering, induction of sterility, salt tolerance and resistance to low temperature etc. e.g., Benzoic acid, salicylic acid, cinnamic acid, caffeic acid, ferulic acid, coumarin, juglone, scopoletin, naringenin, chologenic acid, Maleic hydrazide, TIBA.

11.5 GROWTH RETARDANTS

These are diverse groups of chemicals having common physiological effect of reducing stem growth by inhibiting cell division of the sub-apical meristem. The formation of leaves, flowers and fruits remain unaffected. Growth retardation is principally induced by inhibition of gibberellin biosynthesis between ent-kaurene and ent-kaurenoic acid.

Example, Uniconazole, paclobutrazole (P333, Cultar), triapenthenol, flurpirimidol, inabefide, AMO-1618, CCC, Phosphon-D, B9.

11.6 CONSTRAINTS IN THE USE OF GROWTH REGULATORS (PRAJAPATI ET AL. 2014)

1. Sensitivity of each plant species or cultivars to a given chemical treatment prevents easy prediction of the biological effects.
2. Screening for PGR activities entails high costs and is much difficult. Some synthetic plant growth regulators cause human health hazards.
3. The cost of developing new PGR is very high due to which they are very much costly.
4. It is difficult in identification of proper stage of crop at which the growth regulators should be applied.
5. Lack of support from agricultural researchers in public and private sectors.
6. Lack of basic knowledge of toxicity and mechanism of action.

11.7 PLANT GROWTH REGULATOR APPLICATION METHODS

The following are the methods used for application of plant growth regulators in different crops (Prajapati *et al.* 2014):

1. ***Application in Powder Form***: Plant growth regulators in powder form are dissolved in organic solvent, mixed with moistened charcoal powder, soybean flour or wheat flour and a uniform paste is prepared. The paste is allowed to stand until the solvent evaporates.
2. ***Application in Lanoline Paste***: Most of the roots promoting plant growth regulators are readily soluble in lanoline. A lanoline paste which promotes adventitious roots in plant is made by mixing plant growth regulators in lanoline and allowing it to cool.
3. ***Soaking Method***: Measured quantity of plant growth regulator is dissolved in alcohol, then diluted with distilled water to make required quantity and concentration of solution (20-2000 ppm). Then the cuttings are soaked in solution for 24 hours before planting.

4. ***Aerosol Method***: This method is popular in green houses, where the plant growth regulator solution is released through a small aerosol bottle/cylinder. Liquid gases soon evaporate leaving the plant growth regulator chemicals in the air.
5. ***Spraying method.***
6. ***Root feeding method.***
7. ***Injection of solution into internal tissues.***

11.8 PLANT GROWTH REGULATORS USED IN RIPENING OF FRUITS

Crop	Chemical	Dose	Response
Mango	Ethephon	1000 ppm	Accelerate fruit ripening and improves surface colour
Citrus	Ethrel	1000 ppm	Induce yellow colour within seven days
Banana	Ethrel	1000 ppm	Accelerate ripening by two days
Papaya	Ethephon + NaOH	2000 ppm	Ripening within 24 hours
Sapota	Ethephon + NaOH	500 ppm	Ripening within two days

Source: Dr. P. Jeyakumar, Associate Professor (Crop Physiology), Dept. of Fruit Crops, HC & RI, TNAU, Coimbatore.

11.9 PLANT GROWTH REGULATORS USED IN FLOWERING AND FRUIT SETTING

Crop	Chemical	Dose	Time of spray and number of sprays
Coconut	2,4-D	30 ppm	One month after opening of spathe, through micro sprayer
Banana	2,4-D	25 ppm	Within a week after opening of last bud
Mandarin orange	2,4-D or NAA	20 ppm, 100 ppm	Spray at flowering
Grape (Thompson seedless)	GA_3	25 ppm	Dip cluster at calyptra falling stage
Pineapple	Planofix 10 ppm +2% urea + 0.04% sodium carbonate +20 ppm ethrel	50 ml/plant applied in to the crown 200 to 300 ppm	At 35-40 leaf stage Sprayed during fruiting
Snake gourd Bitter gourd Bottle gourd	Ethrel	100 ppm	4 times 10 -15 days after sowing at weekly intervals
Ribbed gourd Pumpkin	Ethrel	250 ppm	4 times 10-15 days after sowing at weekly intervals

Source: Dr. P. Jeyakumar, Associate Professor (Crop Physiology), Dept. of Fruit Crops, HC & RI, TNAU, Coimbatore.

11.10 USE OF GROWTH REGULATORS IN FRUIT CROPS

The use of plant growth regulators has assumed an integral part of modern fruit production to improve the quality and production of fruits and it has resulted in outstanding achievements in a number of fruit crops with regard to improvement in yield and quality (Jain and Dashora, 2011). The formative effect of growth hormones is gaining its importance for managing canopy, ensuring uniform flowering and enhancing fruit retention and yield under commercial cultivation for perennial fruit trees. There have been numerous reports considering increased yield due to the use of hormones especially in the horticultural sector.

11.10.1 Seed Germination

Plant growth regulators are used to promote early seed germination and improve the germination percentage. Many seeds have natural dormancy which can be overcome by dipping the seeds in auxins, e.g., GA_3 @ 500 ppm solution enhances seed germination in aonla, 500 ppm GA in ber.

11.10.2 Vegetative Propagation

Auxins play an important role in the initiation of roots in cuttings and air layering.

Guava cuttings	:	5000 ppm IBA by quick dip method
Grape cuttings	:	4000 ppm IBA by quick dip method
Pomegranates	:	2000 ppm IBA by quick dip method
Litchi cuttings	:	3000 ppm IBA by quick dip method
Jamun cuttings	:	5000 ppm IBA by quick dip method
Fig cuttings	:	1000 ppm IBA by quick dip method
Air layering-guava	:	3000 ppm IBA by pasting lanoline paste
Air layering-pomegranate	:	3000 ppm IBA by pasting lanoline paste
Air layering-litchi	:	5000 ppm IBA by pasting lanoline paste
Air layering-jamun	:	10000 ppm IBA by pasting lanoline paste
Air layering-tamarind	:	4000 ppm IBA by pasting lanoline paste
Air layering-cashew	:	500 ppm IBA by pasting lanoline paste

Source: Subbaiah *et al.* (2017).

11.10.3 Avocado

Biennial bearing has been reported in avocado. Paclobutrazol is recommended to overcome the concerned problem of biennial bearing in many perennial tree fruits (Adato, 1990). He reported an increase in the yield during the off year with reference to the on year. The additional yield of about 379 and 546% were obtained in two different groves, when cultar (25% PBZ) was sprayed at the stages of inflorescence elongation and incipient anthesis. Lovatt and Salazar-Gracia (2006) reported that application of GA_3 (25 mg/L) at the cauliflower stage of inflorescence development during March significantly increased the 2-year cumulative total yield with commercially valuable large size fruit (178: 325 g/fruit).

11.10.4 Aonla

Chiranjeevi *et al.* (2017) evaluated the influence of growth regulators on germination, seedling growth and vigour attributes of aonla. The seeds pre-soaked with GA_3 200 ppm solution recorded the earliest germination (8.33 days), highest germination percentage (88.88%), maximum seedling height (28.47 cm), seedling stem girth (1.26 cm) and seedling biomass (2.28 g) compared to other treatments. In conclusion, soaking of aonla seeds in GA_3 200 ppm solution for 12 hours followed by 12 hours shade drying will improve the seed germination characters and seedling attributes. The application of gibberellic acid, NAA may have favourably influenced the metabolic activities possibly due to their increased endogenous level which increased the ascorbic acid of aonla fruit. The shelf life of aonla fruits can be increased that might be due to antagonistic effect of GA_3 which inhibited ethylene production and delayed the conversion of starch to sugar (Patel *et al.*, 2017). Ghosh *et al.* (2009) observed that fruit drop in aonla was mainly associated with the hormonal imbalances as NAA 10 ppm sprayed gave maximum fruit retention of 22.3% followed by NAA 20 ppm, which resulted in second best fruit retention (17.1%). Gholap *et al.* (2000) studied the effect of plant growth regulators on seedling growth of aonla and revealed that 200 ppm GA_3 was found significantly superior in respect to seedling height (27.63 cm), stem girth of seedlings (0.86 cm) and number of roots per seedling (24.00) followed by application of thiourea 200 ppm.

11.10.5 Ber

Bhosale (2012) observed that an application GA_3 at 20 ppm significantly increased the plant height at harvest (3.64 m) and plant spread at harvest (3.59 and 3.94 m); minimum plant height (2.43 m) and plant spread (2.68 and 2.86 m) found in treatment 2, 4-D 20 ppm, while treatment 2, 4-D 20 ppm recorded the maximum number of main and subsidiary branches. He also obtained maximum yield kg per plant (23.04 kg/plant), kg per hectare (6382.08 kg/ha) and tonne per hectare (6.38 tonne/ha) with the application of GA_3 at 20 ppm. Foliar application of paclobutrazol (200 ppm) was effective in increasing yield and minimizing fruit drop and fruit cracking in ber (Singh, 2000). Plant growth regulators can significantly change the hormonal status of the plant resulting in good fruit retention and thereby improve the fruit yield and quality. Application of NAA at 25 mg/L gave significantly highest fruit retention (75%) which resulted in highest fruit yield of 120.5 quintals as against 64.7 quintals/ha in control (Ghosh *et al.*, 2009).

11.10.6 Bael

The role of different plant growth regulators on fruit setting, controlling fruit drop and increasing yield and fruit quality of bael have been studied by Kundu and Ghosh (2017) at the Horticultural Research Station, Mondouri of Bidhan Chandra Krishi Viswavidyalaya. Maximum fruit set percentage (51.25%) has been recorded by spraying of NAA at 20 ppm followed by GA at 20 ppm (46.25%) while the opposite trend has been recorded in case of fruit retention with NAA 20 ppm followed by GA (20 ppm) with percentage of 11.25% and 6.75% respectively. Uniyal and Mishra (2015) obtained maximum fruit weight (2.23 kg), fruit length (17.99 cm) and fruit diameter (17.77 cm) with the foliar spray of NAA 30 ppm; while minimum fruit weight (1.99 kg), fruit length (15.00 cm) and fruit diameter (14.87 cm) were recorded in control under *Tarai* conditions of Uttarakhand.

11.10.7 Banana

Beneficial effect of various plant growth regulators have been studied in several banana cultivars after the last hand opening stage of a bunch. Among the PGR's, Gibberellic acid, 2,4-D and CPPU are commonly used in banana, which have been shown to regulate several physiological processes (Jayakumar *et al.*, 2010). Mulagund *et al.* (2015) conducted an experiment on "Influence of post-shooting sprays of sulphate of potash and certain growth regulators on bunch characters and fruit yield of banana cv. Nendran and concluded that the combined foliar sprays of 2 per cent sulphate of potash and 2 ppm Brassinosteroid significantly increased the bunch characters *viz.*, bunch weight (11.35 kg), finger weight (215.40 g), finger length (29.10 cm), pulp weight (180.22 g), pulp to peel ratio (5.13) and total bunch yield (29.38 tonnes/ha) with relatively higher benefit: cost ratio (2.87). Thus, the study clearly indicates that combined post-shoot application of SOP (2%) with 2 ppm Brassinosteroids improves the bunch characters and fruit yield with economically cost viability.

11.10.8 Citrus

The use of plant growth regulators has become an important component in the field of citriculture because of the wide range of potential roles they play in increasing the productivity of crop per unit area. The plant growth regulating compounds actively regulate the growth and development by regulation of the endogenous processes and their exogenous applications have been exploited for modifying the growth response. Plant growth regulators have been used in citrus fruit production for influencing flowering, fruit set and fruit drop and play a major role in fruit growth and abscission. These regulators have also been used to influence fruit quality factors like peel quality and colour, fruit size, juice quality and to improve total soluble solids in different citrus species. Gibberellins reduce the flower production resulting in higher productivity of better-quality fruits. It acts like a thinner agent but it also showed the ability to retain the flowers (Iglesias *et al.* 2007) whereas 2,4-D delay or stimulate the abscission. However, the triazole compounds that inhibit gibberellic acid biosynthesis have been reported to promote inflorescence production (Harty and van Staden,1988). Attempts to promote flowering using growth retardants that are reported to inhibit synthesis of gibberellins *viz.*, CCC and Paclobutrazol have not been able to provide conclusive results as reported by Harty and van Staden (1988). The most widely tested growth regulators used for thinning are ethephon (Guardiola and Lazaro, 1987). Flower drop is caused by the appearance of ethylene—produced auto-catalytically. Early application of auxins increases final fruit size more consistently. According to Randhawa *et al.* (1961) application of 2,4-D (15 and 20 μg/ml) and 2,4,5-T (5 and 10 μg/ml) reduced the fruit drop in Lahore local and Nagpur mandarin. Gibberellins and cytokinins are generally considered to be positive regulators of fruit growth while auxins have been reported to act as stimulators of growth and also as abscission agents (El-Otmani and Oubahou,1996). According to Kaur *et al.* (2000) fruit weight increased with increase in amount of 2,4-D in trees of Kinnow mandarin. The use of 2,4-D as a growth regulator to promote size and to control fruit and leaf drop was reported by Hield *et al.* (1964). NAA at 40 ppm proved to be the best treatment for managing fruit cracking and improving fruit quality in lemon (Sandhu, 2013). Garcia-Luis *et al.* (2001) found that application of growth regulators markedly influenced rind structure, affecting both cell size and thickness of flavedo, as it is relevant to cracking. Moreover, growth regulators play a significant role in peel resistance and plasticity that determine intensity of cracking.

11.10.9 Fig

Fig seeds treated with GA_3 (500-1000 ppm) gave the highest percentages of germination and emergence and also reduced the time to germination and emergence from the seeds in Bursa 'Siyahı' and 'Sarılop' fig cultivars (Caliskan *et al.,* 2012). Quick dip treatment of cuttings with IBA (3000 ppm) increased number of shoots, average shoot diameter, shoot length, number of leaves and leaf area (Kaur *et al.,* 2018). GA_3 treatment advanced fruit maturity by 3-4 weeks in Abbodi and Abiad Aswan fig fruits (El- Mahdy *et al.,* 2015). Singh and Chayrusia (1991) reported that Ethephon (750 ppm) when applied on mature unripe fruits brought about ripening of fruits 5 times more in 12 days as compared to control.

11.10.10 Guava

Application of growth regulators plays an important role in all the stages of growth and development of guava which helps to improve the final yield and quality of produce. The growth regulators are also used to improve the seed germination by breaking seed dormancy, root initiation in cuttings and air layers, regulate proper canopy, flowering and fruit set. Application of GA_3 will improve germination as well reduces the germination period (Kalyani *et al.,* 2014). Sharma *et al.* (1991) reported that 10,000 ppm IBA increased success of air-layers and root quality in guava. Lal *et al.* (2007) reported that application of NAA and IBA will increase the rooting percent in stooling of guava cv. Sardar. Different concentrations of IBA and NAA affect rooting of stooled shoots. Application of IBA at 7500 ppm in lanolin paste during first week of August in cv. Sardar recorded maximum rooting percentage (96.67%) and survival of rooted stooled shoots (75%) after transplanting in the field (Lal *et al.,* 2007). Paclobutrazol is known to inhibit gibberellin biosynthesis and can cause several physiological changes in plants including increased photosynthetic pigments, improved nutrient uptake, senescence retardation and enhanced flowering and seed yields. Paclobutrazol and ethephon may be useful in high density planting as paclobutrazol helps in making the plants dwarf by producing a retarding effect on the growth of tree through inhibition of gibberellin biosynthesis which is a key plant growth promoter. Similarly, ethephon acts as a ripening hormone and it enhances the ripening process along with its growth retardation effect. Crop regulation plays major role in Guava for generating more income to the farmers. Deblossoming of the rainy season crop with certain chemicals, growth regulators and cultural practices can be done to obtain a better winter crop. Deblossming of rainy season crop with foliar application of NAA 800 ppm will help to increase yield of winter crop (88 kg/tree) in cv. Sardar (Tiwari and Lal, 2007). A number of studies show that *in vitro* propagation of guava is successful only when it is supplemented with different combination of plant growth regulators.

11.10.11 Mango

Fruit drop is a major problem resulting in low production and reduction in the income of mango growers in tropical and subtropical regions. Mango growers are facing problems of low fruit set, fruit drop and poor quality in terms of size of fruit. Deficiency of auxins, gibberellins and cytokinins as well as high level of inhibitors appears to be the cause of fruit drop in mango trees (Krisanapook *et al.*, 2000). Plant growth regulators have primitive role in minimizing fruit drop at different stages. Plant growth regulators have potential to enhance productivity of fruits by bringing out a change in nutritional and hormonal status of the plant (Tripathi *et al.*, 2006) Naphthalene acetic acid and CPPU are fruit drop reducing PGR. A period of heavy fruit drop coincides with low auxin levels and exogenous application of auxin, such as NAA, controls the fruit drop in mango (Gofur *et al.*, 1999). Rani and Brahmachari (2004) noticed increased fruit retention with foliar application on GA_3 in different mango cultivars.

11.10.12 Papaya

Meena *et al.* (2012) studied the effect of seed treatment with gibberellic acid on growth parameters of different papaya cultivars *viz.*, Honey Dew, Coorg Honey Dew, Farm Sel: 1 & Hybrid Madhu and reported that 100 ppm GA_3 significantly increased the seedling height (17.83 cm), stem diameter (0.417 cm), number of leaves per plant (10.08), fresh weight (11.54 g) and dry weight of seedling (1.30 g). Papayas are polygamous herbs having plants that produce flowers but do not develop into fruits. Due to this condition, farmers have to plant up to five seedlings per hill if they possess their own seeds or are forced to buy expensive F_1 seeds from seed companies. Plant growth regulators play an important role in the sex expression of plants. Although GA usually promotes maleness in flowers, this is not true for all species. GAs is now known to be essential in many diverse stages of plant growth and development, including germination, leaf expansion, flower induction and development, and fruit and seed growth (Davies, 1995). Papaya genetics, sex determination, and sex-linked characteristics associated with the three sex types have been extensively studied (Ghosh and Sen, 1975; Story, 1953; reviewed by Ming *et al.*, 2007). The application of GA_3 promotes male secondary sexual characteristics in papaya plants but has minimal effect on sex expression and flower development. Exogenous applications of GA_3 on female and hermaphrodite flowers of papaya did not yield any sex reversal phenotype but caused a significant increase in peduncle elongation and inflorescence branch number in all treated plants. An increase in flower number was seen in females but not in hermaphrodites or males. There was an increase in plant height for all treated plants except Sun Up Diminutive mutant, suggesting that the mechanism causing the dwarf phenotype is independent of gibberellins (Han and Murray, 2014). Generally, application of PGRs resulted in increased femaleness expression of plants but does not affect the number of days to flower initiation. However, excessive application of ethrel (200 ppm) may reduce the number of female papaya plants. Moreover, application of IBA at 100 ppm is the most effective type and concentration of PGR that enhance femaleness expression of papaya plants (Higida and Pascual, 2015). Jindal and Singh

(1976) reported that when papaya was treated separately with morphactin (methyl ester of 2-chloro-9-hydroxy fluerene-(9)-carboxylic acid), ethephon (2-chloroethane phosphoric acid) and TIBA (2,3,5-triiodobenzoic acid), the sexual character of the plant became more female. The most striking effect was obtained with morphactin, a low concentration of which resulted in a maximum number of female plants. The increase in femaleness was manifested by a reduction in the number of male plants and an increase in the number of female plants. Hazarika *et al.* (2016) claimed the highest vegetative growth, yield attributing characters and yield with GA_3 (200 ppm) in papaya cv. Red Lady. They also recorded maximum value with respect to different growth parameters, *viz.* plant height, girth and E-W and N-S spread. Treatment with GA_3 (200 ppm) also showed superiority in different yield-attributing characteristics, such as fruit set percentage, number of fruits/plant, fruit length, fruit diameter, fruit circumference, fruit weight, and fruit volume. Quality parameters, such as TSS, acidity, total reducing and non-reducing sugars of fruits, ethrel (400 ppm), exhibited significantly maximum value. According to Pusdekar and Pusdekar (2009), maleic hydrazide (MH) was found to be the most effective regarding enhancement of fruit weight, fruit volume, cavity width and TSS of the fruit in papaya. Highest fruit yield was recorded with MH at all concentrations (200, 400 and 600 ppm). CCC (500 ppm) recorded maximum ascorbic acid content in fruits whereas acidity content (0.12-0.15%) was significantly lower in ethrel.

11.10.13 Jamun

Jamun is commercially propagated through seeds through exploitation of nucellar embryony. Besides soaking in water, several other efforts had been reported for enhancing germination by use of chemicals and growth regulators like gibberellic acid, thiourea, KNO_3 etc. with varied success (Shanmugavelu, 1970). Even though seed propagation is commercial method of propagation of jamun, asexual propagation is commonly recommended to shorten gestation period. Pandey *et al.* (2011) observed that application of IBA at 3000 ppm was found to be most suitable for rooting and growth of jamum cuttings followed by 2500 ppm concentration. Kumar (2007) studied the influence of IBA on rooting of jamun air layering and recorded maximum rooting (100%) in the air layers prepared during June and July months with IBA (7500 ppm) as well as July and August months with IBA (10000 ppm). Hegde *et al.* (2018) claimed that soil application of Paclobutrazole at 1.5 g a.i/plant at preflushing stage resulted in highest number (292) of new flushes per plant which ultimately led to highest fruit number (4217) and fruit yield per plant (47.13 kg). The reduction in plant height also occurred with the use of Paclobutrazole which might be due to its inhibitory effect on gibberellins biosynthesis pathway at the sub-apical meristem, that ultimately reduced cell elongation, rate of cell division and decreased the shoot growth.

11.10.14 Jackfruit

Harshavardhan and Rajasekhar (2012) conducted an experiment to enhance the germination and seedling growth of jack fruit (*Artocarpus heterophyllus* Lam.) by different pre-sowing treatments. Seeds from fully ripened fruits were soaked in distilled water (control), GA_3 at 100 and 200 ppm, NAA at 25 and 50 ppm and KNO_3 at 0.25 and 0.5% for 12 hrs and 24 hrs. GA_3 at 200 ppm for 24 hrs recorded the tallest seedlings with more absolute growth rate and less number of days taken for attaining graftable size.

11.10.15 Custard Apple

Custard apple is an arid fruit crop and hardy in nature, and requires dry climate with mild winter. Growth regulators have been used to increase flowering, fruit set and fruit retention, yield and improve the quality of custard apple. Maximum germination percentage was recorded when seeds were soaked in GA_3. Soaking of custard apple seed for 12 hours recorded the minimum days taken to germinate (24.00 days) and maximum (63.99%) germination percentage (Parmar *et al.,* 2016). According to Choudhury *et al.* (2016), highest number of fruit set per shoot, number of fruit retention per shoot and lowest number of fruit drop per shoot were noted when applied with GA_3 (50 ppm) whereas NAA (200 ppm) treatment enhanced maximum number of flowers per shoot. Prajapati *et al.* (2016) found beneficial effect of GA_3 + NAA (20 + 15 mg/l) for getting number of fruits set/branch (5.20), number of fruits/tree (118.33), average fruit weight (227.67 g) and fruit yield/tree (26.94 kg). Pino (2008) showed that the combination of gibberellins and cytokines was also efficient in the generation of seedless fruits in *cherimoya.* Maximum TSS, reducing sugar and total sugar were found significantly highest and acidity, non-reducing sugar and ascorbic acid were noted lowest with application of 50 ppm GA_3 (Chaudhari *et al.,* 2017).

11.10.16 Pomegranate

Plant growth regulators are reported to play a significant role in pomegranate. Sharma *et al.* (2009) obtained maximum rooting, root number and root length with IBA 500 ppm + Borax 1% both in semi-hard and hard wood cuttings. Hard wood cuttings respond better to the hormonal treatment as compared to semi-hard wood cuttings. Phawa *et al.* (2017) recorded maximum plant height (194.90 cm), spread of plant (194.20 cm), canopy volume (3.81 m^3), internode length (8.07 cm), shoot length (15.57 cm), plant girth (3.24 cm), leaf area (7.37 cm^2), and number of flowers per plant (29.01) with the application of GA_3 at 75 ppm while minimum days to flowering (23.67 days) were recorded under 40 ppm NAA. Choudhari and Desai (1992) investigated the effect of plant growth regulators for flower thinning with Ethrel (250, 500 ppm), NAA (250, 500 ppm), MH (1000 ppm) and Carbaryl (0.7%) applied on trees after removal of male and intermediate flowers and retaining only 50 hermaphrodite flowers. They observed that NAA 500 ppm and 250 ppm affect flower thinning to the extent of 15 per cent and 12 per cent respectively as compared to control (5%). Desai *et al.* (1993) observed higher yield (19.5 kg/tree) with 250 ppm NAA + 0.7% Carbaryl as compared to NAA alone (15.3 kg/tree) or under control (18.1 kg/tree) treatment.

11.11 USE OF GROWTH REGULATORS IN VEGETABLE CROPS

11.11.1 Tomato

Tomato is one of the most popular solanaceous vegetable cultivated round the year in one or another part of the country from temperate to tropical region. Use of growth regulators had improved the production of tomato including other vegetables in respect of better growth and quality (Saha, 2009). According to Pramanik *et al.* (2018), combined application of 40 ppm GA_3 with 25 ppm 4-CPA or 25 ppm NAA as foliar sprays had a stimulatory effect on plant growth, flowering, fruit setting, yield and quality of fruit which was accompanied by increases in endogenous auxin, gibberellins and cytokinin contents in tomato plant. Increased fruit size and setting in tomato due to application of 2, 4-dichlorophenoxy acetic acid (2,4-D), 4-chlorophenoxy acetic acid (4-CPA), and β-naphthoxyacetic acid (β-NAA) has been reported by Gemici *et al.* (2006). Tomato plants treated with a mixture of 4-CPA and GA_3 (Sasaki *et al.*, 2005) showed increased fruit set and proportion of normal fruits compared to plants of the same crop treated with 4-CPA alone. Arvind (2012) reported that foliar application of 15 ppm GA_3 followed by 25 ppm NAA produced superior yield attributing characters and ultimately fruit yield of tomato. The beneficial response of plant growth regulators to increase the yield and quality of solanaceous and other vegetables have been reported by Singh and Singh (1993).The highest fruit yield per plant was recorded in CIPA 20 ppm, 2 4-D 5 ppm and Alar 50 ppm and maximum fruit yield was recorded in CIPA 20 ppm, 2, 4-D 5 ppm and Alar 50 ppm which showed 71.47, 40.1 and 33.2% higher yield over control, respectively (Tiwari and Singh, 2014). GA_3 is one of the important growth stimulating hormones which enhances cell division and cell elongation, thus helps in the growth and development of plants. GA_3 increases the leaf size, stem length and fruit set (Serrani *et al.*, 2007). In order to enhance the ripening of fruits, ethrel 1000 ppm can be sprayed on the plants at the time of initiation of ripening. An early spray may damage the foliage and reduce the size of fruits. Tomato fruit dipping in solution of GA_3 (100-500 ppm) retards the ripening and extends the storage life. Application of GA_3 at the time of flowering elongates the stigmatic position of flower and avoids selfing. Such lines can be used as female line in hybrid seed production programme (Meena, 2015).

11.11.2 Brinjal

In Brinjal, treatment with GA_3 at 50 ppm resulted in higher total number of flowers per plant (38.49), number of fruits per plant (18.56) and fruit yield (1.58 kg/plant and 377 q/ha) and GA3 proved to be the best in improving the morphological, physiological and yield attributing parameters in brinjal (Kropi *et al.*, 2018). Application of 2, 4-D 2.0 ppm at flowering induces parthenocarpy, increases fruit-set, advances fruit maturation and significantly increases the total yield. Soaking of seeds for 24 h in GA_3 at 10–40 ppm improves germination. A significant increase in yield (50%) was obtained by whole plant spray of 2,4-D at 2 ppm at intervals of one week over a period of 60–70 days from commencement of flowering. Higher yield was obtained from plants whose roots were dipped in GA_3 and ascorbic acid each at 250 ppm solution. MH at 100-500 ppm is very effective for induction of male sterility (Meena, 2015). According to Netam and Sharma (2014), the combined application of GA_3, NAA and 2,4-D @ 10 ppm, 20 ppm and 1 ppm had significantly increased plant growth, flowering, quality and yield.

11.11.3 Chilli

According to Meena (2015) application of Planofix (10 ppm) at flowering and three weeks later in chilli cultivars increased the number of branches whereas fruit set in chilli can be improved by application of GA3 (10-100 ppm), NAA (20-200 ppm) and CCC (1000 ppm). MH at 100-500 ppm is very effective for induction of male sterility. Foliar spray of Triacontanol (1-2 ppm) improves the fruit set and reduces the flower and fruit drop at high temperature condition (Meena, 2015).

11.11.4 Potato

Soaking of potato seed tuber in CCC at 500 mg/litre, sodium ascorbate at 100 mg/l, cytozyme at 5% or foliar sprays with ethephon at 400 mg/l, CCC at 25 mg/l or gallic acid at 10-100 mg/l increased tuber yield (Meena,2015). Cutters (1992) reported that gibberellins inhibit and abscisic acid promotes tuber induction in potato. According to Lovell and Booth (1967), exogenously applied gibberellins stops the growth of preformed tubers by promoting the development of new stolons. At the same time, the length of the internodes and height of the plants increase (Sharma *et al.* 1998). The growth inhibitors daminozide and chlormequat chloride promote tuberization and significantly reduce the height of potato plants derived from seed tubers. Exogenous application of gibberellins promotes the breaking of dormancy of tuber sprouts (Van Ittersum and Scholte, 1993). Tuber skin colour and appearance are the most important quality factors considered by consumers in purchasing specialty potatoes. Low dosage foliar application of 2,4-D has been reported to be effective in intensifying skin colour, improving appearance, and reducing oversized tubers in several red potato cultivars (Nelson and Bristol 1975). A mixture of 1-naphthaleneacetic acid (NAA) and benzyl adenine (cytokinins) was shown to increase tuber size and number when applied to potatoes as a foliar spray (Ahmed and Sagar 1981). Post-harvest storage can also potentially influence tuber skin colour and pigment content. Jansen and Flamme (2006) reported no significant changes in anthocyanins content of coloured potatoes during cold storage at 4 °C and 86% humidity for 135 days. However, visual skin colour of red skinned potatoes has been reported to fade during storage (Waterer, 2010).

11.11.5 Cole Crops

Among plant growth regulators, GA_3 and kinetin exhibited beneficial effect in several cole crops (Badawi and Sahhar, 1978).

Cabbage

Cabbage (*Brassica oleracea* var. *capitata* L.) is popular as a winter season vegetable in India. It was found to show a quick growth, early head formation and higher yield when treated with plant growth regulators especially GA_3 and NAA (Dhengle *et al.*,

2008). Foliar application of GA_3 (60 ppm) or NAA (80 ppm) can be recommended to cabbage growers for obtaining better growth and yield of cabbage (Chaurasiya *et al.* 2014). Maximum weight of head (1.72 kg) was obtained with 50 ppm GA_3 as against 0.81 kg under control (Chauhan and Bordia,1971). Badawi and Sahhar (1979) conducted an experiment at the experimental station of the Faculty of Agriculture, Cairo University, Egypt. They sprayed 0, 50, 100 and 200 ppm GA3 and 0, 10, 20 and 40 ppm IBA after 4 and 8 weeks of transplanting to determine the extent of stimulating effect of different concentrations of GA_3 and IBA on cabbage. In most cases, treatments showed a decline in both diameter and height of the edible head. The highest edible head weight (5.21 kg) was obtained with GA_3 (50 ppm) applied 4 weeks after transplanting. Drarmender *et al.* (1996) studied on the effect of GA_3 alone or in combination with NAA (both at 25, 50 or 75 ppm) on the growth of cabbage (cv. Pride of India) in the field at Horticulture Farm S.K.A. College of Agriculture, Jobner, Rajasthan, India during *rabi* (winter) 1993-94. The best growth (plant height, plant spread, number of leaves, leaf area and days to maturity) was observed following treatment with GA_3 (50 ppm) followed by NAA (50 ppm). GA_3 (75 ppm) reduced the mean number of days required to start head formation and the highest chlorophyll content in outer leaves was observed following treatment with NAA at 50 ppm. Thomas (1976) found that N-6 benzyladenine (BA) and N-4 pyridyle N1 phenyle urea had same effect on lateral bud development on Brussels sprout seedlings and both chemicals increased flowering of young plants and produced higher seed yield. However, application of gibberellins improved the plant development and seed yield in cabbage and cauliflower.

Cauliflower

Gibberellic acid stimulates plant growth as it increases the height, number of leaves per plant and the fresh weight of whole plant.GA_3 at 800 ppm concentration was the most appropriate one for floral induction and this concentration also gave the highest total yield in cauliflower (Attallah and Abbas, 2012). Mishra and Singh (1986) revealed that there was significant increase in growth characters, namely, plants height, diameter of stem, number of leaves per plant, weight of plant, curd yield and nitrogen content in the stem and the leaves due to N, B and GA_3 application. However, length of stem was increased only by GA_3 spray. Abdalla *et al.* (1980) conducted an experiment with the cauliflower cultivars and the plants were treated with different concentrations of IBA (5-40 ppm), GA3 (10-80 ppm) or NAA (120-160 ppm) 4 weeks after transplanting and twice more at fortnightly intervals. NAA at 160 ppm gave the highest yield with regard to curd diameter, weight and colour. Similar results were obtained from plants treated with GA_3 at 80 ppm and NAA at 40 ppm. Vijoy *et al.* (2000) treated thirty days old cauliflower (cv. Pant Subhra) seedlings with 50 or 100 ppm GA_3, 5 or 10 ppm IBA, or 100 or 200 ppm NAA at 15 and 30 days of growth. They revealed that GA_3 produced the tallest plants, the largest curds and highest curd yields. Sinha (1973) reported a significantly higher seed yield in cauliflower cultivar snow ball -16 when the plants were sprayed with 250 ppm Ethrel at full bloom stage. Similarly, application of GA_3 at 50, 100 and 250 ppm also increased the seed yield in cauliflower.

Broccoli

Broccoli (*Brassica oleracea* var. *italica*), also known as 'green gobhi', belongs to family *Brassicaceae* and contains high level of vitamins, proteins and minerals beneficial to human health. It is also a rich source of sulphoraphane, a compound associated with reducing the risk of cancer. The application of nitrogen at 2.0%, NAA at 120 ppm and GA_3 at 100 ppm are recommended for better growth, yield and quality of broccoli (Viswakarma *et al.*, 2017). Khalili *et al.* (2008) found that kinetin also affects vitamin C content in broccoli.

11.11.6 Leafy Vegetables

Simao *et al.* (1958) reported that application of GA_3 had increased the leaf size and number of lettuce. Jauhari *et al.* (1960) noted increase in number of leaves in spinach with the use of GA_3. According to Yabuta *et al.* (1981), application of GA_3 had significantly increased marketable weight, petiole length, number of leaves and height of many leafy vegetables but decreased the leaf area.

11.11.7 Tuber and Root Vegetables

Root and tuber crops are the second group of cultivated species after cereals in tropical countries. They are capable to produce higher yield with minimum inputs even in adverse climatic conditions and poor soils. Root vegetables especially cassava, sweet potato, elephant foot yam, taro, tannin, yams, yam bean, arrow root, onion, carrot, radish etc. contribute significantly to human and animal food apart from finding use in various industrial applications. Growth regulators are reported to improve yield of root vegetables in which the underground part is economically important.

Elephant Foot Yam

Nedunchezhiyan *et al.* (2011) reported that among the growth regulators, thiourea was most effective in inducing earliness in first sprouting in elephant foot yam. According to Mukherjee *et al.* (2009) and Bhagavan (2005), foliar spraying of KNO_3 (1–2%) and thiourea (0.5–1.0%) recorded early and increased sprouting of seed corms of elephant foot yam. Punna *et al.* (2018) inferred that chemicals had highly influenced vegetative and yield parameters. The pre-planting soaking of setts in thiourea @ 200 ppm was found better for boosting up the production of elephant foot yam cv. Gajendra.

Colocasia

Colocasia is also known as taro, cocoyam and dasheen in different places and used as an important vegetable in various parts of the tropics. Ud-Deen (2009) reported that different treatments of uniconazole (growth retardant) showed significant influence on plant height, petiole length, number of leaves per plant, weight of leaf and weight of petioles per hill, number and weight of corms and cormels per hill and yield of cormel whereas GA_3 (growth promoter) enhanced foliage growth, flowering and cormel development in colocasia.

Sweet Potato

Abdul Vahab and Mohana Kumaran (1980) obtained highest increase of sweet potato yield with Ethrel (Ethephon) at 450 ppm and 300 ppm whereas tuber girth was increased with CCC applied at higher concentrations of 500 and 1000 ppm. Sarkar (2008) studied the effect of GA_3, CCC and their interactions on yield of sweet potato and concluded that spraying of GA_3 and CCC influenced yield of sweet potato irrespective of concentrations.

Cassava

Remison *et al.* (2002) conducted an experiment on ten cassava cultivars with three growth regulators (GA_3, ABA and IAA) at 5 levels (0, 25, 50, 75 and 100 ppm) to study growth and yield of cassava. They reported that growth regulators increased the tuber yield and dry matter production by foliar application of GA_3 (25 ppm).

Radish

Foliar spray of GA_3 (30, 35 and 45 ppm) recorded maximum plant height, number of leaves per plant, leaf length, width of leaf and fresh weight of leaf in radish (Singh *et al.*, 1989). Mohamed Yassin and Anbu (1996) reported that root length, root girth and root weight were increased by foliar application of CCC at 1000 ppm in radish.

Carrot

Abbas (2011) reported that foliar application of GA_3 decreased root fresh weight and root dry weight in carrot.

11.11.8 Cucurbitaceous Vegetables

Cucumber

The concept that sex expression of cucumber plants may be regulated by a balance between native auxins and gibberellins (Atsmon,1968). Applied auxins, especially a-naphthalene acetic acid, induce femaleness (Galun *et al.* 1965; Ito and Saito,1965), whereas gibberellins induce maleness (Galun *et al.* 1965; Saito and Ito, 1963). Ethrel, which releases ethylene in the presence of plant tissues is remarkably effective in increasing femaleness in cucumber (Rudich *et al.*, 1969; Sims and Gledhill,1969).

Pumpkin

Application of growth regulators, especially IAA 100 mg/l proved to be the best in producing longer nodes, longer vines, larger leaves and higher yield in pumpkin. Spraying with IAA 100 mg/l increased yield and number of seeds per fruit (Ntui *et al.*, 2007). They also suggested that IAA was generally more effective in inhibiting male flowers and increasing the number of female flowers than BAP in pumpkin.

Melons

Spraying of ethrel increased the number of perfect flowers per plant for 7.18 (31.42%), reduced the number of male flowers per plant for 21.47 (17.98%), affected earlier appearance of the first pistillate/perfect flower for 3.68 days, and delayed the appearance of the first staminate flower for 16.07 days in *Cucumis melo* (Girek *et al.* 2008). Thomas (2008) reported that in bitter melon, ethrel was more effective in producing pistillate flowers which formed fruits than GA_3.

Ridge Gourd

According to Vyas *et al.* (2015) the application of Ethrel 200 ppm was found to be the most effective in increasing more number of female flowers (6.13, 30.92 and 18.22 at 45, 60 and 75 DAS, respectively), decreasing the number of male flowers (112.82 and 118.46 at 60 and 75 DAS, respectively) and thereby reducing the male female ratio (3.61, 3.43 and 2.94 at 60, 90 and 120 DAS, respectively) in ridge gourd cv. Pusa Sadabahar. They also claimed that Ethrel (200 ppm) increased the fruit setting (21.64), total number of fruits per plant (20.09), length of marketable fruits (23.73 cm), weight of marketable fruits (331.67 g), thereby fruit yield per plot (20.18 kg) and per hectare (134.50 q) which was at par with Ethrel (100) ppm.

11.12 USE OF GROWTH REGULATORS IN FLOWER CROPS

Commercial floriculture is one of the most profitable agro-industries in the word. Plant growth regulators are used by the commercial growers as a part of cultural practice in most of the ornamental plants. Plant growth regulators encompass quicker impact on vegetative as well as flower yield of flower crops. Use of growth regulators in flower crops must be specific in their action and toxicologically and environmentally safe. The physiological activities of flower crops are regulated by the application of growth regulators and finally affect the growth and flower production.

11.12.1 Chrysanthemum

Sharma *et al.* (1995) reported the effect of foliar application of plant growth regulators on growth and flowering of chrysanthemum cv Move-in-Carvin. They confirmed that plant height was inversely proportional to Maleic Hydrazide (250, 500, 750 and 1000 ppm) concentration and directly proportional to that of NAA (25, 50, 75 and 100 ppm). However, MH and NAA had no effect on plant girth. Gautam *et al.*, (2006) observed the effect of plant growth regulators viz., GA3 (50, 100,150 and 200 ppm), NAA (50, 100, 150 and 200 ppm), Ethrel (>50, 1000, 1250 and 1500 ppm) and B-nine (daminozide) (1000, 1500, 2000 and 2500 ppm) on the growth of chrysanthemum cv. Nilima and inferred that GA_3 at all concentrations and NAA at 100 ppm increased plant height, internal length and basal diameter, while ethrel and B-nine at all concentrations retarded plant height, number of nodes and internodal length.

11.12.2 Gladiolus

Exogenous application of GA_3 at 200 ppm was effective for enhancing growth and flower quality of gladiolus (Sable *et al.* 2015). Dorajeerao and Mokashi (2012) noted that foliar spray of Cycocel (3000 ppm) produced maximum number of flowers per plant, when compared to other concentrations. Hatamzadeh *et al.* (2012) assessed the effect of salicylic acid (SA) on the quality and vase life of cut gladiolus cv. 'Wings Sensation' flowers over four developmental stages (bud stage; half bloom; full bloom; senescence). The flowers were treated in different concentrations of SA (50, 100, 150 and 200 mg/L). The results showed that the SA delayed flower senescence and leakage of ion in petals, as well as decreased fresh weight loss. Faraji and Basaki

(2014) observed the effect of Indole-3-acetic acid (IAA) and benzyl adenine (BA) on growth, flowering and corm production of cut flower gladiolus cv. White Prosperity. The results indicated that IAA (0,100,150, 200 mg/l) and BA (0, 100, 150, 200 mg/l) increased germination rate of gladiolus. Also, onset stalk flower, diameter of floret and bulb wing were affected by IAA and BA. Ram *et al.* (2012) assessed the effect of salicylic acid on growth and flowering of gladiolus. The results showed that the foliar application of 100 ppm salicylic acid increased number of leaves, leaf length, leaf width, number of flowers, emergence of earlier spike and opening of flower. Kumar (2015) assessed the effect of pulsing solutions on postharvest life of cut spikes of gladiolus cv. Peater Pears. Among all the pulsing treatments, treatment (20% Sugar + 200 ppm STS + 200 ppm GA_3) gave maximum vase life, floret size, minimum days to open basal floret, maximum floret longevity, floret opening percentage while with the treatment (20% sucrose + 300 ppm Al_2SO_4 + 200 ppm GA_3), it attained maximum number of floret, floret weight and floret open at a time during the study.

11.12.3 Marigold

Marigold is very popular due to easy to grow and wider adaptability. In India, African marigold flowers are sold in the market as loose for making garland. Flowers are traditionally used for offering in temple, churches and used in festival for beautification of landscape. Gibberellic acid was found to be very effective in manipulating growth and flowering in marigold. Plant height, numbers of primary and secondary branches per plant and plant spread and different flowering as well as yield attributing traits like early flower bud initiation, opening of first flower and maximum duration of flowering and flower yield per plant were found to be maximum in the treatment Gibberellic Acid @ 200 ppm (Anirudh *et al.*,2017). Tiwari *et al.* (2018) assessed the effect of organic and inorganic fertilizers with foliar application of gibberellic acid on productivity, profitability and soil health of Marigold (*Tagetes erecta* L.) cv. Pusa Narangi Gainda. They obtained maximum plant height, plant spread, number of branches per plant, earliest flower bud initiation, days taken to opening of first flower, duration of flowering, length of flower stalk, diameter of flower, number of flowers per plant, weight of flower, flower yield, net return and cost: benefit ratio under the treatments 100% R.D. of NPK (100 kg N, 75 kg P and 75 kg K) + 25% R.D. of vermi-compost (17.9 q/ha) + GA_3 100 ppm followed by 100% R.D. of NPK (100 kg N, 75 kg P and 75 kg K) + 25% R.D. of FYM (50 q/ha) + GA3 100 ppm.

11.12.4 Tuberose

Tuberose (*Polianthes tuberosa* L.), an ornamental bulbous plant native to Mexico, is one of the most important cut flowers in tropical and subtropical areas. The spikes are useful as cut flowers in vase decoration and bouquets; while individual floret is used for making veni, garland, button-holes or crown. It has a delightful fragrance and is the source of tuberose oil. Amin *et al.* (2017) obtained tallest tuberose plant (68.9 cm), longest length of rachis (21.9 cm), highest number of floret/spike (41.2), highest diameter of spike (1.1 cm), maximum weight of single spike (40.1 g) and highest number of spikes per hectare (3.9 lakh) from GA_3 at 300 ppm. Singh (1999) found highest number of florets per spike by using GA_3 at 200 ppm. Pathak *et al.* (1980) found the maximum yield of spikes by treating with GA_3 at 200 ppm.

11.12.5 Gerbera

Gerbera is an important cut flower having single and double flowers. It is used in fresh and dry flower arrangement, floral decoration, exhibition and in high group bouquet. Sangma *et al.* (2017) reported that GA_3 @150 ppm treatment was found to be the best one as compared to others and significantly gave higher yield of gerbera and was effective for enhancing plant growth, flower quality and yield of gerbera.

REFERENCES

Abdalla IM, Helal RM and Zaki ME 1980. Studies on the effect of some growth regulators on yield and quality of cauliflower. *Ann. Agric. Sci.*, **12**: 199-208.

Abdul Vahab M and Mohan Kumaran N 1980. *National seminar on tuber crops production technology. November, Tamil Nadu Agricultural University* (India), pp. 137-141.

Abbas ED 2011. Effect of GA_3 on growth and some physiological characters in carrot plant (*Daucus carota* L.). *Ibn al-haitham J. for Pure and Applied Sci.*, **24** (3): 52-57.

Adato I 1990. Effects of paclobutrazol on avocado (*Persea americana* Mill.) cv. Fuerte. *Scientia Horticulturae,* **45**: 105-110.

Amin MR, Pervin N, Nusrat N, Mehraj H and Jamal Uddin AFM 2017. Effect of plant growth regulators on growth and flowering of tuberose (*Polianthes tuberosa* L.) cv. Single. *J. Biosci. Agric. Res.*, **12**(01): 1016-1020.

Anuradha RW, Sateesh RP, Naveenakumar, Priyanka TK and Kulakarni BS 2017. Effect of growth regulators on vegetative, flowering and flower yield parameters in African marigold cv Culcatta Orange. *Int. J. Pure App. Biosci.*, **5** (5): 636-640.

Atsmon D, Lang A and Light EN 1968. Contents and recovery of gibberellins in monoecious and gynoecious cucumber plants. *Plant Physiol.,* **43**: 806- 810.

Attallah SY and Abbas HS 2012. Effect of gibberllic acid on earliness of cauliflower curd initiation under Assuit conditions. *Assiut J. of Agric. Sci.*, 43: 1:48-56.

Arvind J 2012. Effect of plant growth regulators on different varieties of tomato (*Lycopersicon esculentum* Mill.), *M. Sc. Thesis, R.A.K College of Agriculture Sehore* 466001, (M.P.).

Basra A (Ed.). 2000. *Plant growth regulators in agriculture and horticulture: their role and commercial uses.* CRC Press.

Badawi MA and EL-Sahhar KF 1979. Influence of some growth substances on different characters of cabbage. *Egypt. J. Hort.*, **6** (2): 221-235.

Badawi MA and Sahhar EL 1978. Influence of some growth substances on different characters of cabbage. *Egypt J. Hort.*, **6** (2): 221-285.

Bhosale GH 2012. Effect of plant growth regulators on growth, yield and quality of ber (*Zizyphus mauritiana* Lamk.) cv. Mehrun under Saurashtra region. *JAU (http://krishikosh.egranth.ac.in/handle/1/5810023628).*

Chaurasiya J, Meena ML, Singh HD, Adarsh A and Mishra PK 2014. Effect of GA and NAA on growth and yield of Cabbage (*Brassica oleracea* var. *capitata* L.) cv. Pride of India. *The Bioscan,* **9** (3): 1139-1141.

Davies P 1995. *Plant hormones: Physiology, biochemistry & molecular biology*. Kluwer Academic Publishers, Dordrecht, The Netherlands.

Dharmender K, Gujar KD, Paliwal R and Kumar D 1996. Yield and yield attributes of cabbage as influenced by GA3 and NAA. *Crop Res.* Hisar, **12** (1): 120-122.

Dhengle RP and Bhosale AM 2008. Effect of plant growth regulators on yield of cabbage (*Brassica oleracea* var. *capitata*). *Int. J. Plant Sci.*, **3** (2): 376-378.

Dorajeerao AVD and Mokashi AN 2012. Yield and quality parameters of garland chrysanthemum (*Chrysanthemum coronarium* L.) as influenced by growth regulators/chemicals. *Indian J. Plant Sci.*, **1** (1):16-21.

Gautam SK, Sen NL, Jain MC and Dashora LK 2006. Effect of plant regulators on growth, flowering and yield of chrysanthemum (*Chrysanthemum morifolium* Ram.) cv. Nilima. *Orissa J. Hort.*, **34** (1):36-40.

Prajapati S, Jain PK, Sengupta S K and Tiwari A 2014. Plant growth regulators (plant hormone) in vegetables: their functions and commercial application. *Popular Kheti*, **2** (4): 109-113.

Caliskan O, Mavi K and Polat A 2012. Influences of presowing treatments on the germination and emergence of fig seeds (*Ficus carica* L.). *Acta Scientiarum (Agronomy)*, **34** (3): 293-297.

Chiranjeevi MR, Muralidhara BM, Sneha MK, Shivan and Hongal 2017. Effect of growth regulators and biofertilizers on germination and seedling growth of aonla (*Emblica officinalis* Gaertn). *Int. J. Curr. Microbiol. App. Sci.*, **6**(12): 1320-1326.

Chaudhari JC, Patel KD, Yadav L, Patel UI and Varu DK 2016. Effect of plant growth regulators on flowering, fruit set and yield of custard apple (*Annona squamosa* L.) cv. Sindhan. *Advances in Life Sciences*, **5** (4): 1202-1204.

Chaudhari JC, Yadav L and Varu DK 2017. Effect of plant growth regulators on quality of custard apple (*Annona squamosa* L.) cv. Sindhan. *International Journal of Chemical Studies*, **5** (5): 1036-1037.

Chaudhari SM and Desai UT 1992. Effect of plant growth regulators on flower sex in pomegranate {*Punica granatum* L.). *Indian Journal of Agricultural Sciences*, **63** (1): 34-35.

Desai UT, Ahire GZ, Masalkar SD and Choudhari SM 1993. Crop regulation in pomegranate: II. Effects of growth regulators on fruit set, yield and fruit quality. *Annals of Arid Zone*, 32 (3): 161-164.

El-Otmani M and Ait-Oubahou A 1996. Prolonging citrus fruit shelf life: recent developments and future prospects. *Proc. Int. Soc. Citriculture*, **1**: 59-69.

El-Mahdy TK. El-Akaad MM, Gouda FM and Hussein AS 2015. Effect of some growth regulators application on fruit growth and ripening of Abbodi and Abiad Aswan fig cultivars. *Assiut J. Agric. Sci.*, **46**(2): 107-119.

Fishel FM 2006. Institute of Food and Agricultural Sciences, University of Florida,Document No. PI-102 (*http://edis.ifas.ufl.edu*).

Garcia-Luis A, Duarte AMM, Kanduser M and Guardiola JL 2001. The anatomy of the fruit in relation to the propensity of citrus species to split. *Sci. Hort.*, **87**:33-52.

Gemici M, Türkyilmaz B, Tan K 2006. Effect of 2,4-D and 4-CPA on yield and quality of the tomato (*Lycopersicon esculentum* Mill). *JFS*. **29**:24-32.

Ghosh SN, Bera B, Kundu A and Roy S 2009. Effect of plant growth regulators on fruit retention, yield and physico-chemical characteristics of fruits in ber 'banarasi karka' grown in close spacing. *Acta Horticulture*, **840**:49

Ghosh SN, Bera B and Kundu A 2009. Effect of nutrients and plant growth regulators on fruit retention, yield and physicochemical characteristics in aonla cv. NA-10. *J. Hortl. Sci.*, **4** (2): 164-166.

Ghosh SP and Sen SP 1975. The modification of sex expression in papaya (*Carica papaya* L.). *J. Hort. Sci.*, **50**: 91-96.

Gholap SV, Dod VN, Bhuyar SA and Bharad SG 2000. Effect of plant growth regulators on seed germination and seedling growth in aonla (*Phyllantus emblica* L.). *Crop Res.*, **20** (3): 546-548.

Galun E, Izhar S, and Atsmon D 1965. Determination of relative auxin content in hermaphrodite and andromonoecious *Cucumis sativus* L. *Plant Physiol.*, **40**: 321-326.

Girek Z, Slaven P, Jasmina Z, Tomislav Z, Ugrinovic M and Zdravkovic M 2008. The effect of growth regulators on sex expression in melon (*Cucumis melo* L.). *Crop Breeding and Applied Biotech.*,**13**: 165-171.

Gofur MA, Alam MS, Karim MR, and Ibrahim M 1999. Effect of foliar spray plant hormone on the control of fruit drop of mango (*Mangifera indica* L.) cv. Khirsapat. *Proc. Sixth Intl. Mango Symp., Kasetsart Univ., Chon Buri*, Thailand, 6–9 April 1999, p. 97.

Guardiola J and Lazaro E 1987. The effect of synthetic auxins on fruit growth and anatomical development in Satsuma mandarin. *Scientia Horticulture*, **31**: 119-30.

Hatamzadeh A, Hatami M and Ghasemnezhad M 2012. Efficiency of salicylic acid delay petal senescence and extended quality of cut spikes of *Gladiolus grandiflora* cv. Wings sensation. *African J. Agric. Res.*, **7** (4): 540-545.

Harty AR and Van Staden J 1988. Paclobutrazol and temperature effects on lemon. *Proc 6th Int. Citrus Cong.* **1**: 343-53.

Han J and Murray JE 2014. The effects of gibberellic acid on sex expression and secondary sexual characteristics in papaya. *Hortscience*, **49** (3):378–383.

Harshavardhan A and Rajasekhar M 2012. Effect of pre-sowing seed treatments on seedling growth of jackfruit (*Artocarpus heterophyllus* Lam.). *J. of Res., ANGRAU*, **40** (4): 87-89.

Hazarika TK, Sangma BD, Mandal D, Nautiyal BP and Shukla AC 2016. Effect of plant growth regulators on growth, yield and quality of tissue cultured papaya (*Carica papaya*) cv. Red Lady. *Indian J. Agric. Sci.*, **86** (3): 404–408.

Hegde S, Adiga JD, Honnabyraiah MK, Guruprasad TR, Shivanna M and Halesh GK 2018. Influence of paclobutrazol on growth and yield of jamun *cv.* Chintamani. *International Journal of Current Microbiology & Applied Sciences*, **7** (1): 1590-1599.

Higida CS and Pascual RL 2015. Modifying sex expression of papaya (*Carica Papaya*) through application of plant growth regulators. *Trop. Tech. J.*, **18** (1): 1-9.

Hield HZ, Burns RM and Coggens CW 1964. Pre-harvest use of 2,4-D on citrus. *University Circular*, **528**: 3-10.

Iglesias DJ, Cercos M, Colmenero-Flores JM, Naranjo G, Rios E, Ruiz-Rivero O, Leiso I, Morillon R, Tadeo FR and Talon M 2007. Physiology of Citrus fruiting. *Brazil J. Plant Physiology*, **19**:333-62.

Ito H and Saito T 1956. Factors responsible for the sex expression of Japanese cucumber. m. The role of auxin on the plant growth and sex expression. *J. Hort. Ass. Jap.,* **25**: 101-110.

Jain MC and Dashora LK 2011. Effect of plant growth regulators on physic-chemical characters and yield of guava cv. Sardar under high density planting system. *Indian Journal of Horticulture,* **68**: 259–61.

Jansen G, and Flamme W 2006. Coloured potatoes (*Solanum tuberosum* L.)-Anthocyanin content and tuber quality. *Genetic Resources and Crop Evolution,* **53**: 1321–1331.

Jauhri DS, Singh RD and Dikshit VS 1960. Preliminary studies on the effect of gibberellic acid on growth of spinach (*Spinacia oleracea*). *Curr. Sci.,* **29**: 484- 485.

Jeyakumar P, Ramesh Kumar A and Kumar N 2010. Effect of post shooting spray of potash (SOP) on yield and quality of banana cv. Robusta (AAA- Cavendish). *Research J. Agric. and Biolog. Sci.,* **4**(6): 655-659.

Jindal KK and Singh RN 1976. Modification of flowering pattern and sex expression in *Carica papaya* by morphactin, ethephon and TIBA. *Elsevier Zeitschrift für Pflanzenphysiologie,* **78**(5): 403-410

Kalyani M, Bharad SG and Parameshwar P 2014. Effect of growth regulators on seed germination in guava. *Inter. J. Bio. Sci.,* **5**(2): 81-91.

Kaur N, Monga PK, Thind SK, Thatai SK and Vij VK 2000. The effect of growth regulators on periodical fruit drop in Kinnow mandarin. *Haryana J. Hort. ScI.,* **29**: 39-41.

Kaur A, Kaur K and Kaur A 2018. Role of IBA and PHB on success of cuttings of fig cv. Brown Turkey. *Asian Journal of Science and Technology,* **9**(5): 8237-8241.

Khalili F, Shekarchi M, Mostofi Y, Pirali-Hamedani M and Adib N 2008. Cytokinins affect fermentation product accumulation, vitamin C reservation and quality of stored broccoli in modified atmosphere packages. *Iran J. Med. Plants,* **7** (26): 53-56.

Kiran A, Sanjay S and Chavda JC 2011. Effect of post-harvest treatments on quality of Jamun (*Syzygium cuminii* Skeels) fruits during storage. *Asian Journal of Horticulture,* **6**(2): 297-299.

Krisanapook K, Phavaphutanon L, Kaewladdakorn P and Pickakum A 2000. Studies on fruit growth, levels of GA – like substances and CK- Like substances in fruits of mango cv. Khiew Sawoey. *Acta Horticulturae,* **509**: 694-704.

Kropi J, Gautam BP, Phonglosa A and Kalita CD 2018. Effect of plant growth regulator on growth and fruit yield of brinjal. *Int. J. Agric. Sci.,* **10** (18): 7199-7201.

Kumar V 2007. Studies on air layering and softwood grafting in jamun (S*yzygium cumini*). M. Sc. Thesis, Division of Horticulture, Gandhi Krishi Vignana Kendra, University of Agricultural Sciences, Bangalore.

Kumar M 2015. Effect of pulsing with chemicals on post-harvest quality of gladiolus (*Gladiolus hybridus*) cv. Peater Pears. J. Plant Devel. Sci., 7(3): 293-294

Kundu M and Ghosh SN 2017. Yield and quality improvement in bael (*Aegle marmelos*) by plant growth regulators. *International Journal of Minor Fruits, Medicinal and Aromatic Plants,* **3** (1): 05-08.

Lal SJP, Tiwari P, Awasthi O and Singh G 2007. Effect of IBA and NAA on rooting potential of stooled shoots of guava (*Psidium guajava* L.) cv. Sardar. Proceedings of the 1st International Guava Symposium. *Acta Horticulture,* **735:**193-196.

Lovatt JC and Salaz-Gracia S 2006. Plant growth regulators for avocado production. *In: Proceedings of 33rd PGRSA Annual Meeting.pp.98-107.*

Lovell PH and Booth A 1967. Effects of gibberellic acid on growth, tuber formation and carbohydrate distribution in *Solanum tuberosum*. *New Phytol*. **66**: 525–537.

Meena OP 2015. A Review: Role of plant growth regulators in vegetable production. *International J. Agric. Sci. and Res.*, (IJASR), **5** (5): 71-84.

Meena RR, Jain MC and Mukerjee S 2012. Effect of pre-sowing dip treatment with gibberellic acid on germination and survivability of papaya. *Annals of Plant and Soil Res.*, **5** (1): 120-121.

Ming R, Yu Q and Moore PH 2007. Sex determination in papaya. *Semin. Cell Dev. Biol.* **18**: 401–408.

Mohamed YG and Anbu S 1996. Effect of cycocel on growth and tuberisation of radish. (*Raphanus sativus*). *South Indian Hort*., **44** (1&2): 49-51.

Mulagund J, Kumar S, Soorianathasundaram K and Porika H 2015. Influence of post-shooting sprays of sulphate of potash and certain growth regulators on bunch characters and fruit yield of banana cv. Nendran. *The Bioscan*, **10** (1): 153-159.

Nelson DC and Bristol DW 1975. 2,4-D on potatoes. Agricultural Extension Service. *University of Minnesota and North Dakota State Univ. R*ed River Valley Potato Facts number 5.

Netam LN and Sharma R 2014. Efficacy of plant growth regulators on growth characters and yield attributes in Brinjal (*Solanum melongena* L.) cv. Brinjal 3112. *IOSR Journal of Agriculture and Veterinary Science*,**7** (7): 27-30.

Ntui VO, Uyoh EA, Udensi O and Enok LN 2007. Response of pumpkin (*Cucurbita ficifolia* L.) to some growth regulators. *J. Food, Agric.& Env.*, **5** (2): 211-214.

Pandey J 2011. Effect of indole butyric acid and time intervals on rooting and growth of jamun (*Syzygium cuminii* (L.) Skeels) cuttings. *M.Sc. Thesis, Department of Horticulture, Jawaharlal Nehru Krishi Yashwa Vidyalaya, Jabalpur,* Madhya Pradesh.

Pathak S, Choudhuri MA and Chatterjee SK 1980. Effect of bulb size on growth and flowering of tuberose. *Indian J. Plant. Physiol*., **23:** 47-54.

Parmar RK, Patel MJ, Thakkar RM and Tsomu T 2016. Influence of seed priming treatments on germination and seedling vigour of custard apple (*Annona squamosa* L.) cv. Local. *The Bioscan,* **11**(1): 389-393.

Phawa T, Prasad VM and Rajwade VB 2017. Effect of plant growth regulators on growth and flowering of pomegranate (*Punica granatum* L.) cv. Kandhari in Allahabad Agro-Climatic Conditions. *International J. Current Microbiology and Applied Sci.,* **6** (8): 116-121.

Prajapati S, Jamkar T, Singh OP, Raypuriya, Mandloi R and Jain PK 2015. Plant growth regulators in vegetable production: An overview. *Plant* Archives, **15** (2): 619-626.

Prajapati RD, Laua HN, Solanki PD and Parekh NS 2016. Effect of plant growth regulators on flowering, fruiting, yield and quality parameters of custard apple (*Annona squamosa Linn.*) cv. Local. *Ecology, Environment and Conservation*, **22**: 177-179.

Pramanik K, Pradhan J and Sahoo SK 2018. Role of auxin and gibberellins growth, yield and quality of tomato: A review. *The Pharma Innovation J.* 7(9): 301-305.

Pusdekar GG and Pusdekar MG 2009. Effect of plant growth regulators on flowering and fruit quality in papaya (*Carica papaya*) cv. CO-2. *New Agriculturist*, **20** (1-2): 107-110.

Pino G 2008. Obtaining parthenocarpic fruits of cherimoya (*Annona cherimola* Mill.) Through the use of crescent regulators. *Thesis (MSC. in Agronomy)*, Pontificia Universidad Católica de Valparaíso, Chile. pp. 51.

Punna S, Desai K., Tandel BM and Desai GB 2018. Response of elephant foot yam (*Amorphophallus paeoniifolius* (Dennst.) Nicolso) to different chemicals. *Int. J. Chem. Studies,* **6**(4): 1482-1484.

Ram M, Pal V, Singh MK, Kumar M 2012. Response of different spacing and salicylic acid levels on growth and flowering of gladiolus (*Gladiolus grandiflorus* L). *Hortflora Research Spectrum*, **1**(3): 270-273

Randhawa GS, Jain NL and Sharma BB 1961. Pre harvest drop, size and quality of Jaffa oranges as effected by dipping in aqueous solutions of plant regulators. *Indian J. Hort. Sci.*, **18**: 277-84.

Rani R and Brahmachari VS 2004. Effect of growth substances and calcium compounds on fruit retention, growth and yield of 'Amrapali' mango. *Orissa J. Hort.*, **32**:15–17.

Remison SU, Ewanlen DO and Okaka VB 2002. Evaluation of cassava varieties and effects of growth regulators on vegetative traits and yield. *Trop. Agric. Res and Extension*, **5** (2): 1-2.

Rudich J, Halevy AH and Kedar N 1969. Increase in femaleness of three cucurbits by treatment with Ethrel, an ethylene releasing compound. *Planta*, **86**: 69-76.

Saito T and Ito H 1963. Factors responsible for the sex expression of Japanese cucumber. XIL. Physiological factors associated with the sex expression of flowers.Role of gibberellin. *J. Jap. Soc. Hort. Sci.*, **32**: 278-290.

Sable PB, Ransingh UR and Waskar DP 2015. Effect of Foliar Application of Plant growth regulators on growth and flower quality of Gladiolus Cv. 'H.B. Pitt'. *J. Hort.*, **2**(3): 1-3.

Saha P 2009. Effect of NAA and GA3 on yield and quality of tomato (*Lycopersicon esculentum* Mill). *Environ. & Ecol.*, **27**(3): 1048-1050.

Sandhu S 2013. Improving lemon (*Citrus limon* (L.) Burm.) quality using growth regulators. *J. Hort. Sci.*, **8**(1): 88-90.

Sangma ZN, Singh D and Fatmi U 2017. Effect of plant growth regulators on growth, yield and flower quality of gerbera (*Gerbera jamesonii* L.) cv. Pink Elegance under naturally ventilated polyhouse (NVPH). *Int. J. Curr. Microbiol. App. Sci.*, **6**(10): 468-476.

Serrani JC, Fos M, Atare´s A and Garcı´aMartı´nez JL 2007. Effect of gibberellin and auxin on parthenocarpic fruit growth induction in the cv Micro Tom of tomato. *J. Plant Growth Regul*, **26**: 211-221.

Seema Sarkar CM 2008. Interaction between GA3 and CCC on yield and quality of sweet potato. *Int. J. Plant Sci.*, **3**(2): 477-479.

Sharma N, Anand R and Kumar D 2009. Standardization of pomegranate (*Punica granatum* L.) propagation through cuttings.CAB Abstracts, *Biological Forum*, **1**(1): 75-80.

Shanmugavelu KG 1970. Effect of gibberellic acid on seed germination and development of seedlings of some tree plant species. *Madras Agricultural Journal*, **57**: 311-4.

Sharma HG, Verma LS, Jain V and Tiwary BL 1995. Effect of foliar application of some plant growth regulators on growth and flowering of chrysanthemum var. Move-in-Carvin. *Orissa J. Hort.*, **23** (1/2):61-64.

Sharma, RS et al. 1991. Orissa J. Hort., **19**(1-2): 41-45.

Simao S, Serzedello A and Whitaker N 1958. Effect of gibberellic acid on tomato plants. *Rev. Agric. Pracicaba*, 33.

Sims W L and Gledhill BL 1969. Ethrel effects on sex expression, growth development in pickling cucumbers. *Calif. Agric.*, **23**(2): 4-6.

Singh AK 1999. Response of tuberose growth, flowering and bulb production to plant bioregulators spraying. *Prog. Hort.*, **31**(3-4), 181-183.

Singh DK 2000. Effect of paclobutrazol on yield and quality of different cultivars of ber (*Zizyphus mauritiana*). *Indian J. Agric. Sci.*, **70**: 20–2.

Singh SP and Chaurusia SN 1991. Induction of early ripening of fig *(Ficus carica* L.) by Ethephone. *J. Hill Res.*, **4**(2): 110-112.

Singh DK and Singh RP 1993. Varietal response to 2,4-D chlorophenoxy acetic acid and methods of its application on summer tomato. *Indian j. Agric. Sci.* **63**.5: 276 – 282.

Singh M, Singh RP and Yadav HS 1989. Response of growth regulators and their methods of application on yield of radish (*Raphanus sativus* L.). *Bharatiya Krishi Anusandhana Patrika*. **4** (2): 84-88.

Story WB 1953. Genetics of the papaya. *J. Hered.* **44**:70–78.

Subbaiah KV, Raju GS, Nirmala TV, Reddy AD, Karuna SE and Reddy RVSK 2017. Importance of plant growth regulators on fruit nursery. *Bioquest*,**1** (1) :12-13.

Thomas TD 2008. The effect of in vivo and in vitro applications of ethrel and GA3 on sex expression in bitter melon (*Momordica charantia* L.). *Euphytica*,**164:** 317-323.

Tiwari JP and Lal S 2007. Effect of NAA, flower bud thinning and pruning on crop regulation. *Proceedings of the 1st International Guava Symposium. Acta Horticulture*,**735**: 311-314.

Thomas TH 1976. Comparison of N6-benzyladenine and N-4-pyridyl-N′-phenylurea Effects on Lateral Bud Growth, Flowering and Seed Production of Brussels Sprouts (*Brassica oleracea* var. *gemmifera*). *Physiologia Plantarum.* **38** (1): 35-38.

Tiwari H, Kumar M, Naresh RK, Singh MK, Malik S and Singh SP 2018. Effect of organic and inorganic fertilizers with foliar application of gibberellic acid on productivity, profitability and soil health of marigold (*Tagetes erecta* L.) cv. Pusa Narangi Gainda. *Int. J Agric. Stat. Sci.*,**14** (2): 575-585

Tiwari AK and Singh DK 2014. Use of plant growth regulators in tomato (*Solanum lycopersicum* L.) under tarai conditions of Uttarakhand. *Indian J. Hill Farming*, **27** (2): 8-40.

Tripathi VK and Shukla PK 2006. Effect of plant bioregulator on growth, yield and quality of strawberry cv. Chandar. *J. Asian Hort.*, **2**(4): 260.

Uniyal S and Mishra KK 2015. Effect of plant growth regulators on fruit drop and quality of Bael under Tarai conditions of Uttarakhand. *Indian J. Hort.*, **72** (1): 126-129.

UD-Deen MMD 2009. Effect of plant growth regulators on growth and yield of Mukhi Kachu. *Bangladesh J. Agril. Res.*, **34**(2): 233-238.

Van Ittersum MK and Scholte K 1993. Shortening dormancy of seed potatoes by haulm application of gibberellic acid and storage temperature regimes. *Am. Potato J.*, **70**: 7-19.

Vijoy K, Ray N and Kumar V 2000. Effect of plant growth regulators on cauliflower cv. Plant Subhra. *Orissa J. Hort.*, **28** (1): 65-67.

Vishwakarma S, Bala S, Kumar P, Prakash N, Kumar V, Singh SS and Singh SK 2017. Effect of nitrogen, naphthalene acetic acid and gibberellic acid on growth, yield and quality of broccoli (*Brassica oleracea* var. *italica* L.) cv. 'Sante'. *J. Pharmacognosy and Phytochem (SP1)*: 188-194.

Vyas MN, Leua HN, Jadav RG, Patel HC, Patel AD and Patel AS 2015. Effect of plant growth regulators on growth, flowering and yield of ridge gourd (*Luffa acutangula* roxb l.) cv. Pusa Nasdar. *Ecology, Environment and Conse.*, **21** (1): 377-380.

Waterer D 2010. Influence of growth regulators on skin colour and scab diseases of red-skinned potatoes. *Canadian J. of Plant Sci.*, **90**: 745-753.

Yabuta T, Sumuki Y, Asoc K and Hayashi T 1981. Effect of foliar spray of plant hormones on yield and quality of cabbage. *J. Jap. Soc. Hort. Sci.,* **50** (3): 360-364.

OUTCOMES ASSESSMENT

PART A

Answer the following questions (True or False).

1. The plant growth regulator includes both naturally occurring phytohormones as well as synthetic compounds. (True/False)
2. There are four classical phytohormones. (True/False)
3. Growth inhibitors do not suppress the growth of plant. (True/False)
4. Ethylene induces uniform ripening in vegetables. (True/False)
5. Kinetin does not affect vitamin C content in broccoli. (True/False)

PART B

Answer the following questions.

1. Define plant growth regulators.
2. Who suggested the use of the term 'phytohormones'?
3. Among PGRs, — and — exhibited beneficial effect in several cole crops.
4. "Hormone" is a Greek word derived from —.
5. What are flowering hormones?

PART C

Write a brief note on each of the following.

1. Use of plant growth regulators in mango.
2. Role of plant growth regulators in *Chrysanthemum*.
3. Plant growth regulator application methods.
4. Constraints in the use of growth regulators.
5. Plant growth regulators in brinjal.

Chapter 12

Water Management

12.1 INTRODUCTION

Water is a factor of prime importance in agriculture industry. Water management is one of the largest and most important inputs into an orchard. An efficient water management refers to artificial application of water, *i.e.*, irrigation in crop root zones in case of soil moisture deficit and removal of water, *i.e.*, drainage from the root zone in case of excess so as to provide the crops a most optimum soil moisture regime for best production. It means how best we are in irrigating and dewatering our fields so that plants are protected from stress as well as water logging (Goswami, 2015). Water resources are continuously overburdened with excessive drafts and water sources are shrinking at faster rate. Groundwater levels are falling at the rates faster than ever. Under this type of circumstances, judicious use of water should be given due attention for water conservation and also in increasing water productivity. Area under horticultural plantations is increasing at exponential rate. The unproductive uplands and low productive midlands and lowlands are brought under fruit based horticultural plantations. For a profitable orchard enterprise, a well-planned irrigation system and efficient water management practice are of utmost importance. Horticultural sector, which includes vegetables, flowers and fruits, has enormous potential for development owing to varying agro-climatic conditions in India. There is tremendous scope for increase in productivity and quality of horticultural crops, which is necessary to compete in this modern competitive world market. There are many production technologies available to increase the per unit vegetable production and also to increase its quality. One of the widely acceptable, reliable and cost-effective technologies is pressurized irrigation.

Water is one of the most crucial inputs for growing irrigated vegetables in large scale. The increasing population and industrialization have led to decrease in the allocation of water to agricultural sector. Thus, water has become a very precious and costly input. Pressurized irrigation includes drip, sprinkler, jet and spray. Drip irrigation is the best available technology for the judicious use of water for growing fruit crops in large scale on sustainable basis (Hasan *et al.*, 2004). It is necessary to have the irrigation system in which both the crop water productivity and irrigation efficiency increase considerably. Micro-irrigation has become the most viable and efficient technology in such a situation. It provides several advantages in the context of crop agronomy, water and energy conservation. It has the potential to achieve the crop water productivity to a desired level of 4 tonnes per hectare and simultaneously maintain the irrigation efficiency above 80% (Hasan, 2015).

12.2 WHAT IS IRRIGATION?

Irrigation can be defined as the artificial application of water to the crop plants in the event of shortage of natural rains in order to obtain rapid growth and increased yields.

12.3 ADVANTAGES OF IRRIGATION

1. The grower has many choices of crops and varieties and can go for multiple cropping for cultivation.
2. Crop plants respond to fertilizer and other inputs and thereby productivity is high.
3. Quality of the crop is improved.
4. Higher economic return and employment opportunities. It makes economy drought proof.
5. Development of pisciculture and afforestation. Plantation is raised along the banks of canals and field boundaries.
6. Domestic water supply, hydel power generation at dam site and means of transport where navigation is possible.
7. Prevention of damage through flood.
8. Increase in gross domestic product of the country, revenue, employment, land value, higher wages to farm labour, agro-based industries and groundwater storage.
9. General development of other sectors and development of the country.

12.4 DISADVANTAGES OF IRRIGATION

1. Over-irrigation coupled with poor drainage and seepage in an area where water table is high leads to water logging of the area. Crop yield is drastically reduced.
2. Salinization problems.

12.5 IMPORTANCE OF IRRIGATION

Water is the main constituent of plants. Plants absorb water and nutrients from the soil through their roots and the absorbed water is transferred or lost mainly through the leaves. In tropical countries the loss of water from plant is very high especially in summer. Plants need adequate supply of water for their normal growth, development and production. Where there is shortage of water, particularly during critical stages like flowering and fruiting, there will be drastic reduction in yields. Hence, the necessity of irrigation arises to make up the deficiency of water.

12.6 FACTORS AFFECTING IRRIGATION OF PLANTS

1. Topography and soil characteristics.
2. Kind of plant (root depth, water absorption capacity, growth habit etc.)
3. Weather condition.

12.7 WHEN TO IRRIGATE?

The time when a plant needs irrigation can only be judged by a keen observing eye. The plants need water when their new leaves begin to show a wilting appearance. A little before the trees show the sign of wilting, some of the broad-leaved weeds in orchards begin to show distress symptoms.

12.8 HOW MUCH TO IRRIGATE?

If water supply is limited, only a light irrigation can be given at a time with higher frequency of irrigation. If water is available in plenty, the irrigation may be heavy with longer intervals between successive irrigations. However, inadequate irrigation reduces the growth and fruiting of the trees while over irrigation serves no useful purpose, and it may even prove to be harmful. It may create water logging, the nutrients may get leached, fruits may become watery and develop poor quality. Plants which have suffered drought should not be given liberal doses of irrigations all at once. That may result in the splitting of fruits and even the splitting of bark, branches and trunk.

12.9 METHODS OF IRRIGATION

Irrigation water can be applied to crop lands using one of the following irrigation methods:

1. Traditional/Surface Irrigation
   (a) Flooding Method
   (b) Furrow Method
   (c) Basins Method
   (d) Modified/ring basin
   (e) Pitcher Irrigation
   (f) Funnel Irrigation
2. Sub-Surface Irrigation
3. Advanced/Pressurized Methods
   (a) Sprinkler Irrigation
   (b) Trickle (Drip) Irrigation

12.9.1 Surface Irrigation

In this method water flows from the ditch directly to the field without much control on either side of the flow. It covers the entire field and moves almost unguided.

Flooding

This system is very simple and easy to lay out. When the land is flat, the entire area is flooded by letting in water from one end. This system is commonly practiced in canal or tank bed areas. It is an easier method and permits the use of bullock drawn implements in the orchards. But in this, there is wastage of water and leads to soil erosion also. It encourages growth of weeds and spread of diseases like gummosis in citrus and collar-rot in papaya. This system is not suitable for arid region where water availability is scarce. Flood irrigation is an ancient method of irrigating crops (Figure 12.1). It is still one of the most commonly used methods of irrigation used today.

Fig. 12.1 Flood Irrigation

Furrow System

Unlike the flood system, here the entire land surface is not covered with irrigation water. The furrows are opened in the entire orchard at 5' or less apart, depending upon the age of the trees. Water is let in these furrows from the main channel. In young orchards, two furrows on each side of the rows are generally made. It is suited to such lands which have a moderate slope to the extent of 1-2% if the water is to run freely and reach the ends of the furrows (Figure 12.2). This system is practiced in newly planted orchards.

Fig. 12.2 Furrow Irrigation

Basin System

In this method circular basins are provided around the trunk of the tree. The basins are inter connected in series and are fed through the main channel running perpendicular to the tree rows (Figure 12.3). When compared to flooding, this system minimizes the loss of water. In this system of irrigation, the water close to trunk, may bring about certain diseases like gummosis and nutrients are likely to be washed from one basin to the other. There is water saving in this system as only limited area is wetted.

Fig.12.3 Basin method

Ring Basin System

This is an improvement over basin system. In this system, a ring is formed close and around the tree and water is let into the basin. This method is recommended for citrus trees as in this system the water is not allowed to touch the bark of the tree thereby reducing the chances of collar-rot to which these trees are often susceptible. The size of the ring will increase as the tree grows. In this system, the spread of diseases like collar-rot etc., are prevented. However, it involves more labour and capital and it does not permit uniform distribution of water throughout the bed or basin as in the basin system of irrigation.

Pitcher Irrigation

This system is very suitable for those areas where water scarcity exists. The pitcher filled with water buried in the periphery of individual tree where feeding roots are confined. It is similar to drip irrigation but less expensive to install. The pitchers are the round earthen containers used in rural areas for water storage, ranging from 10 to 20 litres in capacity (Figure 12.4). This kind of irrigation is ideal for saplings, promoting deep root growth. Soluble fertilizers can also be mixed with water and applied through the pitcher. If the water used for irrigation has high salinity, the pitcher location should be changed in every 3 years (Goswami,2015).

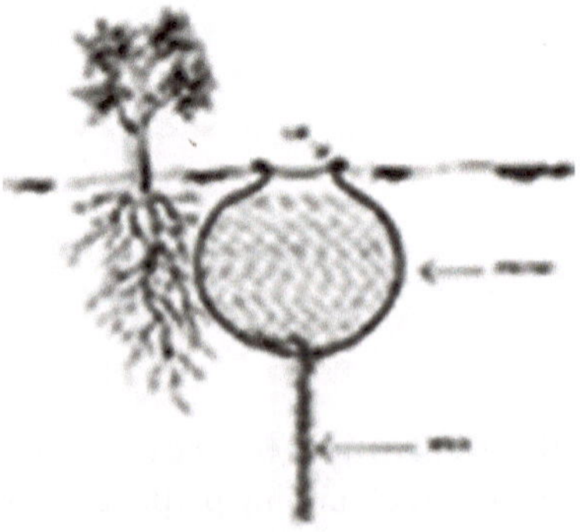

Fig. 12.4 Pitcher Irrigation

Funnel System

In this system, a funnel of about 1 litre capacity is fixed along the side of the plant. The water is applied in the funnel. As the funnel is fixed in the side of the plant, the applied water reaches directly to the feeding roots and the plant survives with the use of very little water (Figure 12.5). This system in is useful only for newly planted plants which require less amount of water (Singh, 2018).

Fig. 12.5 Funnel Irrigation

12.9.2 Sub-Surface Irrigation

Subsurface irrigation, also known as sub irrigation, involve irrigation to crops by applying water from beneath the soil surface either by constructing trenches or installing underground perforated pipe lines or tile lines. Water is discharged into trenches and allowed to stand during the whole period of irrigation for lateral and upward movement of water by capillarity to the soil between trenches. Underground perforated pipe or tiles in which water is forced, trickle out water through perforations in pipes or gaps in between the tiles. Water moves laterally and upward to moist the root zone soil under capillary tensions. Pipelines remain filled with water during the period of irrigation. The upper layers of soil remain relatively dry owing to constant evaporation while lower layers remain moist. This method consists of conducting water in number of furrows or ditches underground in perforated pipe lines, until sufficient water is taken into the soil so as to retain the water table near the root zone. Moisture reaches the plant roots through capillary action. In limited situations, this may be a very desirable system of irrigation. In general, however, it must be used with great caution because of the danger of water logging and salt accumulation. If the substrata are slowly permeable, practically no water move through, water added may stand in soil sufficiently for long time resulting an injury to plant roots due to poor aeration. Where irrigation water or the sub soil contains appreciable amounts of salt, sub soil irrigation is usually not advisable. Land must be carefully levelled for successful sub soil irrigations so that raising the water table will wet all parts of the field equally.

The conditions which are favourable for sub irrigation are as follows:

1. Impervious subsoil at a depth of 2 meters or more.
2. A very permeable subsoil.
3. A permeable loam or sandy loam surface soil,

4. Uniform topographic conditions, and
5. Moderate ground slopes.

12.9.3 Advanced/Pressurized Methods

Overhead or Aerial Irrigation or Sprinkler Irrigation

Sprinkler irrigation may have definite economic advantages in developing new land that has never been irrigated, particularly where the land is rough uneven or the soil is too much porous, shallow or highly erodible. Sprinkling is the method of applying water in the form of a spray which is somewhat similar to rain. In this method, water is sprayed into the air and allowed to fall on the soil surface in a uniform pattern at a rate less than the infiltration rate of the soil (Figure 12.6). It is quite useful where only small streams are available such as irrigation wells of small capacity. It is also helpful in irrigating at the seedling stage when furrowing is difficult and flooding leads of crusting of soil. Fertilizer materials may be evenly applied by this method. This is usually done by drawing liquid fertilizer solution slowly into the pipe. It has several disadvantages like high initial cost, difficult to work in windy locations, trouble from clogging of nozzles, interference in pollination processes and requirement of more labour while removing or resetting. The loss of water by seepage and evaporation are avoided. In general, this method is best adopted for areas where ordinarily surface method is inefficient. This is an ideal system for hilly regions where other system cannot be used. The main difference between sprinkler systems and drip systems of irrigation is the wetting of a larger soil volume by the spray or jet emitters. This occurs by virtue of the water being distributed over a larger area of soil but the drip systems apply water to the one point and rely on the soil properties for distribution of the water. The wetting of a larger surface of soil is important on sandy soils where little lateral movement occurs within the soil and on some clay soils where cracking of the soil is severe. The wetting of a larger soil volume should result in bigger trees but not necessary more productive trees.

Fig. 12.6 Sprinkler Irrigation

Sprinkler Irrigation for Different Horticultural Systems

i. **Ground cover landscape irrigation:** They are particularly useful in irrigating ground covers. The objective in designing a superior landscape irrigation system is to apply the same amount of water over the area being irrigated within the same window of time; this concept is called distribution uniformity. It is a key element for a high-quality irrigation design. Poor uniformity results in over watering in some areas and under watering in other areas. The amount of water is applied in terms of depth per unit area and time to the surface. The amount of water applied must not exceed the 'infiltration rate' and will vary greatly based on soil type and degree of compaction. Different areas of the landscape may well have different infiltration rates and water holding capabilities, therefore will have differing water requirements. Beds are different than sod, for example. For this reason, a basic rule of landscape irrigation design is to never water beds and sod at the same time (in the same zone). It is impossible to do both things well simultaneously.

ii. **Bedding plants:** Bedding plants are typically grown on benches or on the floor with walk isles between the blocks of plants. Installation required for designing a system includes the width of other tables or blocks, length of the tables or blocks and the height of the plant material at maturity. This scenario usually calls for an overhead device being installed on lengths of tubing called drops. Sprinklers that produce small droplets are installed on the drops and spaced appropriately for the performance of the sprinkler.

iii. **Propagation Systems:** Propagation systems are similar in all respects to bedding plants except that the irrigation device is most often some type of mister or fogger. In this application it is desired only to keep a small amount of moisture on the surface of what is being grown. The media is not saturated. This is where cycles of five seconds on, five minutes off from six o' clock in the morning until six o'clock at night is followed. Since the droplet size of the devices available for this application is so small great care must be given to proper spacing of the devices. Typically, the diameter of a mister would be limited to 150 cm (5 feet). If there is any wind movement in the greenhouse it will greatly affect the pattern of the system.

iv. **Plantation crops:** Irrigation is found to increase yield in mature tea. In young tea, yield increases due to sprinkler irrigation when compared to sub-soil irrigation. Among different methods of irrigation, sprinkler irrigation is suitable for both young and mature tea

Drip or Trickle System

This is the most recent method of irrigating the plants. Drip irrigation is a low labour intensive and highly efficient system of irrigation, which is also amenable to use in difficult situations and problematic soils, even with poor quality water. Irrigation water savings ranging from 36-79% can be affected by adopting a suitable drip irrigation system. Drip irrigation or low volume irrigation is designed to supply filtered water directly to the root zone of the plant so as to maintain the soil moisture near to field capacity level for most of the time (Figure 12.7). Water and fertilizer saving around 25 and 30 percent respectively through drip fertigation system over traditional irrigation system was reported for various fruit crops for Delhi region (Hasan *et al.*, 2006). It is usually practiced for high value crops, especially in greenhouses and glasshouses. There will be an installation of pipe lines with nozzles very close to the soil. The nozzle is fitted in such a way that the water is dripped almost at the root zone of the plants. The water is allowed to move in pipes under very low or no pressure and it drops at regular interval. This system is very important for economic use of scarcely available water. It offers many advantages but also suffers from certain disadvantages. Its advantages include, economic use of water, reduced weed growth, no seepage, land levelling not required, labour cost is less and frost protection through application of hot water. However, disadvantages include high initial cost, requires clean water, high skill in maintenance and operation of equipment and poor water distribution when pressure is low and intercultural cultivation is blocked. Drip irrigation system comprises main line, sub mains, laterals, valves (to control the flow), drippers or emitters (to supply water to the plants), pressure gauges, water meters, filters to reduce clogging of the emitters (to remove all debris, sand and clay), pumps, fertilizer tanks, vacuum breakers, and pressure regulators. The drippers are designed to supply water at the desired rate directly to the soil. Low pressure heads at the emitters are considered adequate as the soil capillary forces causes the emitted water to spread laterally and vertically. Flow is controlled manually or set to automatically either to deliver desired amount of water for a predetermined time or to supply water whenever soil moisture decreases to a predetermined amount.

Fig. 12.7 Drip Irrigation

Drip Irrigation for Different Horticultural Systems

1. **Nursery irrigation to field grown nursery stock**: The length of the row will determine the size of tubing required. The objective is to evenly wet the row and manage the water in the root zone. Water that move out of this target zone is lost to the plant and represents a wasted resource. In most cases, it is difficult to apply all of the water needed in a day without losing water from the root zone. Hence water has to be applied more than once a day and as many times as necessary to supply the total amount required. It is called pulse irrigation. Fertilizer and other materials can be applied with safe limits and precautions.

2. **Container Irrigation**: Irrigation of containers is different from field grown crops in that it has a confined space above ground subject to temperature extremes, artificial growing medium such as pine bark, excessive air movement around containers, and the force of gravity. Problem is, it takes the same amount of water to grow a crop in a container as it does in the field: however, there is no reservoir to store the water except for the medium, which is usually very coarse. If too much water is applied at one time, or in only one place in the container, the water will quickly channel through the media and out of the pot.
3. **Greenhouse Irrigation**: There are many variations of greenhouse irrigation system layouts. The critical thing is to establish what crops will be grown. The total volume of water required is dependent on what is being irrigated. Irrigation in containers generally required relatively low flow rates per greenhouse, while propagation and bedding plants generally require overhead irrigation with comparatively higher flow rates per house.

Control of greenhouse irrigation:
Since all greenhouse irrigation applications require high frequency irrigation, automation of these types of systems using an irrigation controller is the rule rather than the exception.

The Drum Kit Drip Irrigation System
Drum kits are the small drip irrigation system that operates on gravity from a tank placed at 1 to 3 metres high. It is a close piped gravity system for growing vegetables, flowers and other horticultural crops on flat or minor slope land. It does not necessarily need any external power for normal operation (Figure 12.8). Drum kit is suitable for small farming areas. The typical system sizes range from 100–1000 m^2. Drum kit is a very low-pressure gravity fed system. It requires no external power, electricity for its operation. It is ideal for rural conditions and small-scale agricultural production in rural areas with water shortage and limited water supply. It can be installed in open fields, greenhouses, low-tunnels and in the kitchen gardens in back yard of houses (Mali and Kumar, 2010).

Fig. 12.8 Drum Kit Drip Irrigation System

Advantages:

Drum kit has all the advantages that other power operated drip irrigation has, like saving of water, better yield of crop, saving in labour, suitable for poor soil, reduced weed growth, convenient for cultural practices and improved efficiency of fertilizer use. Besides this it has several other advantages which are listed below:

1. It is highly suitable for small and marginal farmers cultivating smallcr arcas.
2. There is no requirement of electricity for its operation.
3. The tank can be filled with manually operated treadle pump or hand pump or whenever electricity is available pump can be used to pump the water from longer distance.
4. Existing tanks can be utilized in place of drums which will reduce the cost of installing a new system.
5. System is simple and its press fit joints makes it farmer friendly.
6. It can prove to be an entry point for large drip irrigation system when farmer himself realizes the benefits accruing from drum kit drip irrigation.

12.10 CALCULATION OF WATER REQUIREMENT OF THE PLANT

The water requirement (WR) of a particular horticultural crop is calculated by the following formula:

WR (Litres/day) = Pan evaporation (mm/day) × Row to row spacing (m) × Plant to plant spacing (m) × (crop factor) × (pan factor) × (% welted area)

(Crop factor for horticultural plants is generally taken as 0.6, pan factor 0.7 and welted area 0.2 to 0.3)

12.11 IRRIGATION REQUIREMENT

The field irrigation requirement (IR) of crops refers to water requirement of crops exclusive of effective rainfall and contribution from soil profile and it may be given as follows:

$$IR = WR - (ER + S)$$

Where, IR: Irrigation requirement, WR: Water requirement, ER: Effective rainfall, S: Soil moisture contribution.

Irrigation requirement depends upon the irrigation need of individual crop based on crop area and losses in the farm water distribution system etc. All the quantities are usually expressed in terms of water depth per unit of land area (ha/cm) or unit of depth (cm).

12.12 IRRIGATION SCHEDULING

Irrigation scheduling is defined as frequency with which water is to be applied based on crop need and nature of the soil. It may be defined as scientific management techniques of allocating irrigation water based on the individual crop water requirement under different soil and climatic condition, with an aim to achieve maximum crop production per unit of water applied over a unit area in unit time.

Excess irrigation is harmful because it (a) wastes water below root zone, (b) results in loss of fertilizer nutrients, (c) causes water stagnation and salinity, (d) causes poor aeration and (e) finally, it damages the crops.

12.13 CRITICAL STAGES OF CROP GROWTH

Optimal soil moisture for plant growth varies with the stage of crop growth. Certain periods during the crop growth and development are most sensitive to soil moisture stress compared to other. The stage at which the water stress causes severe yield reduction is also known as critical stage of water requirement. These periods are known as moisture sensitive periods. The term critical period is commonly used to define the stage of growth when plants are most sensitive to shortage of water.

Table 12.1 Critical soil moisture periods of crops

Crop	Available soil moisture (%)	Critical soil moisture stage of crops
Chillies	50	Tenth leaf to flowering and fruit development and after periodical harvests
Potato	65	Stolon formation, tuberization and tuber enlargement
Onion	60	Bulb formation and bulb enlargement
Tomato	60	Flowering and fruit development and each harvest
Peas	40	Flowering and pod development
Cabbage	60	Head formation and enlargement
Cauliflower	70	Curd formation and enlargement
Brinjal	50	Flowering and fruit development and after each harvest
Cucumber	50	Flowering and fruit development
Okra	45	Flowering and pod development
Leaf vegetables	70	Entire crop duration
Fruit crops	For fruit crops 50% available soil moisture is taken as critical limit	
Citrus	Flowering, fruit setting and fruit development	
Banana	In the early hours of vegetative period, flowering and fruit formation	
Mango	Commencement of fruiting to maturity	
Pine apple	Vegetative growth	
Grape	Vegetative growth	
Guava	Fruit growth and development	
Ber	A drought resistant plant but irrigation is required during fruit growth	
Coconut	Nursery stage, root enlargement	
Coffee	Flowering and fruit development	

Source: ecoursesonline.iasri.res.in

12.14 QUALITY OF IRRIGATION WATER

Irrigation water must not have direct or indirect adverse effects on the health of human beings, animals and plants. The irrigation water must not damage the soil and not endanger the quality of surface and ground waters with which it comes into contact. The presence of toxic substances in irrigation water may threaten the vegetation besides unbecoming the suitability of soil for future cultivation. The various types of impurities, which make the water unfit for irrigation, are classified as:

1. Total concentration of soluble salts in water.
2. Proportion of sodium ions to other ions.
3. Concentration of potentially toxic elements present in water.
4. Bacterial contamination.
5. Sediment concentration in water.

12.15 WATER USE EFFICIENCY (WUE)

It is defined as the yield of marketable crop produced per unit of water used in evapotranspiration. A higher level of return is possible by enhancing the profit and reducing the water utilized.

This may be achieved by the following way:

1. Selection of high value crop and high yielding cultivars.
2. Optimum plant population.
3. Optimum production packages for higher yield.
4. Suitable method of irrigation for minimizing irrigation water requirement.
5. Scheduling of irrigation by following scientific principles.

12.16 WATER MANAGEMENT IN PROBLEM SOILS

The soil environment is not always supporting a higher productivity in relations with water management. The soils may have or develop problems when water availability exceeds the requirement and saturates or stagnates. In arid and semi-arid regions soils are affected due to rise in water table and accumulation of salts leading to unfavourable soil-water-plant relationships and decrease in crop productivity.

The following table gives an idea of the risk of salinization:

<table>
<tr><th>Salt concentration of the irrigation water in g/l</th><th>Soil salinization risk</th><th>Restriction on use</th></tr>
<tr><td>Less than 0.5 g/l</td><td>No risk</td><td>No constraint</td></tr>
<tr><td>0.5-2 g/l</td><td>Slight to moderate risk</td><td rowspan="2">It should be used with appropriate water management practices. Not generally advised.</td></tr>
<tr><td>More than 2 g/l</td><td>High risk</td></tr>
</table>

Source: ecoursesonline.iasri.res.in

12.16.1 Soil Salinity

The salt concentration in the water extracted from a saturated soil (called saturation extract) defines the salinity of the soil. If this water contains less than 3 grams of salt per litre, the soil is said to be non-saline. If the salt concentration of the saturation extract contains more than 12 g/l, the soil is said to be highly saline.

Salt concentration of the soil in g/l	water (saturation extract) in millimhos/cm	Salinity
0 -3	0-4.5	Non saline
3 -6	4.5-9.0	Slightly saline
6 -12	9.0-18.0	Medium saline
more than 12	more than 18	Highly saline

Source: ecoursesonline.iasri.res.in

12.16.2 Crops and Saline Soils

Most crops do not grow well on saline soils that contain salts. Some salts are toxic to plants when present in high concentration. The highly tolerant crops can withstand a salt concentration of the saturation extract up to 10 g/l. The moderately tolerant crops can withstand salt concentration up to 5 g/l. The limit of the sensitive group is about 2.5 g/l. Some plants are more tolerant to a high salt concentration than others. Some tolerance level of crops to salt is mentioned below:

Highly tolerant	Moderately tolerant	Sensitive
Date palm	Tomato	Peas
Sugar beet	Potatoes	Beans
Asparagus	Carrot	Pear
Spinach	Onion	Apple
	Cucumber	Orange
	Pomegranate	Prune
	Fig	Almond
	Olive	Apricot
	Grape	Peach

Source: ecoursesonline.iasri.res.in and Singh J (2018).

12.16.3 Improvement of Saline Soils

Improvement of saline soils means reduction of the salt concentration of the soil to an intensity that is not injurious to the crops. More water is applied to the field than is required for crop growth. This supplementary water infiltrates into the soil and percolates through the root zone. During percolation, the water washes the salts out of the root zone. This washing process is called leaching. Proper water management execution and adequate drainage helps in improvement of salty soils, and also in prevention of salinization. A sodic soil may develop when the concentration of sodium salts is high comparative to other types of salt. Sodic soils are characterized by a poor soil structure and they have a low penetration rate. Thus, sodic soils adversely affect plant's growth. Improvement of sodic soils can be done through application of gypsum and the calcium replacing sodium can be leached from the root zone by irrigation water.

12.17 MANAGEMENT OF POOR-QUALITY WATER

For a sustained crop production, the poor-quality water can be managed in following ways:

1. Mixing with good quality water.
2. Flooding with good quality water once or twice to wash out salts beyond root zone.
3. Gypsum mixing with water to reduce sodium hazards and also to improve soil structure.
4. Providing drainage to remove salts.
5. Growing salt tolerant crops.
6. In acidic soils, lime application should be adequate and excessive leaching should be avoided.
7. Adoption of drip irrigation for possible crop is also recommended to overcome soil physical and chemical problems.
8. Moderate application of FYM and green manuring.

12.18 DRAINAGE

It is the process of removal of excess water as free or gravitational water from the surface and the sub surface of farm lands with a view to avoid water logging and create favourable soil conditions for optimum plant growth.

Need for drainage

It is generally assumed that in arid region drainage is not necessary and water logging is not a problem. Even in arid region due to over irrigation and seepage from reservoirs, canals etc., drainage becomes necessary. Irrigation and drainage are complementary practices in arid region to have optimum soil water balance.

REFERENCES

Goswami AK 2015. Methods of irrigation in fruit crops. *In: Fundamental of Fruit production,* edited by Usha K, Thakre M, Goswami AK and Nayan Deepak, Division of Fruits and Horticultural Technology, Indian Agricultural Research Institute, New Delhi, pp.70-78.

Hasan M, Singh AK and Sirohi NPS 2004. Performance and evaluation of different irrigation scheduling methods for peach through efficient fertigation system network. *Acta Hort.,* **662**: 193-197.

Hasan M and Sirohi NPS 2006. Irrigation and fertigation scheduling for peach and citrus crops. *Journal of Agricultural Engineering,* **43**(4): 43-46.

Hasan M 2015.Water and Nutrient Management for Fruit Crops. *In: Fundamental of Fruit production,* edited by Usha K, Thakre M, Goswami AK and Nayan Deepak, Division of Fruits and Horticultural Technology, Indian Agricultural Research Institute, New Delhi, pp.79-87.

Mali SS and Kumar S 2010. *Water management practices in Horticultural crops.* https://www.researchgate.net/publication/305180928

Singh J 2018. Water management. *In: Fundamentals of Horticulture.* Edited by Jitendra Singh, Kalyani Publishers.Pp.131-144.

http://ecoursesonline.iasri.res.in/mod/page/view.php.

OUTCOMES ASSESSMENT

PART A

Answer the following questions (True or False).

1. An efficient water management refers to natural application of water. (True/False)
2. Gypsum mixing with water to reduce sodium hazards. (True/False)
3. Irrigation requirement and water requirement are synonyms.(True/False)
4. Improvement of saline soils means reduction of the salt concentration. (True/False)
5. Date palm is moderately tolerant to salinity. (True/False)

PART B

Answer the following questions.

1. What is drainage?
2. Define water use efficiency.
3. What do you mean by drip irrigation?
4. Irrigation scheduling is defined as ....
5. Why irrigation is required?

PART C

Write a brief note on each of the following.

1. Sprinkler irrigation.
2. Drip irrigation.
3. Critical stages of crop growth.
4. Surface irrigation.
5. Drum kit drip irrigation system.

Chapter 13

Weed Management

13.1 INTRODUCTION

Weed is defined as an unwanted plant growing at a place where it is not desired. Weeds are unwanted and undeserved plants that interfere with the utilization of the land and water resources and thus adversely affect crop production and human welfare. Thus, a plant out of its place or a plant growing where it is not desired at that time is a weed. Weed has no species, but name was suggested as a useless and harmful plant that persistently grows where it is quite unwanted. ***According to Jethro Tull, the father of Weed Science, weed is defined as the unwanted, undesirable plant, growing out of their proper place, which interfere with the utilization of natural resources, prolific, persistent, competitive, harmful and even poisonous in nature and can grow in adverse climatic conditions.*** Of the total losses caused by pests, weeds have a major share (30%). They reduce the crop yield and decrease the quality of produce and hence reduce the market value of the turnout. Weeds are a major problem to horticultural production as they take advantage of their initial slow growth rate. Weeds reduce crop yield, lower their quality, act as alternate host of many pests/diseases and also have allelopathic effects on crops. Weeds also interfere with vital farm operations like weeding, fertilizer application, herbicide application and harvesting. Both annual and perennial weeds are common in their actions reducing the productivity of fruit trees. Weeds also provide shelter to various pathogens by becoming an alternate or collateral host of invaded crops by a number of fungal, bacterial and viral diseases.

The worst weeds found in the world are *Cyperus rotundus, Cynodon dactylon, Echinochloa crusgalli, Echinochloa colona, Eleusine indica, Sorghum halepense, Imperata cylinderica, Eichhornia crassipes, Portulaca oleracia, Chenopodium album, Digitaria sanguinalis, Convolvulus arvensis, Avena fatua, Amaranthus hybridus, Amaranthus spinosus, Cyperus esculentus, Paspalum cojugatum, Rottboelia exaltata* (Rana and Rana, 2016). All of these weeds are found in India. It has been reported that about 36-42% losses may occur due to inadequate management of weeds in apple (El-Metwally and Hafez, 2007). Therefore, management of weeds in all agro-ecosystems is crucial to sustain our crop productivity and to ensure the food security to the rapidly increasing population.

13.2 CROP-WEED COMPETITION

Competition is fight between two or more organisms for a limited resource (water, nutrient, light and space) that is necessary for growth. Competition between crop plants and weeds is most severe when they have similar vegetative habit and common demand for available growth factors. Weeds appear much more adapted to agro-ecosystems than our crop plants. Without interference by man, weeds would easily wipe out the crop plants. Competition from weeds is the most important of all biological factors that reduce crop yield. This occurs primarily because weeds use resources that would otherwise be available to the crop. If crop plants occupy the soil and are vigorous, weeds are excluded or retarded in their growth. On the other hand, if crop stand is thin or lacks vigour, weeds will flourish. Any environmental condition or any intervention that promotes the growth of crop plants tends to diminish the ill effects of weeds. Weeds compete with the crop for nutrients, moisture, solar energy, space and CO_2. *The period at which maximum crop-weed competition occurs is called **critical period**.*

Critical period of crop weed competition in *Kharif* vegetables

Crop	DAS	Crop	DAS
Tomato	20-40	French bean	20-30
Brinjal	20-40	Cucumber	30-45
Chilli	20-40	Bitter gourd	30-45
Capsicum	20-40	Colocasia	30-45
Okra	20-30	Potato	30-45
Carrot	20-60	Cauliflower	30-45
Onion	30-75	Peas	30-45
Garlic	30-75	Cabbage	30-45
French bean	20-30		

Source: Rana and Rana, 2016.

13.3 WEED MANAGEMENT IN FRUIT ORCHARD

Our country is in the midst of the few countries of the world where all types of fruits are grown by virtue of having diverse agro-climatic conditions. As such, we cultivate temperate, subtropical and tropical fruits with the same intensity (Singh *et al.* 2016). Fruits have a great potential but orchards are generally, infested with various types of annual, biennial and perennial weeds which compete with the fruit plants for nutrients, space, moisture, light and adversely affect growth, vigour, fruit set, flower initiation, yield, fruit quality and winter hardiness (Majek *et al.*, 1993) and thereby directly reducing the productivity of fruit trees. Weeds also provide shelter to various pathogens by becoming an alternate or collateral host of invaded crops by a number of fungal, bacterial and viral diseases. Weeds are a major constraint to the profitability of arable cropping causing the highest potential losses (34%) of all crop pests (Bullock and Murphy 1986). Additionally, weeds also interfere with the fruit picking, tree pruning and other horticultural operations. Further, weeds in orchards may lead to more severe frost bite in fruits because of reduced radiation from the soil. Thus, weeds pose a serious menace in horticulture and considerably reduce the yield of fruit crops.

13.3.1 Major Weeds in Fruit Orchards

A number of weeds have been reported in fruit orchards and some of the prominent weeds are listed in Table 13.1.

Table 13.1 Common Weeds Found in Fruit Orchard

S. No.	Scientific name	Family
A. Grasses		
1.	*Dactyloctenium aegyptium*	Poaceae
2.	*Perotis indica*	Poaceae
3.	*Heteropogon contortus*	Poaceae
4.	*Cynodon dactylon*	Poaceae
5.	*Chloris barbata*	Poaceae
B. Broad leaved		
1.	*Aristolochia bracteata*	Aristolochiaceae
2.	*Abutilon indicum*	Malvaceae
3.	*Celosia argentea*	Amaranthaceae
4.	*Hibiscus micranthus*	Malvaceae
5.	*Leucas aspera*	Labiatae
6.	*Leucas utricaefolia*	Labiatae
7.	*Ocimum canum*	Labiatae
8.	*Oldenlandia umbellate*	Rubiaceae
9.	*Phyllanthus maderaspatensis*	Euphorbiaceae
10.	*Trichodesma indicum*	Boraginaceae
11.	*Tridax procumbens*	Asteraceae
12.	*Cyanotis culculata*	Commelinaceae
13.	*Tephrosia purpurea*	Leguminosae
14.	*Mimosa pudica*	Mimosoideae
15.	*Vicoa indica*	Asteraceae
16.	*Gompherena decumbens*	Amaranthaceae
C. Sedges		
1.	*Cyperus rotundus*	Cyperaceae

Source: Sivmurugan *et al.* (2011)

Table13.2 Weed Hosts for Insect-Pests, Viral Diseases and Nematodes in Fruit Crops

Fruit Crop	Common name	Scientific name	Weed hosts
A. Insect- pests			
Citrus	**Fruit sucking moth**	*Othereis futlonica*	*Tinospora cordifolia, Cocculus hirsutus*
		Othereis materna	*Cocculus pendulus, Quisqualis indica, Acbysantbes* spp.
Cashew nut	**Bark feeder**	*Indarbela tetraonis*	*Pettophorum femtgineana*

Fruit Crop	Common name	Scientific name	Weed hosts
B. Viral diseases and their vectors			
Banana	**Mosaic and heart rot**	*Aphis qossypii, Myzus persicae*	*Achyranthes aspera, Amaranthus polygamus, A. spinosus, Datura fistula, D. metel, D. stramonium, Corchorus acutangulus, Pbysallsl floridana, Dianthus barbatus, Petunia axillaries, Xantbium strumarium, Celosia argentea, Digera arvensis, Chenopodium album, C. murale, Medicago sativa, Solatium nigrum, Euphorbia geniculate.*
	Bunchy top	*Petalonia nigronervosa*	*Colocasia sp.*
Papaya	**Leaf curl, Ring spot**	*Bemisia tabaci, Aphis gossypii*	*Zinnia elegans, Datura stramonium, Chenopodium quinoa*
Coconut	**Root wilt, Cadang Cadang**	*Strphanitis typipus, Aphis gossypii*	*Chenopodium amaranticob, Borreiria articularis, Commelina bengalensis*
C. Nematodes and related diseases			
Citrus	**Slow decline**	*Tilenchtdus, Semipenitrans*	*Andropogon rbizomatus, Panicum sp.*
	Lance nematode	*Haplolaimus indicus*	*Cynodon dactylon*
Banana	**Rhizome root rot, Black head, Toppling, Necrotic lesion, Die back**	*Radopholus similis, Helicotylenchus multicinctus*	Some grasses, *Cynodon dactylon* and broad-leaved weeds
Strawberry	**Root lesion**	*Pratylenchus penetrans*	*Chenopodium album, Capsella bursa pastoris, Cirsium awense, Erigeron strigosus, Lamhm amplexicaule, Plantago major, Polygonum persicaria, Senicio vulgaris, Sonchus arvensis, Stellaria media*
Many fruits	**Root lesion**	*Pratylenchus* spp.	*Cynodon dactylon, Tribulusi* spp. *Echinocloa crusgalli, Amaranthus* spp., *Digitaria* spp. *Sorghum halepense*

Source: (Sharma, 2006)

13.3.2 Methods of Weed Management in Fruit Crops

Weed control has been observed as one of the most important practice in crop production because good weed control will ensure maximum yield and high quality of farm produce. Since most horticultural crops are very slow in growth especially at the early stages of their establishment, it becomes imperative to begin weed control early enough in order to ensure high yields and quality (Njoroge,1999). Weed management is an important aspect in horticultural crop production. Weeds reduce crop yield and quality as well as increase cost of production. Weed management is a system approach whereby whole land use planning is done in advance to minimize the invasion of weeds in aggressive forms and give crop plants a very strong competitive advantage over the later. The systems approach is called integrated weed management (IWM).

Weed management refers to the overall approach of minimizing weed infestation so as to prevent yield losses of fruit crops. The first step in effective weed management is the accurate identification which in turn will help in a basic understanding of the weeds' life cycle. Correct identification can be an important step in making sure that new weeds can be eradicated before they become established. There are four methods of weed control, namely, mechanical, cultural, biological and chemical.

(a) *Mechanical*: In this method, weeds are controlled mainly by hand weeding, digging etc. Hand weeding is effective against annuals and biennials. In fruit nurseries, hand weeding is the most commonly and widely used method of weed control.

(b) *Cultural*: Cultural methods rely upon weed control by modification of cultural practices like selection of crop, use of compost or manure, allowing the land to fallow, mulching etc.

(c) *Biological*: In biological methods, there is reduction in weed infestation by direct or indirect action of biological entity such as plants, parasites, predators and pathogens. Natural enemies of plants like insects and disease-causing organisms are used. *Lantana camara* weed has been successfully controlled by torcid moth.

(d) *Chemical*: In this method, chemicals are used to kill the weeds. The chemicals employed for weed control are commonly referred to as weedicides or herbicides. Only such chemicals are used for which the standing crop has a degree of resistance or tolerance. Several types of chemicals may be used on the basis of their effect on weeds. There are two types of herbicides, one which is sprayed before the germination of weed seed, called as 'pre-emergent' herbicide, and the other which is sprayed on the weed when they are 6-8 inches tall, called as 'post-emergent' herbicides. Some commonly used herbicides are enumerated as under:

Pre-Emergent Herbicides: Alachlor, butachlor, atrazine, oxyfluorfen, diuron, pendimethalin, fluchloralin etc.

Post-Emergent Herbicides: Gramaxone, fernoxone and glyphosate

Calculation of Herbicidal Doses:

The containers of all commercial herbicides have label expressing the amount of active chemical contained in a particular product either in percentage of active ingredient (a.i) or acid equivalent (a.e). All the recommendations are made on these bases. The following formula may be used to calculate the quantity of a commercial product required to give a specific dose of active ingredient.

$$\text{Weight of herbicide required} = \frac{\text{Dose of a.i. required} \times 100}{\text{Per cent a.i. in a commercial product}}$$

Suppose, a grower wants to spray a weed infested field @1.0 kg a.i./ha with an herbicide containing 50% active ingredient. Then the quantity of herbicide required will be calculated as under:

$$\text{Amount of herbicide required} = \frac{1 \times 100}{50} = 2.0 \text{ kg}$$

Preparation of The Herbicide Mixture:

Herbicides may be in powder or solution form. Weighed herbicide powder may be taken in a plastic beaker and make a fine paste by adding small quantity of water. Mix this paste in the remaining water (1000 litres per hectare) with constant stirring.

How to spray the herbicides?

Pour the prepared herbicide solution into the sprayer tank using a sieve or muslin cloth to avoid clogging of the spray nozzle with dust particles. Use WFN-40 or VLV-200 nozzle for the spray. Apply uniformly to the soil as pre-emergent spray or on weeds if it is a post-emergent spray. Normally, few drops of detergents like 'teepol' or 'sandovit' are added to post-emergent spray solution to increase its efficiency. Use 1200 litres of water per hectare for a post-emergent spray. For pre-emergent spray, prepare the land thoroughly without any clods. Remove all extraneous material like dry grass, nuts and stolons before giving the spray.

Different Methods of Herbicide Application:

Several types of herbicides may be used on the basis of their effect on weeds. Broadly, they have been classified into two types *viz*., selective and non-selective. Selective herbicides kill special types of weeds on the other hand, non-selective herbicides kill all types of plants. These weed controlling chemicals are applied mainly through following two methods:

(a) *Foliar Application*: After working out the quantity of herbicide, the spray solution is prepared and applied to the weeds and other undesirable vegetation with the sprayer. These chemicals kill the weeds either by contact action or get translocated into the various parts of the plants through xylem and phloem and damage the plants. The examples of contact herbicides are Diquat, Paraquat and Nitrofen whereas herbicides like 2,4-D, 2,4,5-T, Dalapon, MH etc. are translocated into the plant system and kill the plants.

(b) *Soil Application*: Herbicides can also be applied through the soil before or after sowing/transplanting of the crop. Atrazine, Propazine, Nitrofen are suitable for soil application. Further, they have been named as soil fumigants or soil sterilants. The fumes of certain herbicides like Methyl bromide, Chloropicrin kill the plants and effect can persist up to one month. On the other hand, soil sterilants make the soil unable to support any vegetation. All types of plants fail to grow with the use of these sterilants, namely, Diuron, Monuron Atrazine, Simazine etc.

13.3.3 Weed Management in Fruit Crops

The weeds found in most of the fruit orchard like *Cynodon dactylon* (Bermuda grass) and *Cyperus rotundus* (Motha) act as alternate hosts of grass hopper; *Parthenium hysterophorus* (congress grass) for mealy bugs, which attack mango, guava, pear and many other fruit trees.

Mango

Three to four ploughing (machine or bullock drawn) are enough to control weed in mango in a year. Cover cropping is another practice followed in mango orchard to suppress the growth of weeds, to bring additional income to the grower until the trees begin to bear and improve the health of trees if the intercrops grown are of right type. The recommended intercrops in mango orchards for are bottle gourd, bitter gourd, onion, chillies, cowpea, black gram and green gram. Paraquat (3.0 kg a.i./ha) as post-emergence; Diuron at 6.67: 8.9 kg/ha as pre-emergence application control weeds, effectively. Application of Atrazine or Diuron at 2.0 kg a.i./ha as preemergence followed by Paraquat at 3.0 kg a.i./ha as post-emergence have been found effective for controlling both monocot and dicot weeds in one to five years old mango orchard. Patel *et al.* (2010) claimed that pre-emergence treatment of Atrazine 2 kg a.i./ha was found superior to minimize weed population, with minimum dry weight of weeds and higher weed control efficiency in mango seedling nursery.

Banana

The major monocot weeds of banana fields are *Cyperus rotundus, Cynodon dactylon, Digitaria marginata, and Eleusine aegyptium*, and the dicot weed are *Euphorbia* spp, *Polygonum plebejum, Portulaca oleracea, and Mimosa pudica.* Banana crop has superficial root system and thus heavy machinery for its cultivation must be avoided. The methods of weed control for banana are hand weeding, mulching, cover cropping and by application of herbicides. Inter-cropping with ginger, cowpea, colocasia and tapioca in various cultivars of banana has been reported. Herbicide has been used for weed control in banana. In banana good control of both grassy and broad leaf weeds for 4-5 months is reported with their pre-emergence spray of diuron at 2.0 kg a.i./ha. Paraquat at 1.8 kg a.i./ha was used as a post-emergence spray on 6-9 inches tall weeds in banana and controlled them for two months. Herbicide application was found to be more economical than hand weeding (6 hand weeding per year were necessary in banana crop). Pre-emergence application of Atrazine or Diuron at 1.5 to 2 kg/ha soon after planting would keep the ground weed free for 3 to 5 months. As post-emergence, Paraquat or Dalapon or 2,4–D or Glyphosate can be used. (Chinnusamy et al. 2009 and Kabeer & Nair, 2009). Weed population is reduced in banana plots when *Nelsonia canescens* is used as a cover crop (Fongod *et al.* 2010).

Citrus

Grasses that cause problems to citrus orchards are generally perennials and reproduce by seeds/rhizomes, e.g., Bermuda and torpedo grass, or by vegetative methods, *i.e.*, by tubers or bulbs. These methods include hoeing, mechanical cultivation, mowing, slashing and burning. Mechanical cultivation is convenient in inter-row area. In citrus orchards weed control is carried out by 2 or3 cross ploughing. While resorting to manual weeding, tillage should not be deeper than 10 to 15 cm. The Spray of glyphosate 41 SL @ 1.6 litres per acre as post-emergence spray during the second fortnight of March followed by glyphosate 41 SL or gramaxone 24 WSC @ 1.2 litre per acre during second fortnight of July in 200 litres of water is effective for weed management in citrus orchards (Kaur and Rattanpal, 2017). Bromacil and diuron combination @ 2.0 kg a.i./ha is found to be superior to single spray and has got weed control up to 4 months.

Papaya

Being a shallow rooted plant, weeds compete with papaya plants for nutrition. Therefore, regular weeding is essential for vigorous plant growth. Weeding is usually done by hoeing or tilling. Use of herbicides like Diuron @2.0-2.5 kg a.i/ha or Glyphosate @1-2 kg a.i./ha were effective in controlling the weeds in papaya orchards. Care should be taken while spraying so as to avoid spilling of the herbicides on papaya leaves. Effective use of paraquat and glyphosate in papaya is dependent on avoidance of spray contact to green bark and foliage. Pre-emergence herbicide tolerance is dependent on papaya age, size and maturity, and soil type. Only one herbicide, oryzalin is shown to be tolerated by papaya immediately after transplanting. Herbicides such as diuron and oxyfluorfen with a broader spectrum of weed control generally injure young papaya. However, they can be effectively used if the initial application of these herbicides is delayed until papaya attains certain size or maturity indices (Nishimoto, 1997).

Grape

Atrazine or Simazine or Diuron at 2.0 kg a.i./ha as pre-emergent followed six months by two post-emergent sprays of Paraquat at 2.0 kg a.i./ha at three monthly intervals for grape cv. Bangalore Blue. Oxyflurofen at 1.0 kg a.i./ha as pre-emergent spray and Glyphosate at 3.0 kg a.i./ha as post-emergent spray proved very effective for controlling weeds for four and six months respectively in grape. (Rana and Rana, 2016).

Guava

The problematic weeds of guava fields are *Cynodon dactylon,Cyperus rotundus, Bidens pilosa, Tridax procumbens, Acanthospermum hispidum and Lagasca mollis. Cynodon dactylon* caused 50 per cent reduction in plant height of guava. The weed in guava orchard can be managed by following hand weeding, mulching and by application of 2,4-D alone at 3.0 kg/ha or 2,4-D (2.0 kg/ha) in combination with gramoxone (1.5 l/ha). Manual weeding and herbicide combination of Glyphosate (0.5 kg a.i./ha) + 2, 4-D (1 kg a.i./ha) were valuable practices in guava (cv. L-49) weed management (Maji *et al.* 2008).

Pineapple

The common weeds in pineapple plantations are Hariyali and nutgrass. Weeding is important from the economic point of view. Hand weeding especially in closely spaced crop is cumbersome and uneconomic. Therefore, chemical control of weed is advisable. Pineapple Research Station, Vellanikkara, Thrissur (Kerala), recommended application of Diuron at 3 kg/ha or combination of Diuron (1.5 kg/ha) + Bromacil (2.0 kg/ha) as pre-emergence spray. Pre-emergence herbicides like Simazine or Atrazine or Diuron 3 to 4 kg/ha application 4: 6 weeks before planting and ploughing the field 2 weeks before the leaves of the crop began to curl down prevent weed establishment. Weed growth emerging later is treated with post-emergence herbicides such as Glyphosate 0.5 kg/ha or Dalapan 3 to 4 kg/ha or 2,4: D 0.5 to 1 kg/ha depending upon the predominant weed species. Dalapan spray drift can cause chlorosis on outer leaves and necrosis on leaf tips when pineapple plants are young. It should be applied carefully any time from 6 months.

Apple

Apple (*Malus domestica*) grows mostly in the temperate regions of Jammu & Kashmir, Himachal Pradesh and Uttarakhand and to small extent in Nilgiris Hills of Tamil Nadu. *Rosa moschata, Rosa eglantaria, Rubus* sp. and *Berberis* sp. are some of the predominant weeds of apple orchards. The weeds can be controlled by applying mulching with black polythene and oat leaves in apple basins, growing intercrops (mustard, wheat, oat, barley). Chemically, weeds can be managed with the spray of gramoxone at 500 ppm + Ansar 529 at 1000 ppm when applied as post-emergent spray. Gramoxone (3.75 kg/ha) was found to be the most efficient post- emergent herbicides out of a number of herbicides tested in apple orchards. An application of Aminotriazole (5.0 kg/ha) and Simazine (5.0 kg/ha) as pre-emergent spray resulted in the elimination of broad leaf weeds and increases in the density of grasses (Rana and Rana, 2016).

Table 13.3 Herbicides for Weed Control in Some Fruit Crops

Fruit crops	Application	Dose
Mango	Paraquat	3.0 Kg a.i./ha
	Atrazine or Diuron	2.0 Kg a.i./ha
Banana	Paraquat	1.8 Kg a.i./ha
Grape	Atrazine or Simazine	2.0 Kg a.i./ha
	Paraquat	2.0 Kg a.i./ha
	Glyphosate	3.0 Kg a.i./ha
Papaya	Alachoor or Butachlor	2.0 Kg a.i./ha
	Diuron or Ametryn	4.0 Kg a.i./ha
Sweet Orange	Bromacil + Diuron	2.0 Kg a.i./ha each
Mandarin	Monouron or Diuron	4.0 Kg a.i./ha
Guava	2,4-D	3.0 Kg a.i./ha
	2,4-D + Granxone	2.0 Kg a.i. +1.5lha
Sapota	Bromacil + Diuron	2.0 Kg a.i./ha each
Pineapple	Bromacil + Diuron	2.0 Kg a.i./ha each
	Glyphosate	1.5 Kg a.i./ha

Fruit crops	Application	Dose
Ber	Oxyfluofon	0.6 Kg a.i./ha
	Oxadiazon	1.5 Kg a.i./ha
Pomegranate	Simazine + oryzalin	2.0 Kg a.i./ha each
Anola	Aminotriazole	4.0 Kg a.i./ha
	Paraquat or Diquat	0.5-1.00 Kg a.i./ha

(Maji *et al.* 2004 and Makhija and Singh 2011)

13.4 WEED MANAGEMENT IN VEGETABLE CROPS

13.4.1 Tomato

Marana *et al.* (1986) estimated the critical period of weed competition to be 30-40 days after sowing and hence, weeds should be removed for 40-50 days after sowing. They further noted that the presence of weeds reduced fruit yield by 70% depending on stage and duration of competition. For the control of annual grasses, pre-planting soil incorporation of Trifluralin 3 to 5 kg/ha or Nitralin 3 to 5 kg/ha or Diphenamid 2 to 4 kg/ha is best. Generally, a pre-emergence herbicide followed by one hand weeding in the later period of crop growth gives complete weed control. According to Kumar *et al.* (2015) application of pendimethalin @1.5 kg a.i. per ha and fluchloralin @ 1.0 kg a.i. per ha being equally effective gave maximum control of weed by lowering the weed density, weed dry matter accumulation and thereby recording higher weed control efficiency of 97.5 and 98.5%, respectively. Effective weed control and increase in yields with application of fluchloralin @1.0 kg a.i. per ha (Behera and Singh, 1999) and metribuzin (Hesammi, 2013) have been reported.

13.4.2 Brinjal

Yield reduction due to weed competition in brinjal is in the range of 49 to 90% (Reddy *et al.* 2000). Leela (1982) reported that 45% annual loss in brinjal would be due to weed menace. Pendimethalin @ 1.0 kg/ha as pre-sowing incorporated spray + one hand weeding at 6 weeks after transplanting or Pre-emergence application of Oxyfluorfen @ 0.5 kg/ha at 3 days after transplanting followed by one hand weeding at 30 days after transplanting. Post emergence application of Quizalofop-p-ethyl @ 1.00 kg/ha. The integrated weed management treatments *viz.*, black polythene sheet mulching followed by fluchloralin 1.5 kg/ha + sugarcane trash mulching at 18 DAT favoured the growth and yield of brinjal cv. Annamalai. However, Sharma and Singh (2012) reported that pendimethalin (Extra) (0.64 kg/ha) pre-transplanting + one hand weeding at 40 DAT + pendimethalin (Extra) (0.64 kg/ha) at 45 DAT as postemergence resulted in better performance followed by pendimethalin (1.0 kg/ha) pre-transplanting + one hand weeding at 45 DAT with respect to growth and yield parameters due to effective weed management in brinjal.

13.4.3 Chilli

The extent of reduction in fruit yield of chilli has been reported to be in the range of 60-70% depending on the intensity and persistence of weed density in standing crop (Khan *et al.* 2012). Pre-emergence application of Pendimethalin 1.00 kg/ha or Oxyfluorfen 0.15 kg/ha was supplemented with one hand weeding at 30 days after transplanting. According to Shil and Adhikary (2014), application of oxyfluorfen @ 200 g/ha, recorded lower weed biomass (6.01 and 17.57 g/m2) and higher WCE (81.7% and 75.0%) at 30 and 60 DAT under Nadia, West Bengal condition. Gare *et al.* (2015) claimed that Pre-emergence application of Pendimethalin (30 EC) @ 0.825 kg a.i./ha within two days after transplanting along with two hoeings + one hand weeding recorded minimum weed density, weed biomass and weed index in chilli under rainfed condition in Maharashtra.

13.4.4 Onion

The major weed flora of onion in vertisols of Dharwad composed of *Cyperus rotundus, Cynodon dactylon, Setaria glauca, Digitaria sanguinalis, Panicum isachne, Commelina benghalensis* among monocots while, the predominant dicots were *Parthenium hysterophorus, Phyllanthus niruri, Hibiscus panduriformis, Trianthema monogyna, Eclipta alba, Amaranthus viridis, Portulaca oleraceae, Cocculus hirsutus* and *Corchorus trilocularis.* (Nandagouda *et al.*1996). Onion germinates and grows relatively slow and hence weed competition is more critical up to 40 days after sowing. Onion exhibits greater susceptibility to weed competition as compared to other crops due to its inherent characteristics such as their slow growth, small stature, shallow roots and lack of dense foliage. Non-availability of labourers during critical period of crop makes hand weeding difficult leading to heavy yield losses. Spraying of pre-emergence herbicides keeps the crop in weed free conditions during the early stages. At later stage, second flush of weeds will affect the bulb formation. The higher onion bulb yield of 38.3 t per ha due to lesser weed population and weed growth from initial crop growth as compared to weedy check was obtained by Patel *et al.* (2011). Ranpise and Patil (2001) observed that preemergence application of oxyfluorfen at 0.4 kg/ha in onion recorded maximum yield (242.2 q/ha) followed by oxyfluorfen 0.2 kg/ha (233.3 q/ha) as compared to the lower yield under control (50 q/ha) due to maximum weed intensity. Panse *et al.* (2014) observed that application of Oxyfluorfen 23.5% EC before planting + Quizalofop ethyl 5% EC at 30 days after transplanting recorded highest weed control efficiency (89.86%) and higher marketable bulb yield (287.21 q/ha) with benefit : cost benefit ratio of 1: 2.30.

13.4.5 Garlic

Weed infestation in garlic is one of the major factors for loss in yield and bulb loss to the tune of 30-60% (Lawande *et al.*, 2009). Weed reduces the bulb yield to the extent of 40 to 80% (Verma and Singh, 1996). The application of Oxyfluorfen 0.150 kg/ha + Quizalofop ethyl 0.050 kg/ha as post-emergence found effective in controlling weeds and increasing bulb yield of garlic crop (Siddhu *et al.*, 2018).

13.4.6 Okra

Pre-emergence herbicide – Pendimethalin @ 1.00 kg/ha. Post emergence herbicide – Fluazifopbutyl @ 0.25 kg/ha + hand weeding on 30 DAS. Pre-emergence application of Oxyfluorfen at 0.15 kg/ha (or) Fluchloralin at 1.00 kg/ha (or) Metolachlor at 0.75 kg/ha followed by one hand weeding on 30 DAS.

13.4.7 Pea

Pre-emergence application of Alachlor 1 to 2 kg/ha or pre-plant soil application of EPTC 2 to 3 kg/ha. Dalapan is applied as post emergence when peas are 5 to 15 cm tall but not within 25 days of harvest. MCPA and MCPB are applied when broad-leaved weeds are at 10 to 15 cm tall.

13.4.8 Tuber and Root Crops

Potato

The yield reduction due to weeds in potato is estimated to be as high as 10 to 80% (Lal and Gupta, 1984). So, control of weeds in the initial stages appears imperative as it plays an important role in maximizing the tuber production. Mukherjee *et al.* (2012) concluded that the application of metribuzin @ 0.30 kg/ha (early post-emergence) or pendimethalin @ 0.60 kg/ha (pre-emergence) treatment in ridge planted potato followed by earthing up at 45 DAP were effective for controlling weeds, getting higher production and profitability. Rice straw mulching significantly increased the tuber yield over the chemical control. According to Hoogar *et al.* (2017), higher weed control efficiency (73.3%), lower weed index (5.9%), higher number of tubers (4.30) and tuber yield (19.77 t/ha) at harvest were recorded by application of Fenoxaprop–p-ethyl 54 g a.i./ha as early post emergent at 20 DAT.

Cassava

Weed infestation is one of the major constraints in cassava growing areas and weeding is the major labour consuming activity. Cassava has a slow initial growth phase and requires wider spacing (90 cm × 90 cm) to accommodate later growth. Its early growth period (1-4 months) provides space for weeds to flourish. Tuber yield losses in cassava due to weeds may go up to 100% (Ambe *et al.*, 1992). Smother crops such as beans, cowpea, maize, groundnut, melon growing along with cassava during its initial growth period up to 90 days was found effective in controlling weeds (Mohamed Amanullah *et al.*, 2006). Integrated use of cowpea and pre-emergence application of alachlor or chloramben or fluometuron or mixture of fluometuron and chloramben effectively controlled the weeds in cassava (Akobundu, 1980).

***Yam** (Dioscorea **spp.**)*

It is a long duration crop which requires frequent and effective weeding for high tuber yield. The critical period of crop weed competition appears to be between 3 and 6 MAP (Roy Chowdhury and Ravi, 1994) because it is during this period yam plants develop maximum leaf area and total dry matter production. Yam was more affected by weeds between 10 and 22 weeks than at other times (Unamma *et al.*, 1980). No herbicide has been recommended for cultivation of yams. Pre-emergence herbicides like oxyfluorfen and pendimethalin can be successfully used for weed control, as the crop takes 15-20 days to emerge (Nedunchezhiyan *et al.* 2013).

***Sweet Potato** (Ipomoea batatas)*

In sweet potato, the common weed affecting the crop are *Celosia argentia, Digitaria sanguinalis, Cleome viscosa* and *Cyperus rotundus.* Yield reduction of 91% was found in sweet potato due to weed competition. Critical period of crop-weed competition was found to occur from 30 to 45 days after planting under Bhubaneswar conditions (Nedunzhiyan *et al.*,1998). Pre-emergence application of isoproturon @ 1kg/ha effectively controlled the weeds in sweet potato (Nedunchezhiyan and Satapathy, 2002).

Colocasia esculenta

The effect of weed interference in *Colocasia* prevents the development of optimum leaf area, which in turn affects the production of necessary assimilates for tuber bulking. The presence of weeds throughout the crop growth period reduced yield of taro by 60% (Nedunzhiyan *et al.*, 1996). Nedunchezhiyan *et al.* (2003) reported that pre-emergence application of isoproturon @ 1 kg/ha controlled the weeds effectively in taro. Nitrofen has been claimed to have excellent weed control when it is applied through irrigation water (Plucknett and de la Pena, 1971).

***Elephant Foot Yam** (Amorphophallus paeoniifolius)*

Elephant foot yam is susceptible to weed growth throughout the crop growth period because of less coverage of field by the leaf canopy. Often, weeds germinate and grow much earlier than the crop establishes because of delay in sprouting of planted corms. Weed growth is maximum in unweeded fields (AICRP, 2004). The critical period of crop-weed competition is between 1-5 months after planting as the major crop growth and corm bulking occurs during this period. In India, grasses (*Echinochloa colonum, Cynodon indica, Cynodon dactylon, Eleusine indica and Dactyloctenium aegyptium*), sedges (*Cyperus rotundus, Fimbristylis miliaceae*) and broad-leaved weeds (*Eclipta alba, Amaranthus viridis, Euphorbia hirta, Amaranthus spinosus, Commelina banghalensis, Corchorus acutangulus, Phyllanthus niruri* and *Cleome viscosa*) were observed (AICRP, 2004). Straw mulch at the time of planting followed by herbicides (pendimethalin, glyphosate, oxyfluorfen @ 1 kg/ha) could effectively reduce weed population as well as dry weed biomass as compared to control (AICRP,2006).

Carrot

Preplant incorporation of Fluchloralin @ 1.00 kg/ha or pre-emergence Pretilachlor @ 0.5 or Metribuzin @ 1.0 kg/ha controls annual weeds.

Radish

Application of pre-emergence Metolachlor @ 1.0-2.0 kg/ha or Alachlor @ 1.5-2.0 kg/ha or Isoproturon @ 1.0-1.25 kg/ha or Pendimethalin @1.00 kg/ha or Fluchloralin @ 1.00 kg/ha.

13.4.9 Cole Crops

Pre-emergence application of Fluchloralin @1.00 kg/ha or Pendimethalin @ 1.0 kg/ha along with one hand weeding at 40 days after planting.

Table 13.4 Selective pre-emergence and early post-emergence herbicides for vegetable seedbeds

Herbicide	Dose (kg a.i./ha)	Crop
A. Pre-emergence		
Trifluralin	1.0	Fenugreek
Clomazone	0.18: 0.27	Pepper, cucumber
DCPA	6.0: 7.5	Onion, cole crops, lettuce
Metribuzin	0.15: 0.5	Tomato
Napropamide	1.0: 2.0	Tomato, pepper, eggplant
Pendimethalin	1.0: 1.6, 1.0: 2.5	Onion, garlic, lettuce, fenugreek
Imezethapyr	0.055	Fenugreek
Propachlor	5.2: 6.5	Onion, cole crop
B. Post-emergence (crops with at least 3 leaves)		
Clomazone	0.27 -0.36	Pepper
Ioxinil	0.36	Onion, garlic, leek
Linuron	0.5: 1.0	Asparagus, carrots
Metribuzin	0.075: 0.150	Tomato
Oxifluorfen	0.18: 0.24	Onion, garlic
Imezethapyr	0.100	French bean
Rimsulfuron	0.0075 -0.015	Tomato

Table 13.5 Selective herbicides for weed control in vegetable crops

Herbicide	Dose kg a.i./ha	Treatment moment	Weeds	Crops
Alachlor	2.4	Post	Grasses	Brassica crops, onion
Benfluralin	1.17-1.71	PPI	Grasses	Lettuce, garlic
Bensulide	5.5-7.2	Pre	Grasses	Cucurbits
Bentazon	0.75-1	Post	Only Dicots	Green peas, beans
Chlorthal-dimetil (DCPA)	5.25-9.00	PP/Pre/Post	Mainly Grasses	Onion, lettuce, cole, tomato, green beans
Clomazone	0.18-0.54	PP/Post	Mainly Grasses	Pepper, green peas
Clomazone	0.18-0.27	Pre	Mainly Grasses	epper, cucumber, squash, pumpkin
Clopyralid	0.70-0.92	Post	Only Dicots	Asparagus
Diuron	0.4-2.4	Post	Mainly Dicots	Asparagus
Ethalfluralin	0.8-1.7	PP	Mainly Grasses	Tomato, pepper, beans, squash
Halosulfuron	24-48 (g)	Pre/Post	Mainly Dicots	Squash, cucumber
Ioxinil	0.36-0.60	Post	Only Dicots	Onion, leek, garlic
Isoxaben	0.1-0.12	PPI	Only Dicots	Onion, garlic
Linuron	0.50-1.25	Pre	Mainly Dicots	Carrot, artichoke, asparagus, faba bean

Herbicide	Dose kg a.i./ha	Treatment moment	Weeds	Crops
Metabenztiazuron	1.75-2.45	Pre/Post	Mainly Dicots	Onion, garlic, faba bean, peas
Metribuzin	0.35-0.52	PP/Post	Grass and Dicots	Tomato, asparagus
Metribuzin	0.10-0.35	Pre/Post	Grass and Dicots	Tomato, carrots, peas
Napropamide	1.57-2.02	PP/Post	Mainly Grasses	Tomato, pepper, artichoke
Naptalam-Na	2.16-2.88	Pre	Mainly Dicots	Melon and cucurbits
Oxifluorfen	0.36-0.48	Pre/Post	Mainly Dicots	Onion, garlic, cole crops
Oxifluorfen	0.24-0.48	PP	Mainly Dicots	Tomato, pepper
Pendimethalin	1.32-1.65	PP/PPI	Grass and Dicots	Artichoke, cole, lettuce, leek, pepper, tomato, onion, green peas
Pendimethalin	0.66-0.99	Pre	Mainly Grasses	Onion
Pendimethalin	0.66-1.65	Post	Grass and Dicots	Onion
Phenmedipham	0.55-1	Pre/Post	Mainly Dicots	Beets, spinach
Piridate	0.22-0.33	Post	D	Brassica crops
Prometryne	0.50-1.50	Pre/Post	Mainly Dicots	Artichoke, celery peas, pepper, tomato, carrot
Pronamide	0.70-1.50	Pre/Post	Mainly Grasses	Chicory, lettuce, endive
Propachlor	4.5	Pre	Mainly Grasses	Brassica crops, onion
Rimsulfuron	7.5-15(g)	Post	Grass and Dicots	Tomato
Trifluralin	0.59-1.44	PPI	Mainly Grasses	Beans, carrots, celery, cole crops, artichoke, onion, pepper, tomato

Source: Zaragoza, C. 2001.

(PP: Pre-plantation, PPI: pre-plant incorporated, Pre: pre-emergence, Post: post-emergence)

13.5 WEED MANAGEMENT IN PLANTATION CROPS

13.5.1 Tea

Although herbicides are now used over 60% of the area under tea, manual methods like sickling and mulching are extensively used wherever labour is available than by chemical method. Cheeling removes the above ground weed growth and prepares the ground for pre-emergence herbicide application. Sickling is done to remove the tall growth of perennial weeds and cut it back to the ground level for a follow up application of foliage applied herbicides on the regrowth. Pre-emergence application of Oxyfluorfen @ 0.40 kg/ha, foliage application of Paraquat (8 ml/l) + 2,4: D (6 g/l) or Glyphosate (1.5 ml/l) depending on the weed spectrum and this would keep tea weed-free for the rest PE application of Oxyfluorfen at 0.40 kg/ha of the year.

Nursery: Weed control in clonal nursery is done by (2 to 3 weeks before planting cuttings) application of Simazine @ 2 kg/ha. After 6 months Simazine is applied once again at the same dose.

Young Tea (Until 3 years): Application of Simazine @ 1.5 to 2 kg/ha or Paraquat @ 0.3-0.4 kg/ha at pre-emergence and 2,4 -D @ 0.5 to 1 kg/ha at post-emergence.

Matured Tea (Above 3 years): Simazine @ 1.5 to 2 kg/ha or Diuron @ 2 kg/ha as pre-emergence herbicides. Paraquat or 2,4-D or Paraquat + MSMA or Glyphosate are applied as post-emergence.

(***Source:*** www.agritech.tnau.ac.in/agriculture/agri_weedmgt_horticrops)

13.5.2 Coffee

Weeds compete with coffee for moisture and plant nutrients, while some perennial grasses and sedges produce root exudates that are toxic to coffee. Examples of perennial weeds in coffee are *Digitaria abbisinica, Commelina benghalensis, Cyperus rotundus, Cynondon dactylon, Pennisetum clandestinum* (Coffee Research Foundation, 2003). Weed control is particularly difficult (and expensive) in unshaded coffee. Control by herbicides has the advantages of zero tillage, including no damage to superficial feeder roots of the coffee. The most commonly practiced weed control methods include manual, mechanical cultivation and use of herbicides or integrated weed management (IWM). At post-emergence spray of 8 ml/l of Paraquat or Glyphosate @1.5 ml/litre are used to control many perennial weeds. Sequential application of Dalapan 1 to 2 weeks later by Paraquat and subsequently by Dalapan controls perennial grasses effectively.

13.5.3 Cashewnut

The productivity of cashew plantations in India is quite low due to many factors including the problem of weeds. Therefore, the main task after establishment of cashew plantation is to control excessive weed growth which might otherwise severely deplete soil moisture and compete for nutrients along with cashew. Their effect becomes more severe in young seedling stage where the roots of the weeds and the cashew are in the same soil layer. It has been reported that the failure of most cashew programmes in West Africa was due to competition from local grasses (Ohler, 1988). In the mature plantations, weed growth could also impede nut picking in situations where nuts are collected from fallen apples. Normally the cashew plantations are severely infested with a wide variety of annual and perennial weeds, which may reduce nut yield considerably. Based on the severity of weed infestation the yield reduction is observed as high as 60-70% in some cases. Abdul Salam *et al.* (1993) reported that application of glyphosate @ 0.8 kg/ha (single spray) or glyphosate @ 0.4 kg/ha + 2, 4-D @ 1.0 kg/ha (single spray) or paraquat @ 0.4 kg/ha + 2, 4-D @ 1.0 kg/ha (single spray) during August can effectively control the weed growth in cashew plantations. It was also found that chemical weed control is economical compared to the common method of sickle weeding. Abraham and Abraham (1999) found that use of glyphosate @ 1.2 kg/ha or 2, 4-D @ 1 kg/ha controlled Mikania infestation effectively.

13.6 SPICES

13.6.1 Coriander/Cumin/Fenugreek

Pre-planting incorporation of Fluchloralin @1.00 kg/ha. (or) PE application of Pendimethalin @ 1.00 kg/ha or Quizalofop @ 1.00 kg/ha as post-emergence supplemented with one hand weeding. (Source: www.agritech.tnau.ac.in/agriculture/agri_weedmgt_horticrops)

13.7 WEED MANAGEMENT IN FLOWER CROPS

13.7.1 Rose

Weed management is a major problem in rose cultivation under this situation, because of perennial growth habit of rose weeds compete with crop for various growth factors such as nutrient, moisture, light, space, etc. and severely reduced crop growth, yield and quality (Calalo *et al.*2010). Manual weed control is expensive, time consuming and laborious. However, pre-emergence use of herbicide has been reported to be effective in reducing the weed growth and density in rose (Singh *et al.*2005). However, Jadhav *et al.* (2018) claimed that application of Pendimethalin as pre-emergence herbicide @ 1.0 kg a.i./ha. followed by post-emergence application of ethoxysulfuron @ 20.0 g a.i./ha. at 30 DAS and hand weeding at 30 days interval starting from 75 days thereafter significantly minimized the weed biomass in rose and produced maximum plant height (61.30 cm), bud length (4.40 cm), flower diameter (5.70 cm), flower stem length (53.40 cm), number of flowers/plant (26.06) and vase life (5.60 days) in open cultivated rose.

13.7.2 Chrysanthemum

Weeds are of special significance when the chrysanthemum is grown in the field for commercial production. Kumar *et al.* (2017) reported minimum weed count (35.50), fresh weight (22.78 g), and dry weight (7.51 g) of weeds/m^2 with the treatment of pendimethalin @ 1.0 kg a.i./ha at 75 days after transplanting in chrysanthemum under Tarai condition of Uttarakhand.

13.7.3 Gladiolus

In India, gladiolus has established itself as a commercial proposition. In the modern agriculture, the weed control is becoming essential for higher yield of gladiolus. Employing labour increases cost of cultivation and affects successful commercial flower production. Integrated weed management is effective, economic and eco-friendly approach in improving and sustaining the agricultural productivity (Foy 1993).Kumar *et al.* (2012) revealed that there is significant enhancement in spike yield with 2 hand weedings at 20 and 40 days after transplanting (6.05 t/ha) and pendimethalin 2 kg/ha + 1 hand weeding (5.79 t/ha), both of which were superior to weedy check (3.25 t/ha). The highest weed control efficiency (78.2%) was also achieved with 2 hand weedings, followed by pendimethalin + hand weeding (76.9%). Application of pendimethalin along with hand weeding proved to be economical.

13.7.4 Tuberose

It is a vertical growing, non-spreading plant but still it gets tougher competition by weeds. Manual weeding is time consuming and costly. Hence, chemical weed control is one of the alternative methods to control weeds. Weeds cause irreparable damage to crops by competing for water, nutrients, light and space, besides acting as alternate hosts to a number of pathogens and insect-pests (Shalini and Patil, 2006). Rathod and Venugopal (2017) reported maximum flower yield in treatment weed free check (4.18 t/ha) followed by pendimethalin 30 EC @ 1 kg a.i./ha (3.67 t/ha) and alachlor 50 EC @ 1.5 kg a.i/ha (3.54t/ha) in tuberose cv Prajwal.

Table13.6 Weed Management for Different Horticultural Crops

Crop	Chemical Name	Dose/ha	Trade Name	Dose/ha	Time of Application
Weed Management for Other Horticultural Crops					
Cole Crops (Cabbage and Cauliflower)	Pendimethalin + Hand weeding	1.0 kg	Stomp 30% EC	3.3 L	3 DAT + 40 DAT
	Fluchloralin + Hand weeding	1.0 kg	Basalin 45% EC	2.2 L	3 DAT + 40 DAT
Carrot	Fluchloralin	1.0 kg	Basalin 45% EC	2.2 L	3 DAT
	Pritilachlor	0.50 kg	Rift 50% EC	1.0 L	3 DAT
Beans	Fluchloralin + Hand weeding	1.0 kg	Basalin 45% EC	2.2 L	3 DAT + 30 DAT
Radish	Fluchloralin + Hand weeding	1.0 kg	Basalin 45% EC	2.2 L	3 DAT + 30 DAT
	Pendimethalin + Hand weeding	1.0 kg	Stomp 30% EC	3.3 L	3 DAT + 30 DAT
	Alachlor + Hand weeding	1.5 kg	Lasso 50% EC	3.0 L	3 DAT + 30 DAT
	Metolachlor	1.0 kg	Dual 50% EC	2.0 L	3 DAT + 30 DAT
Peas	Alachlor	1.0 kg	Lasso 50% EC	2.0 L	3 DAT
Cucurbits	Oxyfluorfen	250 g	Goal 23.5% EC	1 L	3 DAS
Cumin/Fengreek/ Coriander	Pendimethalin + Hand weeding	1.0 kg	Stomp 30% EC	3.3 L	3 DAS + 40 DAS
	Fluchloralin	1.0 kg	Basalin 45% EC	2.2 L	Preplant incorporation
Turmeric	Pendimethalin + Hand weeding	1.0 kg	Stomp 30% EC	3.3 L	3 DAS + 45 DAS
	Alachlor + Hand weeding	1.0 kg	Lasso 50% EC	2.0 L	3 DAS + 45 DAS
Tea	Oxyfluorfen	400 g	Goal 23.5% EC	1.7 L	3 DAS
Coconut	Paraquat	1.0 kg	Gramaxone 24% SL	4.0 L	2-3 leaf stage of weeds

(Contd.)

Crop	Chemical Name	Dose/ha	Trade Name	Dose/ha	Time of Application
	Glyphosate+ Ammonium sulphate	10 ml (Commercial product) + 20 g/lit of water	Roundup 41% SL	10 ml (Commercial product)	pre flowering stage
Problematic weed management					
Cynodon dactylon & *Cyperus rotundus*	Glyphosate+ AGF activator	4.1 g + 2 ml/ lit of water	Roundup 41% SL	10 ml (Commercial product)	pre flowering stage
Under non crop situation,	Glyphosate + Ammonium sulphate	2.05 kg + 2%	Roundup 41% SL	5.0 L + 2%	pre flowering stage
Parasitic weed management					
Sriga management in Sugarcane	Atrazine + Hand weeding + earthing up combined + 2,4-D sodium salt + urea + trash mulch	1.0 kg+ 5 g/ litre (0.5%) + urea 20 g/ litre (2%)	Atratop 50% WP + Fernoxone 80% WP	2.0 kg + 5 g/litre (Commercial product)	3 DAP + 45 DAP + 60 DAP + 90 DAP + 150 DAP
Cuscuta	Hand removal & Paraquat	0.80 kg	Gramoxone 24% SL	3.3 L	20 DAS
	Pendiemthalin	1.0 kg	Stomp 30% EC	3.3 L	3 DAT
Solanum elaeagnifolium	Glyphosate alone or in combination with 2-4, D Na salt	1.75 kg +2.0 kg/ha	Roundup 41% SL + Fernoxone 80% WP	4 L + 2.5 kg	25 DAT
	Glyphosate + 2,4-D + soap solution	4.1 g + 6 g + 2 ml/lit of water	Roundup 41% SL + Fernoxone 80% WP	10 ml + 6 g (Commercial product) + 2 ml/lit of water	25 DAT
Water hyacinth	Glyphosate + Ammonium sulphate + surfactant	4.1 g + 20 g + 1 ml/lit of water	Roundup 41% SL	10 ml + 20 g (Commercial product) + 2 ml/lit of water	25 DAT

Source: AICRP- Weed Control, TNAU, Coimbatore
http://www.agritech.tnau.ac.in/agriculture/agri_weedmgt_horticrops

13.8 PRECAUTIONS FOR SAFE USE OF HERBICIDES

1. Spray herbicides at recommended concentration and time
2. Avoid spray drift of herbicide because this can cause injury to the susceptible plants growing in the vicinity especially in the nursery area.
3. Clean the sprayer with warm water every time after use. If possible, sprayer used for herbicides should not be used for any other kind of spray.
4. Avoid inhaling vapours, dust or spray mist. Irrigate at regular intervals soon after spray of pre-emergent herbicides like atrazine, diuron, etc.; this is necessary to

carry the herbicides to the lower layers of soil where weed seeds and nuts are germinating.

5. Irrigate immediately either before or after the spray of fluchloralin, as it is highly volatile compound.
6. Do not disturb the soil at least for a month in case of pre-emergent spray.
7. Post-emergent herbicides should not be sprayed on crop plants as it is contact in nature and kills those parts on which it falls. These herbicides must not be sprayed on a rainy day. At least a gap of 8-10 hours is required, between one spray and rain for effective control of weeds.

Important Weeds Found in Different Horticultural Crops

Cynodon dactylon

Cyperus rotundus

Sorghum halepense

Parthenium hysterophorus

Echinochloa crusgalli

Digitaria sanguinalis

Echinocloa muricata

Cyperus rotundus

Eleusine indica

Commelina benghalensis

Weed flora in Rabi

Poa anuua

Anagallis arvensis

Chenopodium album

Vicia sativa

Source: *Rana and Rana. (2016)*

REFERENCES

Abdul Salam M, Pushpalatha PB, Suma A and Abraham CT 1993. Chemical weed control in cashew plantations. *J. Plantation Crops,* **21** (1): 54-56.

Abraham M and Abraham CT 1999. Control of Mikania micrantha with herbicide. In: Proceeding of 11th Kerala Science Congress, Kasaragod, India. pp. 68-70.

AICRP 2004. Integrated weed management in elephant foot yam. In: All India Co-ordinated Research Project on Tuber Crops, Annual Report 2003-2004, Central Tuber Crops Research Institute, Thiruvananthapuram, Kerala, pp. 59-62.

AICRP 2006. Integrated weed management in elephant foot yam. In: All India Co-ordinated Research Project on Tuber Crops, Annual Report 2005-2006, Central Tuber Crops Research Institute, Thiruvananthapuram, Kerala, pp. 89-91.

Akobundu IO 1980. Weed control in cassava cultivations in the sub humid tropics. *Trop. Pest Mgmt*, **26**: 420-426.

Ambe JT, Agboola AA and Hahn SK 1992. Studies on weeding frequency in cassava in Cameroon. *Trop. Pest Mgmt*, **38**: 302-304.

Behera B and Singh GS 1999. Studies on weed management in monsoon season crop of tomato. *Indian J. Weed Sci.*, **31**:67-70.

Calalo JMG, Basch G and Mario de Carvalho 2010. Weed management in no till winter wheat (*Triticum aestivum* L.). *Crop Protection*,**29**:1-6.

Chinnusamy C, Prabhakaran NK, Janaki P and Govindarajan K 2009. Compendium on Weed Science Research in Tamil Nadu (25 years). AICPR on Weed Control, Department of Agronomy, TNAU, Coimbatore: 641 003.

CRF 2003 (Coffee Research Foundation), Weed Control in Coffee. Technical Circular No. 502.

El-Metwally IM and Hafez OM 2007. *Arab Univ. J. Agril. Sci.*, Ain Shams University, Cairo 15(1): 157-166.

Fongod AGN, Focho DA, Mih M, Fonge BA and Lang PS 2010. Weed management in Banana production: The use of *Nesonia canescens* spreng as a non-leguminous cover crop. *African J. Env. Sci. and Tech.*, 4:167-73.

Foy CL 1993. Perspective on integrated weed management for sustainable agriculture, *In: International Symposium at* HAU, Hisar, pp. 5–15.

Gare B N, Raundal PU and Burli AV 2015. Integrated weed management in rainfed chilli (*Capsicum annum* L.). *Karnataka J. Agric. Sci.*, **28** (2): 164-167.

Jadhav SB, Katwate SM, Kakade DS and Pawar BG 2018. Effect of pre and post emergence herbicides in rose weed management. *Int. J. Chem. Studies*, **6**(5): 1583-1585

Kabeer KAA and Nair VJ 2009. Flora of Tamil Nadu- Grasses. Botanical Survey of India, Kolkatta.

Kaur N and Rattanpal HS 2017.Weed management in citrus orchards. *Rashtriya Krishi*,**12** (1) :122.

Khan A, Muhammad S, Hussain Z and Khattak AM 2012. Effect of different weed control methods on weeds and yield of chillies (*Capsicum annuum* L.). *Pakistan J. Weed Sci. Res,* **18** (1): 71-78.

Hoogar R, Jayaramaiah R and Pramod G, Bhairappanavar ST and Tambat B 2017. Effect of Weed Management Practices on Weed Density, Weed Control. Efficiency, Weed Index and Yield of Potato (*Solanum tuberosum* L.). *Int. J. Curr. Microbiol. App. Sci.*, **6** (12): 493-499.

Kumar A, Manuja S, Singh J and Chaudhary DR 2015. Integrated weed management in tomato (*Lycopersicon esculentum* Mill) under dry temperate climate of western Himalayas. *J. Crop and Weed*, **11**(1):165-167.

Kumar A, Sharma BC and Kumar J 2012. Integrated weed management in gladiolus. *Indian J. Weed Sci.,* **44** (3): 181-182.

Kumar A, Kumari M, Ghosh S, Tewari T and Bhardwaj SB 2017. Effect of Weed Management Practices in Chrysanthemum (*Dendranthema grandiflora* T.) Under Tarai Conditions of Uttarakhand. *Int. J. Curr. Microbiol. App. Sci*., **6** (8): 3028-3034

Lawande KE, Khar A, Mahajan V, Srinivas PS, Sankar V and Singh RP 2009. Onion and Garlic research in India. *J. Hort. Sci*., **4** (2): 91-119.

Leela D 1982. A decade of weed research in horticultural crop. *Pesticide.* **15**: 3-8.

Maji S, Das BC and Bandyopadhaya P 2008. Studies on weed management practices in Guava cv L-49. *J Crop and Weed*, **4**:52-56.

Makhija M and Singh S 2011. Weed management studies in aonla. *Haryana J Hort. Sci*., **39**:185-87.

Majek BA, Neary PE and Polk DF 1993. Smooth pigweed interference in newly planted peach trees. *J. Production Agric*., 6: 244-246.

Mukherjee PK, Rahaman S, Maity SK and Sinha B 2012. Weed management practices in potato (*Solanum tuberosum* L.). *J. Crop and Weed*, **8** (1): 178-180.

Mohamed Amanullah M, Alagesan A, Vaiyapuri K, Sathyamoorthi K and Pazhanivelan S 2006. Effect of intercropping and organic manures on weed control and performance of cassava (*Manihot esculenta* Crantz). *Indian J. Agron*., **5**: 589-594.

Nadagouda BT, Kudrikeri CB, Salakinkop SR, Hunshal CS and Patil SL 1996. Integrated weed management in drill sown onion (*Allium cepa* L.). Farming Syst., **13** (3&4): 22-27.

Njoroge JM 1999. *17th East African Biennial Weed Science Conference Proceedings*, pp: 65-71

Nedunchezhiyan M, Ravindran CS and Ravi, V 2013. Weed Management in Root and Tuber Crops in India: Critical Analysis. *J. Root Crops*, **39** (2): 13-20.

Nedunzhiyan M Varma SP and Ray RC 1998. Estimation of critical period of crop-weed competition in sweet potato (*Ipomoea batatas* L.). *Adv. Hort. Sci*., **12**: 101-104.

Nedunchezhiyan M and Satapathy BS 2002. Effect of weed management practices on weed dynamics in sweet potato. *J. Root Crops*, **28**(2): 73-77.

Nedunchezhiyan M and Satapathy BS 2003. Effect of weed management practices on root development in taro (*Colocasia esculenta* (L.) Schott). *J. Root Crops*, **29** (1): 60-64.

Nishimoto RK 1997. Herbicide options for weed control in papaya. *Integrated Pest Management Reviews*, **2**: 109–111.

Ohler JG 1988. *Cashew*, Koninklijk Instituut voor de Tropen, Amsterdam, ISBN90-6832-074-2, 1988 (originally printed in 1979), pp. 260.

Plucknet DL and de la Pena RS 1971. Taro production in Hawaii. *World Crops*, **23**: 244-249.

Patel BN, Patel NN and Raj VC 2010. Influence of different weed management practices on growth of mango seedling. *The Asian J. Hort*., **5** (1): 89-92.

Panse R, Gupta A, Jain PK, Sasode DS and Sharma S 2014. Efficacy of different herbicides against weed flora in onion (*Allium cepa* L.). *Journal of Crop and Weed*, **10** (1): 163-166.

Patel TU, Patel CL, Patel DD, Thanki JD, Patel PS and Jat RA 2011. Effect of weed and Fertilizer management on weed control and productivity of onion (*Allium cepa*). *Indian J. Agron*., **56**: 267-272.

Ranpise SA and Patil BT 2001. Effect of herbicides on weed intensity and yield of summer onion cv. N-2-4-1. *Pestol.,* **25**: 59-60.

Rana SS and Rana MC 2016. Principles and Practices of Weed Management. Department of Agronomy, College of Agriculture, CSK Himachal Pradesh Krishi Vishvavidyalaya, Palampur,138 pages. (DOI:10.13140/RG.2.2.33785.47207)

www.hillagic.ac.in/edu/coa/agronomy/lect.)

Rathod A and Venugopal CK 2017. Weed management studies in tuberose (*Polianthes tuberosa* L.) cv. Prajwal. *J. Farm Sci.,* **30** (1): 100-103.

Reddy CN, Reddy MD and Devi MP 2000. Efficiency of various herbicides on weed control and yield of brinjal. *Indian J. Weed Sci.,* **32** (3&4): 150-152.

Sivamurugan AP, Nanthakumar S, Sganthi A and Davidson SJ 2011. Weed management in fruit crops. *Rashtriya Krishi,* **6** (2): 52-52.

Sharma RR 2006. Fruit Production Problems and Solutions. International book distributing Co., Lucknow, India.

Sharma KG and Singh AP 2012. Weed management practices on growth and yield of winter season brinjal under Chhattisgarh plain conditions. *Indian J. Weed Sci.*, **44** (1): 18-20.

Shil S and Adhikary P 2014. Weed management in transplanted chilli. *Indian J. Weed Sci.*, **46** (3): 261-263.

Sha K and Karuppaiah P 2005. Integrated Weed Management in Brinjal (*Solanum melongena* L.). *Indian J. Weed Sci.*, **37** (I & 2): 137-138.

Siddhu GM, Patil BT, Bachkar CB and Handal BB 2018. Weed management in garlic (*Allium sativum* L.). *J. Pharmacology and Phytochemistry*,**7** (1): 1440-1444.

Singh N, Sharma DP and Chand H 2016. Impact of climate change on apple production in India: a review. *Current World Environment*, **11**(1): 251-259.

Singh AK and Kavita K 2005. Effect of herbicides and mulching on growth and flowering parameters in rose. *J. Ornamental Hort.*, **8**(1): 49-52.

Verma SK and Singh T 1996. Weed control in *kharif* onion (*Allium cepa* L.). *Ind. J. Weed Sci.,* **28** (1-2): 48-51.

Zaragoza C 2001. Uso de herbicidas en cultivos hortícolas. *En Uso de Herbicidas en la Agricultura del Siglo XXI,* editado por De Prado, R. & Jarrín, J., Cap.15. Servicio de Publicaciones. Universidad de Córdoba, Spain. pp.169-182.

http://www.agritech.tnau.ac.in/agriculture/agri_weedmgt_horticrops.

OUTCOMES ASSESSMENT

PART A

Answer the following questions (True or False).

1. Oryzalin is shown to be tolerated by papaya immediately after transplanting. (True/False)
2. Hand weeding is effective against perennials. (True/False)
3. Weed is an unwanted plant growing at a place where it is not desired. (True/ False)
4. Weeds do not interfere with vital farm operations. (True/False)
5. Glyphosate is a pre-emergence herbicide. (True/False)

PART B

Answer the following questions.

1. Who is the father of weed science?
2. What is a weed?
3. Pendimethalin is a pre-/post-emergence herbicide.
4. Critical period of crop-weed competition in tomato is — to — DAS.
5. Cite one example of post-emergence herbicide.

PART C

Write a brief note on each of the following.

1. Precautions for safe use of herbicides.
2. Weed management in tomato and mango.
3. Crop-weed competition.
4. Methods of weed management.
5. Different methods of herbicide application.

Chapter 14

Fertility Management in Horticultural Crops

14.1 INTRODUCTION

India is conferred with a varied agro-climatic condition which are highly favourable for growing a large number of horticultural crops such as fruits, vegetables, root tubers, ornamental, aromatic plants, medicinal herbs, spices and plantation crops like coconut, areca nut, cashew and cocoa (Ministry of Information and Broadcasting, GOI, 2008). India is next only to China in area and production of vegetables and occupies prime position in the production of cauliflower, second in onion and third in cabbage in the world. After achieving food security, it is being increasingly felt that India needs to achieve nutritional security for betterment of its population. In achieving nutritional security fruits and vegetables play an important role (Ganeshmurthy *et al.*2011). Although we have achieved good production levels of fruits and vegetables, the present production of fruits and vegetables may not be enough to meet out the demand. To feed the ever-increasing population in a sustainable way a tremendous effort will have to be made to increase the production of fruits and vegetables both horizontally and vertically through a rational and balanced use of production inputs, specially the mineral fertilizers. Soil fertility is a key factor for successful crop production, and it is a measure of capacity of soil to supply plant nutrients. The importance of soil fertility and fertilizer management is being increasingly recognized in all countries recently to meet the demand for food and other agricultural raw materials. Intensive use of fertilizer, intensive cropping with high yielding cultivars have increased the food production and reduced the food shortage, but it has also brought in numerous problems of soil fertility, soil and water pollution. On the other hand, fast depletion of nutrients due to over exploitation, a wide-spread deficiency of N, P, K and S coupled with micro-nutrients deficiencies especially Zn and boron has been noticed in many soils. Further deforestation, shifting cultivation, burning of trees, bushes, grasses and cow dung, soil erosion, soil degradation, nutrient losses, excessive fertilizer application, leaching losses etc., have aggravated the depletion of soil fertility status. It is being realized that the future of Indian agriculture is closely related to scientific management of soil fertility along with judicious and efficient use of fertilizers. Soil fertility and nutrient management is one of the important factors that have a direct impact on crop yield and quality. Soil test reports help in determining soil organic matter, pH, electrical conductivity, cation exchange capacity, and levels of important macronutrients (phosphorus, potassium, calcium, magnesium) and micronutrients (boron, zinc, etc.). These reports also help estimating lime or sulphur application rates to increase or reduce soil pH, respectively. Maintaining a soil pH between 6.0 and 7.0 is recommended for most crop rotations that include vegetable crops.

14.2 SOIL FERTILITY

It is defined as the inherent capacity of a soil to supply available nutrients to plants in an adequate amount and in suitable proportions to maintain growth and development (Table 14.1). It is measure of nutrient status of soil which decides growth and yield of corp. *Soil productivity means the crop producing capacity of a soil which is measured in terms of yield.* Productivity is a very broad term and fertility is only one of the factors that determine the crop yields. Soil, climate, pests, disease, genetic potential of crop and man's management are the main factors governing land productivity, as measured by the yield of crop. To be productive, soil must contain all the 13 essential nutrients required by the plants.

14.3 NUTRITION

Nutrition is the supply and absorption of chemical compounds needed for growth and metabolism of an organism. Plants absorb or utilize more than 90 nutrient elements from the soil and other sources during their growth and development and about 64 nutrients have been identified in plants by their tissue analysis. But all are not essential for their growth and development. They require only 17 elements/nutrients. These 17 have been recognized as essential elements. They are:

1. Carbon (C)
2. Hydrogen (H)
3. Oxygen (O)
4. Nitrogen (N)
5. Phosphorous (P)
6. Potassium (K)
7. Calcium (Ca)
8. Magnesium (Mg),
9. Sulphur (S)
10. Iron (Fe)
11. Manganese (Mn)
12. Zinc (Zn)
13. Copper (Cu)
14. Boron(B)
15. Molybdenum (Mo)
16. Chlorine (Cl)
17. Nickel (Ni)

Source: ecoursesonline.iasri.res.in

Crops such as asparagus, cole crops, garlic, onions and spinach are crops that are sensitive to low pH, requiring pH maintenance above 6.5. Managing optimum soil nutrient levels is the key to maintaining a sustainable and productive vegetable production enterprise. Before a fertilization program can be planned it is important to know the cropping and soil fertilization history of the field. To maintain productivity of trees in long run and maintain the sustainability of tree production capacity, application of nutrient should be based on actual requirement and availability of nutrient in the soil. Nutrients are required for flowering and fruiting besides maintaining proper growth and vigour for producing high yields in following years.

14.3.1 Classification of Essential Nutrients

Essential nutrients are classified into two major groups based on relative utilization or absorption by the plants and based on their biochemical behaviour and physiological functions.

Based on relative utilization or absorption by the plants:

(a) Macro or Major Nutrients and (b) Micronutrients

Further Macronutrients are classified into two types

1. Primary Nutrients: Nitrogen, Phosphorus and Potassium. These three elements are also called as fertilizer elements.
2. Secondary Nutrients: Calcium, Magnesium and Sulphur.

14.3.2 Criteria of Essentiality of Plant Nutrients

This concept was given by Arnon and Stout (1939) and they considered 16 elements essential for plant nutrition. For an element to be regarded as an essential nutrient, it must satisfy the following criteria:

- A deficiency of an essential nutrient element makes it impossible for the plant to complete the vegetative or reproductive stage of its life cycle.
- The deficiency of an element is very specific to the element in question and deficiency can be corrected only by supplying that element.
- The element must directly be involved in the nutrition and metabolism of the plant and have a direct influence on plant apart from its possible effects in correcting some micro-biological or chemical conditions of the soil or other culture medium.

14.4 USE OF MANURES, FERTILIZERS AND BIO-FERTILIZERS

Fertilizers supply plant food and help to increase the yield of different crops through the improvement in the soil fertility. Manures not only supply plant nutrients but also improve the soil physical environment influencing plant growth. Bio-fertilizers are about to make a significant contribution towards the development of strategies for productivity improvement which do not lead to an exponential rise in the consumption of non-renewable forms of energy. Plant roots absorb the nutrients from the soil solution in the ionic (inorganic charged) form. Larger molecules can also be absorbed by roots, but their rate of absorption is slow. Thus, if a fertilizer (organic or inorganic) is applied, it must first be broken down to its simplest forms to be used efficiently by plants. The amount of plant nutrients in a fertilizer programme can be reduced if organic manures are used. The nutrient content of manures varies with type of livestock and methods used in storage, handling and application. Generally, the suggested rates of N, P_2O_5, and K_2O can be reduced by 5.5, 2.2 and 5.5 kg for each tonne of wet manure applied per hectare respectively. Therefore, it is essential to evolve and adopt a strategy of integrated nutrient supply by using a judicious mix of chemical fertilizers, organic manures and bio-fertilizers in relation to their importance, possible reactions in soils etc.

Table 14.1 Source of plant nutrients in soil

Nutrient	Source	Nutrient	Source
C	Carbamate	Fe	Pyrites, Magnetite
N	Organic matter	B	Tourmaline
P	Apatite (Calcium phosphate minerals), Fe or Al phosphates, organic matter	Cu	Chalcopyrite, Olivine, Hornblende, Augite, Biotite
K	Micas, Feldspars	Mn	Manganite, Pyrolusite, Olivine, Hornblende, Augite
Ca	Dolomite, Calcite, Apatite, Calcium carbonate, Gypsum, Augite	Mo	Olivine
Mg	Dolomite, Muscovite, Biotite, Olivine, Augite, Hornblende	Zn	Sphalerite, Olivine, Hornblende, Augite, Biotite
S	Pyrites, Gypsum. Organic matter, Olivine, Hornblende, Augite	Cl	Apatite

Source: Rana, S.S. (2011)

14.4.1 Manures

Manures are plant and animal wastes that are used as sources of plant nutrients. They release nutrients after their decomposition. Manures can be grouped into **bulky organic manures** and **concentrated organic manures** based on concentration of the nutrients.

Bulky Organic Manures

Bulky organic manures contain small percentage of nutrients and they are applied in huge quantities. **Farmyard manure** (FYM), **compost** and **green manure** are the most important and widely used bulky organic manures. Use of bulky organic manures have several advantages: (1) they supply plant nutrients including micronutrients, (2) they improve soil physical properties like structure, water holding capacity etc., (3) they increase the availability of nutrients, (4) carbon dioxide released during decomposition acts as a CO_2 fertilizer.

Farmyard Manure

Farmyard manure refers to the decomposed mixture of dung and urine of farm animals along with litter and left-over material from roughages or fodder fed to the cattle. On an average well decomposed farmyard manure contains 0.5% N, 0.2% P_2O_5 and 0.5% K_2O. The present method of preparing farmyard manure by the farmers is defective. Urine, which is wasted, contains 1% N and 1.35% K. Nitrogen present in urine is mostly in the form of urea which is subjected to volatilization losses. Even during storage, nutrients are lost due to leaching and volatilization. However, it is practically impossible to avoid losses altogether, but can be reduced by following improved method of preparation of farmyard manure. Vegetable crops like potato, tomato, sweet potato, carrot, radish, onion etc., respond well to the farmyard manure. The other responsive crops are sugarcane, rice, Napier grass and orchard crops like oranges, banana, mango and plantation crop like coconut. The entire amount of nutrients present in farmyard manure is not available immediately.

Compost

A mass of rotted organic matter made from waste is called compost. The compost made from farm waste like sugarcane trash, paddy straw, weeds and other plants and other waste is called farm compost. The average nutrient content of farm compost is 0.5% N, 0.15% P_2O_5 and 0.5% K_2O.

Green Manure

Green undecomposed plant material used as manure is called green manure. It is obtained in two ways: by growing green manure crops or by collecting green leaf (along with twigs) from plants grown in wastelands, field bunds and forest. Green manuring is growing in the field plants usually belonging to leguminous family and incorporating into the soil after enough growth. *The plants that are grown for green manure are known as green manure crops.* The most important green manure crops are Sunnhemp, Dhanicha, Cluster beans and *Sesbania rostrata.* Application to the field, green leaves and twigs of trees, shrubs and herbs collected from elsewhere is known as green-leaf manuring. The important plant species useful for green-leaf manure are neem, mahua, wild indigo, *Glyricidia*, Karanji (*Pongamia glabra*), *Calotropis, avise* (*Sesbania grandiflora*), subabul and other shrubs.

Vermicompost

The earthworm soil casts are richer in available plant nutrients (nitrate nitrogen, exchangeable Ca, Mg, K and P) and organic matter. Earthworm eats on fungal mycelia. Earthworms convert farm waste and organic residues into high quality compost. For this, *Eisenia foetida, Perionyx exacavatus, Eudrillus euginiae* and *Lumbrius rubellus* are important. These species can be cultured on organic wastes and dung. The technique of culturing them is called **vermiculture** and using them for decomposing residue to make compost is called **vermicomposting**. About 1000 adult earthworms can convert 5 kg waste into compost per day. The earthworm assimilates 5-10% of the substrate and rest passes through the alimentary canal and is excreted as cast. Earthworm cast contains nutrients, vitamins, hormones and antibiotics.

Sheep and Goat Manure

The dropping of sheep and goats contains higher nutrients than farmyard manure and compost. On an average, the manure contains 3% N, 1% P_2O_5 and 2% K_2O.

Night Soil

Night soil is human excreta, both solid and liquid. It is richer in N, P and K than farmyard manure and compost. Night soil contains on an average 5.5% N, 4.0% P_2O_5 and 2.0% K_2O.

Concentrated Organic Manures

Concentrated organic manures have higher nutrient content than bulky organic manure. The important concentrated organic manures are **oilcakes**, **blood meal**, **fish manure** etc. These are also known as organic nitrogen fertilizer. Before their organic nitrogen is used by the crops, it is converted through bacterial action into readily usable ammoniacal nitrogen and nitrate nitrogen. These organic fertilizers are, therefore, relatively slow acting, but they supply available nitrogen for a longer period.

Oil Cake

After oil is extracted from oilseeds, the remaining solid portion is dried as cake which can be used as manure. The oilcakes are of two types: (i) Edible oilcakes which can be safely fed to livestock *e.g.,* groundnut cake, coconut cake etc., and (ii) Non-edible oilcakes which are not fit for feeding livestock *e.g.,* castor cake, neem cake, mahua cake etc. Both edible and non-edible oilcakes can be used as manures (Table 14.2). However, edible oil cakes are fed to cattle and non-edible oil cakes are used as manures especially for horticultural crops. Nutrients present in oilcakes, after mineralization, are made available to crops 7 to 10 days after application. Oilcakes need to be well powdered before application for even distribution and quicker decomposition.

Table 14.2 Average nutrient contents of oil cakes

Oil cakes	**Nutrient content (%)**		
Non-edible oil cakes	N	P_2O_5	K_2O
Castor cake	4.3	1.8	1.3
Cotton seed cake (un-decorticated)	3.9	1.8	1.6
Karanj cake	3.9	0.9	1.2
Mahua cake	2.5	0.8	1.2
Safflower cake (un-decorticated)	4.9	1.4	1.2
Edible oil cakes			
Coconut cake	3.0	1.9	1.8
Cotton seed cake (Decorticated)	6.4	2.9	2.2
Groundnut cake	7.3	1.5	1.3
Linseed cake	4.9	1.4	1.3
Niger cake	4.7	1.8	1.3
Rape seed cake	5.2	1.8	1.2
Safflower cake (Decorticated)	7.9	2.2	1.9
Sesamum cake	6.2	2.0	1.2

Source: Rana, S.S. (2011)

14.4.2 Fertilizers

Fertilizers are defined as materials having definite chemical composition with a high analytical value that supply essential plant nutrients in available form. They are usually manufactured by industries and sold with a trade name. They are commonly synthetic in nature and called as **chemical fertilizers/inorganic fertilizers/ commercial fertilizers** other than lime and gypsum. Most of the chemical fertilizers are inorganic in nature. The only exception to this is urea and calcium cyanamide ($CaCN_2$), the solid organic N fertilizer. Presently fertilizers have become an integral part of agricultural economy as they increase the fertility of soils and enable them to support high yields (Table 14.3 to 14.7).

Classification of Inorganic Fertilizers

(A) Based on number of nutrients present, fertilizers are classified into following:

1. Straight fertilizers
2. Complex fertilizers
3. Mixed Fertilizers or Fertilizer Mixtures

(B) Classification of fertilizers based on plant nutrient element

1. Nitrogenous fertilizers
2. Phosphatic fertilizers
3. Potassic fertilizers
4. Secondary nutrient fertilizers
5. Micronutrient fertilizers.

Methods of Fertilizer Application

Two major methods are used to apply fertilizers

(A) Application of fertilizers in solid form

1. Broadcast:
   (a) Broadcasting at sowing/planting or before sowing/planting time
   (b) Broadcasting at later stages of crop growth i.e., top dressing
2. Placement:
   (a) Plough sole placement
   (b) Deep placement
   (c) Sub-soil placement
3. Localized placement:
   (a) Drilling
   (b) Band or hill placement:
      (i) Hill placement
      (ii) Row placement
      (iii) Side dressing
   (c) Pellet application
   (d) Basin application

(B) Application of fertilizers in liquid forms

(a) Starter solution: mixed with ratio of 1:2:1 or 1:1:2 of NPK respectively for vegetable seedlings at the time of transplanting

(b) Foliar application or spray fertilization: Spraying the nutrient solution with appropriate concentration over the foliage

(c) Direct application to the soil or plant (injection to soil or plant)

(d) Fertigation: Application fertilizer through irrigation water (Sprinkler or drip method)

Application of Fertilizers in Solid Form

Solid fertilizers of straight or complex or mixed NPK fertilizers are directly applied to soil by adopting different methods of application based on the type of crop, season, age of the crop, type of fertilizer and their concentrations.

1. **Broadcasting:** It is a cheapest method and consists of spreading the fertilizers uniformly over the entire floor of orchard, where the fruit plants have grown to large size and their roots have occupied the entire space in the orchard. It is a process where the fertilizer is spread over the entire soil area evenly and uniformly. This may be done before the land is ploughed, before planting or while the crop is standing. It may be of two types:

   (a) **Broadcasting at planting or sowing**: During sowing time, fertilizers are uniformly spread over the soil and mixed properly.

   (b) **Broadcasting as top dressing**: Followed for closely spaced crops. It may be made at the time of critical crop growth period. It is essential to note that the fertilizers are not to be applied when the leaves of plants are wet. This may create injury to the leaves or direct fall of fertilizer granules like urea or KCl on leaves leads to leaf scorching.

2. **Placement:** Fertilizers are placed in the soils irrespective of the position of seed, seedling or growing plants before sowing or after sowing the crops. The following methods are most common:

   (a) **Plough-sole placement:** It is applied in a continuous band at the bottom of the furrow during the process of ploughing.

   (b) **Deep placement:** It is specially followed in paddy fields for placement of N fertilizers ($(NH_4)_2SO_4$ and urea) in the reduced zone. It prevents loss of ammonia and makes the nutrient easily available to crop. The fertilizer is applied under the plough furrow in the dry soil before flooding the land and making ready for transplanting.

   (c) **Sub-soil placement:** It is recommended in humid and sub humid regions where sub-soils are strongly acid. P and K fertilizers are used.

3. **Localized placement:** It is application of fertilizers into the soil close to the seed or plants in bands or in pockets. Only small quantities of fertilizers are applied. It reduces fixation of P and K.

(a) Drilling

(b) Band or hill placement: i) Hill placement, ii) Row placement, iii) Side dressing

(c) Pellet application

(d) Basin application

(a) **Drilling**: In this method, the fertilizer is applied at the time of sowing by means of a seed-cum-fertilizer drill. This places fertilizer and the seed in the same row but at different depths.

(b) **Band placement**: If refers to the placement of fertilizer in bands. When the plants are placed 3 feet or more on both sides, fertilizers are placed close to plants in bands on one or both sides of the plant. It is practiced for N and P Fertilizers to orange, banana, papaya, apple, pear, coconut, cashew nut and other fruit trees.

(i) **Hill placement**: It is practiced for the application of fertilizers in orchards. In this method, fertilizers are placed close to the plant in bands on one or both sides of the plant. The length and depth of the band varies with the nature of the crop.

(ii) **Row placement**: When the crops like sugarcane, potato, maize, cereals etc., are sown close together in rows, the fertilizer is applied in continuous bands on one or both sides of the row, which is known as row placement.

(iii) **Side dressing**: It refers to the spread of fertilizer in between the rows and around the plants. The common methods of side-dressing are: placement of nitrogenous fertilizers by hand in between the rows of crops like maize, sugarcane, cotton etc., to apply additional doses of nitrogen to the growing crops and placement of fertilizers around the trees like mango, apple, grapes, papaya etc.

(c) **Pellet application**: It refers to the placement of nitrogenous fertilizer in the form of pellets 2.5 to 5 cm deep between the rows of the paddy crop. The fertilizer is mixed with the soil in the ratio of 1:10 and made small pellets of convenient size to deposit in the mud of paddy fields

(d) **Basin application**: In this method, the manure and fertilizers are applied in the basins of the fruit trees. This method is applied when the plants are young and/or basins have been prepared.

(*Source:* http://agritech.tnau.ac.in/agriculture)

Application of Fertilizers in Liquid Form

(i) Starter solution: Fertilizer solutions are prepared by dissolving the fertilizers in water at the ratio of 1:2:1 or 1:1:2 of N: P_2O_5: K_2O and applied to young vegetable plants at the time of transplanting. It is used in place of watering which helps in better establishment and early growth of plants. Only small amount of fertilizer is needed.

(ii) Foliar application: Spraying over the leaves of growing plants with suitable fertilizer solutions of lower concentrations (recommended) to supply one or combination of nutrients. Most of the micronutrient fertilizers (Zinc or Boron) are applied as foliar spray and urea is also used for foliar spray.

(iii) Direct application of Liquid fertilizers to soils: Liquid anhydrous ammonia or N, P, K fertilizers are applied to soils (10 cm depth) with the help of special equipment. Plant injury or wastage of ammonia is very much limited.

(iv) Fertigation: It is the process of direct application of liquid fertilizers through irrigation water. Straight and mixed fertilizers of N P K which are easily soluble in irrigation water can be applied through furrow or drip irrigation or sprinkler irrigation system.

Source: ecoursesonline.iasri.res.in

Table 14.3 Nutrient content of fertilizers

S. No.	Material	Total Nitrogen (N)	Ammoni-acal nitrogen (N)	Nitrogen Nitrate (N)	Nitrogen in form of urea (amide) (N)	Neutral ammoni-um citrate soluble phosphate (P_2O_5)	Water soluble phosphate (P_2O_5)	Water soluble potash (K_2O)
I. Nitrogenous fertilisers								
1.	Ammonium Sulphate	20.6	20.6					
2.	Ammonium chloride	25.0	25.0					
3.	Calcium Ammonium Nitrate	25.0	12.5	12.5				
4.	Calcium Ammonium Nitrate	26.0	13.0	13.0				
5.	Calcium Nitrate	15.5	1.1	14.4				
6.	Urea	46.0			46.0			
II. Phosphatic fertilisers								
7.	SSP 14% SSP 16%						14.0 16.0	
8.	Rock Phosphate (powder/ granular)					18.0		
III. Potassic fertilizers								
9.	Potassium chloride (powder/ granular)							60.0
10.	Potassium Sulphate							50.0

S. No.	Material	Total Nitrogen (N)	Ammoniacal nitrogen (N)	Nitrogen Nitrate (N)	Nitrogen in form of urea (amide) (N)	Neutral ammonium citrate soluble phosphate (P_2O_5)	Water soluble phosphate (P_2O_5)	Water soluble potash (K_2O)
11.	Potassium Magnesium Sulphate							22.0
12.	Potassium Schoenite							23.0
IV. Complex fertilizers								
13.	Ammonium Phosphate							
	11-52-0	11.0	11.0			52.0	44.2	
	1846-0	18.0	15.5		2.5(max)	46.0	41.0	
14.	Ammonium Phosphate Sulphate							
	16-20-0	16.0	16.0			20.0	19.5	
	20-20-0	20.0	18.0		2.0	20.0	17.0	
	18-9-0	18.0	18.0			9.0	8.5	
15.	Ammonium Phosphate Sulphate Nitrate 20-20-0	20.0	17.0	3.0		20.0	17.0	
16.	Nitrophosphate 20-20-0 23-23-0	20.0 23.0	10.0 11.5	10.0 11.5		23.0 23.0	12.0 18.5	
17.	Ammonium Nitrate Phosphate 23-23-0	23.0	13.0	10.0		23.0	20.5	
18.	Urea Ammonium Phosphate 28-28-0 24-24-0 20-20-0	28.0 24.0 20.0	9.0 7.5 6.4		16.5	28.0 24.0 20.0	25.2 20.4 17.0	
19.	Potassium Nitrate (crystalline) (13-0-45)	13.0		13.0				45.0
20.	Mono Potassium Phosphate (0-52-34)						52.0	34.0

(Contd.)

S. No.	Material	Total Nitrogen (N)	Ammoniacal nitrogen (N)	Nitrogen Nitrate (N)	Nitrogen in form of urea (amide) (N)	Neutral ammonium citrate soluble phosphate (P_2O_5)	Water soluble phosphate (P_2O_5)	Water soluble potash (K_2O)
21.	NPK fertilizers							
	15-15-15	15.0	7.5	7.5	3.0	15.0	4.0	15.0
	10-26-26	10.0	7.0		3.0	26.0	22.1	26.0
	12-32-16	12.0	9.0		15.0	32.0	27.2	16.0
	22-22-11	22.0	7.0			22.0	18.7	11.0
	14-35-14		14.0			35.0	29.0	14.0
	17-17-17	17.0	5.0			17.0	14.5	17.0
	14-28-14	14.0	8.0			28.0	23.8	14.0
	19-19-19	19.0	5.6			19.0	16.2	19.0
	17-17-17	17.0	8.5			17.0	13.6	17.0
	20-10-10	20.0	3.9	8.5		10.0	8.5	10.0

Source: http://fert.nic.in/aboutfert/nutrient_content.asp

Table 14.4 Nutrient content of liquid fertilizers

(Per cent by wt. min.)

S. No.	Material	Total Nitrogen (N)	Ammoniacal nitrogen (N)	Nitrogen Nitrate (N)	Nitrogen in form of urea (amide) (N)	Neutral ammonium citrate soluble phosphate (P_2O_5)	Water soluble phosphate (P_2O_5)	Water soluble potash (K_2O)
1	Urea Ammonium Nitrate	32.0	7.7	7.7	16.6 (max)			
2	Superphosphoric Acid					70.0	18.9	0.5 (max)
3	Ammonium Polyphosphate	10.0	10.0			34.0	22.1	0.5 (max)

Source: http://fert.nic.in/aboutfert/nutrient_content.asp

Table 14.5 Nutrient content of mix fertilizers

Mix Fertilizers	Nitrogen (%) by wt.	Water soluble P_2O_5 (%) by wt.	Water soluble K^2O (%) by wt.	Ca	Mg	S
DAP	18	46				
Potassium Nitrate	13		45			
NPK (15:15:15)	15	15	15			
NPK (10:26:26)	10	26	26			
NPK (12:32:16)	12	32	16			
NPK (20:10:10)	20	10	10			

Mix Fertilizers	Nitrogen (%) by wt.	Water soluble P_2O_5 (%) by wt.	Water soluble K^2O (%) by wt.	Ca	Mg	S
MANURES						
Cattle Manure	1.5	1.5	1.2	1.1	0.3	
Poultry Manure						
a) Broiler litter	3.0	3.0	2.0	1.8	0.4	0.3
b) Hen-litter	1.8	2.8	1.4			
Sheep manure	0.6	0.3	0.2			
Sewage sludge	5.0	6.0	0.5	3.0	1.0	1.0

Source: Usha *et al.*, (2015).

Table 14.6 Percent of micronutrients in commercial products

Commercial product	Content (%)	Commercial product	Content (%)
Cu EDTA	13 Cu	Ferrous sulphate	20% Fe
Copper sulphate	35 Cu	Fe EDTA	12% Fe
Borax	11% B	Manganese sulphate	24% Mn
Boric acid	17% B	Ammonium molybdate	54% Mo
Calcium borate	10% B	Sodium molybdate	46% Mo
Sodium tetraborate	14% B	Zinc chelate	14% Zn
Magnesium borate	21% B	Zinc sulphate	36% Zn
Basic slag	13% Fe; 3% Mn		

Source: Usha *et al.*, (2015).

Table 14.7 Factor for calculating quantity of different fertilizers

Fertilizers	Factor	Fertilizers	Factor
Nitrogenous fertilizers		**Phosphatic fertilizers**	
Urea	2.17	Single super phosphate	6.25
Calcium Ammonium Nitrate	3.85	Rock phosphate	5.56
Ammonium Nitrate	2.94	Bone meal	5.00
Ammonium chloride	4.0	**Potassic fertilizers**	
Calcium Nitrate	6.45	Potassium sulphate	1.92
Ammonium sulphate	4.84	Potassium chloride	1.66
		Potassium Magnesium Sulphate	4.55

Source: Usha *et al.*, (2015).

14.4.3 Biofertilizers

Biofertilizers are inputs containing useful microorganisms. They are capable of mobilizing plant nutrients from unavailable forms to available forms through biological processes. They are cost-effective and inexpensive sources of plant nutrients. Biofertilizers are grouped under three distinct categories *viz.*, N-fixers, phosphate solubilizers and mycorrhizal fungi. Nitrogen-fixing organisms associated with horticultural crops are *Rhizobium* sp. associated with legume vegetables and free-living microorganisms belonging to *Azotobacter* sp. and *Azospirillum* sp. which live in association with root systems of non-leguminous horticultural crops. Several soil bacteria particularly those belonging to the genera *Pseudomonas* and *Bacillus* and fungi belonging to the genera *Penicillium* and *Aspergillus* possess the ability to convert insoluble P in soil into soluble forms. The *vesicular arbuscular mycorrhizae* (VAM) are formed by the fungi belonging to the genera *Glomus, Gigaspora, Acaulospora* and *Sclerocystis* which produce vesicles and arbuscules inside the root system. Arbuscles are highly branched fungal hyphae and the vesicles are bulbous swellings of these fungal hyphae. These fungi make nutrients available to the host plants. Chemical fertilizers and biofertilizers should not be applied together as the microorganisms in biofertilizers may get killed.

14.5 LEAF SAMPLES

Leaf analysis is an important tool to estimate nutritional requirement of horticultural crops. Leaf nutrient content can be obtained by analysing leaf tissues at proper growth stage (Table 14.8).

Table14.8 Plant tissues sampling guideline for horticultural crops

S. No.	Crop	Index tissues	Growth stage
Fruit crops			
1.	Mango	Leaves and petiole	4 to 7 months old leaves from middle of the shoot
2.	Banana	Petiole of 3rd open leaf from apex	Bud differentiation stage
3.	Citrus	3 to 5-month-old leaf from new flush. 1st leaf of the shoot	June
4.	Papaya	6th petiole from the apex	6 months after planting
5.	Guava	3rd pair of recently matured leaves	Bloom stage
6.	Grape	5th petiole from base	Bud differentiation stage for yield forecast. petiole
7.	Ber	16th leaf from apex from secondary or tertiary shoot	Two months after pruning
8.	Pomegranate	8th leaf from apex	Bud differentiation. In April for February crop and August for June crop
9.	Sapota	10th leaf from apex	September
10.	Custard apple	5th leaf from apex	2 months after new growth
11.	Fig	Fully expanded leaves, mid shoot current growth	July-August
12.	Phalsa	4th leaf from apex	One month after pruning
Plantation crops			
1.	Coconut	Pinnal leaf from each side of 4th leaf	

S. No.	Crop	Index tissues	Growth stage
2.	Oil palm	Middle 1/3rd minus midrib of 3 upper and 3 lower leaflets from 17 frond of mature trees and 3rd frond of young trees	
3.	Coffee	3rd or 4th pair of leaf from apex of lateral shoots	
4.	Tea	Third leaf from tip of young shoots	
5.	Clove	10th to 12th leaves from tip of non- fruiting shoot	End of blooming period
Vegetable Crops			
1.	Beans	Upper most recent fully developed trifoliate leaves	
2.	Cabbage	Wrapper leaf	2-3 months old
3.	Carrot	Most recent fully matured leaf	Mid-grown
4.	Cauliflower	Most recent fully matured leaf	At heading
5.	Peas	Most recent fully developed leaflet	First bloom
6.	Cluster bean	First fully developed leave	
7.	Cucumber	5th leaf from tip	Flower bud (start) to small fruit
8.	Brinjal	Leaf blades with midribs minus petioles from most recent fully developed leaf	
9.	Garlic	Most recent fully matured leaf	Pre-bulb
10.	Onion	Top-no white portions	1/3 to ½ grown
11.	Tomato	Leaves adjacent to inflorescence	Mid bloom
Ornamental Crops			
1.	Jasmine	Most recent fully developed leaves	
2.	Chrysanthemum	4th leaf from tip, omit unfurled	Bud burst
3.	Hibiscus	Most recent fully developed whole leaves	
4.	Lilly	Most recent fully developed leaf	
5.	Rose	Most recent fully developed compound 5th leaflet leaf	Flower bud pea size

Source: Bhargava and Chadha (1993) and Raghupathi *et al.*, (2014).

14.6 MANURING AND FERTILIZER APPLICATION IN FRUIT CROPS

The manure and fertilizer requirements of fruit crops are summarized below (Table 14.9 and 14.10):

Table 14.9 Manuring and fertilizer application in important fruit crops

Crop	Time of application
Mango	The nutritional requirement of mango varies with the region, soil type and age. A dose of 73g N, 18g P_2O_5 and 68g K_2O_5/year of age from first to tenth year and thereafter a dose of 730g N, 180g P_2O_5 and 680g K_2O should be applied in 2 split doses during June-July and October respectively. Spraying of zinc sulphate (0.3%) during February, March and May is recommended to correct the zinc deficiency. Spraying of Borax (0.5%) after fruit set twice at monthly intervals and 0.5% manganese sulphate after blooming corrects boron and manganese deficiencies respectively. Organic manures and phosphatic fertilizers should be applied immediately after harvest, whereas ammonium sulphate should be given before flowering.
Banana	Banana is a heavy feeder crop that requires very large quantity of nutrients for growth and yield, which accounts for 20-30% of the total cost of production. For normal plant growth and development 100-250g of N/plant is advised depending on nutrient status of soil and cultivar. Urea is commonly used as a source of N. It should be applied in 3-4 splits. Application of 150g N in vegetative phase and 50g N in reproductive phase enhances the yield and delays the leaf senescence. Application of 25% N as farmyard manure and 1kg neem cake is beneficial. Phosphorus is applied in single dose at the time of planting and quantity of P_2O_5 depends upon soil type and varies from 20-40g/plant. Potassium is indispensable in banana nutrition due to its role in vital functions. It is not stored, and its availability is influenced by temperature. Thus, continuous supply is required to be assured at finger-filling stage. Application of K (100g) in 2 splits during vegetative phase and 100g in 2 splits during reproductive phase is recommended. Application of 200-300g K_2O is recommended depending upon cultivar. Invariably, plantains require higher K than other group of cultivars. Muriate of potash is invariably used as source of K. But in soils with pH above 7.5, potassium sulphate is advantageous. Among micronutrients, Zn, Fe, B, Cu and Mn play an important role in normal growth and development of banana. The application of Zn (0.1%), B (0.005%) and Mn (0.1%) improves yield.
Mandarin Orange	Application of 20-25kg farmyard manure together with 0.4kg calcium ammonium nitrate is recommended at the time of planting. A mixture of 90g each of N, P, K/tree may be applied in first year after planting. This dose may be gradually increased to 450g each of N and P, and 900g of K/tree in the seventh year and kept constant thereafter, The micronutrients should be supplied through foliar spraying (multiplex micronutrient mixture @3 ml/l) during November and March. Normally young plants are manured once a year, while bearing plants more than once. Total amount of P and K fertilizers is applied at one time, while N fertilizers are applied in 2 or 3 split doses. In north-western India, manures are added twice a year, once during June and another after harvesting in December-January. In central India, mandarins are manured during December for ambe bahar and in May for mrig bahar. However, for Coorg mandarin, manuring thrice (March-April, June-July and September-October) is highly beneficial. For soil application, first 30-40cm area around the tree trunk should be given hoeing and fertilizers should be applied in the form of ring. They should be mixed thoroughly. The trees should be irrigated immediately

Crop	Time of application				
Sweet orange	State	Nitrogen (g)	Phosphorus (g)	Potassium (g)	Farmyard manure (Kg)
	Andhra Pradesh	1,500	350	400	-
	Karnataka	550	370	550	30
	Maharashtra	1,000	100	200	24-30
	Tamil Nadu	400	200	300	30
Guava	For guava-growing regions of the country, different fertilizer schedules—600g N, 400g K in northern region; 260g N, 320g P and 260g K in eastern region; 900g N, 600g P and 600g K in southern region and 600g N, 300g P and 300g K/ plant/year in western region—have been recommended. In north India, fertilizer is given in the first week of May for rainy season crop and in first week of July for winter season crop. In West Bengal, fertilizers are applied in 2 equal split doses, one in January and the other in August. At Bangalore, full K and 70% N are applied in June and full P and 30% N in September. Since 48% of feeder roots of guava are found in the surface soil up to 25cm depth, the fertilizer should be placed in 25cm trenches 1m away from the trunk for better uptake. Bronzing can be reduced by improving the soil pH and treating the soil with N, P, K and Zn at 200, 80, 150 and 80g/year respectively, or fortnightly foliar spraying of these nutrients each at 2% for 4 months.				
Papaya	Papaya is a heavy feeder and needs heavy doses of manures and fertilizers. Apart from the basal dose of manures applied in the pits, 200-250g each of N, P_2O_5 and K_2O are recommended for getting high yield. Application of 200g N is optimum for fruit yield but papain yield increases with increase in N up to 300g. A dose of 250g N, 250g P and 500g K/plant is recommended for papaya Coorg Honey Dew under Bangalore conditions, while 200g each of N, P and K in split doses in the first, third, fifth and seventh month is recommended for papaya Co 1 under Coimbatore conditions. Deficiency of lime and B has often been observed in papaya orchards. Spraying of 0.5% zinc sulphate (twice) and one spray of Borax (0.1%) may be done depending upon the nutrient status of soil.				

Source: Chadha, K.L (2001)

Table14.10 Nutritional Requirement of full grown up (other fruit) trees/plant/year

Crop	Manures (Kg)	N:P:K (g)	Time of application
Ber	50	750: 150: 150	Half in July and half in October
Date Palm	50	1200: 350: 300	March-April
Pomegranate	50	625 :250: 250	Dec-Jan: Ambe Bahar May-June: Mrig Bahar
Aonla	50	800:300: 400	Half N, half P and full K in Jan-Feb and rest half N and half P in Sep
Phalsa	10	100: 50: 25	After pruning
Fig	25	900: 150: 200	February-March
Guava	40	360: 180: 180	Half N, Full P and full K in July-August, and half N in October

Source: More and Sharma (2008)

14.7 INTEGRATED PLANT NUTRIENT MANAGEMENT

Nutrient management is the science and art directed to link soil, crop, weather, and hydrologic factors with cultural, irrigation, and soil and water conservation practices to achieve the goals of optimizing nutrient use efficiency, yields, crop quality, and economic returns, while reducing off-site transport of nutrients that may impact the environment. Nutrient management, particularly in horticultural crops, is the skilful task of matching a specific field, soil, climate, and crop management conditions to rate, source, timing, and place of nutrient application (Mikkelsen, 2011). *Integrated nutrient management refers to the use of chemical fertilizers in conjunction with organic manures, green manures, crop residues and legumes in a cropping system and locally available resources with objectives of sustaining high yield and ensuring environmental safety.* The objectives of integrated plant nutrient management are: (i) to decrease inorganic fertilizer requirement; (ii) to re-establish the organic matter in soil and to increase nutrient use efficiency; (iii) to continue quality in terms of physical, chemical and biological properties of soil and (iv) to conserve the nutrient balance between the supplied nutrient and nutrient removed by plant and to improve soil health and productivity on sustainable basis. The selection of appropriate nutrient management practices for individual horticulture farms needs to be tailored to the specific conditions existing at a site. The science and art of nutrient management has progressed immensely during the last decade because of the integration of new developments in the areas of computers, simulation models, GIS, GPS, and other in situ sensors. Technology will continue to advance and help integrate these multiple factors for agricultural sustainability and profitability (Ganeshmurthy *et al*., 2016).

14.8 MANAGEMENT OF SOIL ACIDITY

The fertility status of acid soils is very poor and under strongly to moderately acidic soils, the plant growth and development is affected to a great extent. The crops grown on such problematic soils do not give remunerative returns. The addition of lime raises the soil pH, thereby eliminating most major problems of acid soils. Cultivation of acid tolerant crops/cultivars is also possible option.

14.9 MANAGEMENT OF CALCAREOUS SOIL

Calcareous soils are those that contain enough free calcium carbonate ($CaCO_3$). The pH of calcareous soil is > 7.0 and regarded as an alkaline (basic) soil. Fertilizer management in calcareous soils is different from that of non-calcareous soils because of the effect of soil pH on soil nutrient availability and chemical reactions that affect the loss or fixation of some nutrients. The presence of $CaCO_3$ directly or indirectly affects the chemistry and availability of nitrogen (N), Phosphorus (P), Magnesium (Mg), Potassium (K), Manganese (Mn), Zinc (Zn) and iron (Fe). The availability of copper (Cu) also is affected. Application of acid forming fertilizers such as ammonium sulphate and urea fertilizers, sulphur compounds, organic manures and green manures is considered as effective measures to reduce the pH of soil to neutral pH value.

14.10 MANAGEMENT OF SALINE SOIL

The saline soils have EC > 4 Ds/m, pH < 8.2 and the ESP is < 15. They are also known as white alkali soils. In areas where only saline irrigation water is available or when shallow saline water table prevails and soil permeability is low, the following cultural practices are adopted:

1. Growing of highly tolerant crops like barley, sugar beet, date palm and tolerant crops like tapioca, mustard, coconut, spinach, amaranthus, pomegranate, guava, ber etc.
2. Crop should be raised by transplanting seedlings (especially vegetables, flowers, fruit trees) than germinating the seeds.
3. Furrow or drip irrigation, sub-surface irrigation systems and sprinkler irrigation should be followed.

14.11 MANAGEMENT OF ALKALI SOIL

Alkali soils are those that contain measurable amounts of soluble salts mostly as carbonates and bicarbonates of sodium. Alkali soils have pH > 8.2 and ESP > 15. The EC of alkali soils is variable but normally <4 Ds/m. *The alkali soils are also known as sodic. They are also called as black alkali soils.* It is mainly because of high pH and Na_2CO_3, the finely decomposed organic matter is dissolved along with the water that imparts a dark black or brown colour to the soil. The alkali soil can be ameliorated by broadcasting and incorporating of Gypsum or pyrite in the soil. Organic amendments like straw, groundnut and safflower hulls, FYM, compost, green manures, tree leaves and saw dust, and cultural practices like land levelling and shaping, maintaining plant population, green manuring with *Sesbania*, dhanicha, subabul and proper drainage and irrigation methods may ameliorate alkali soil.

REFERENCES

Bhargava BS and Chadha KL 1993. Leaf nutrient guide. In Advances in Fruit Crops, Malhotra Publishers, New Delhi, India.

Chadha KL 2001. Hand Book of Horticulture, Directorate of Information and Publication of Agriculture, ICAR.

Ganeshamurthy AN, Satisha GC and Patil P 2011. Potassium nutrition on yield and quality of fruit crops with special emphasis on banana and grapes. *Karnataka J. Agric. Sci.*, **24** (1):29-38.

Ganeshamurthy AN, Kalaivanan D and Selvakumar G 2016. Nutrient management in horticultural crops. *Indian J. Fert.*, **11** (12) :30-42.

Mikkelsen RL 2011. The "4R" nutrient stewardship framework for horticulture. *Horttechnology*, **21**(6): 658-662 (2011).

Ministry of Information and Broadcasting, Government of India 2008. Soochna Bhawan, CGO Complex Lodhi Road, New Delhi, *India a reference manual*, pp. 64 (2008).

More TA and Sharma BD 2008. Paper on "Recent technologies for developing arid and semi-arid horticulture in India." presented in the 4th Indian Horticulture Congress held at OUAT, Bhubaneswar, 2008.

Raghupathi HB, Ganeshamurthy AN and Ravishankar H 2014. Comparison of DRIS ratio norms of selected fruit crops. *Indian J. Hort.,* **71**(2): 168-175.

Singh J 2018. *Basic Horticulture*. Kalyani Publishers, Ludhiana.

Rana SS 2011. Principles and Practice of Soil Fertility and Nutrient Management. Compendium of Lectures (Theory cum Practical), Department of Agronomy, Forages and Grassland Management, COA, CSK HPKV, Palampur-176062.

Usha K, Thakre M, Goswami AK and Nayan DG 2015. Nutrient assessment in fruit crops. In: Fundamental of Fruit Production, Division of Fruits and Horticultural Technology, Indian Agricultural Research Institute, New Delhi., pp. 209-212.

e-resources:

ecoursesonline.iasri.res.in

http://fert.nic.in/aboutfert/nutrient_content.asp

http://agritech.tnau.ac.in/agriculture

OUTCOMES ASSESSMENT

PART A

Answer the following questions (True or False).

1. The technique of culturing earthworms is called vermi composting. (True/False).
2. Leaf analysis is an important tool to estimate nutritional requirement. (True/False)
3. Concentrated organic manures have lower nutrient content than bulky ones. (True/False)
4. Borax contains 11% boron. (True/False)
5. DAP contains 46% nitrogen.

PART B

Answer the following questions.

1. What is fertigation?
2. Define fertilizers.
3. Chemical fertilizers are also called — and — fertilizers.
4. What are biofertilizers?
5. The alkali soils are also known as — soils.

PART C

Write a brief note on each of the following.

1. Discuss integrated plant nutrient management.
2. What are the methods of fertilizer application?
3. Classification of essential nutrients.

Chapter 15

Cropping Systems

15.1 INTRODUCTION

Cropping system is defined as a cropping pattern followed on a farm and its interaction with farm resources, other farm enterprises and production technology. In other words, cropping system refers to a combination of crops in time and space. When annual crops are considered, a cropping system usually means the combination of crops within a given year (Willey *et al.*, 1989). This cropping system might provide insurance against crop failure by reducing disease (Fininsa and Yuen, 2002) and insect incidence (Gahukar, 1989) or against unstable market prices by planting two or more crops under intercropping and thus, reducing the risk of unexpected changeable prices. Intercropping in scientific way should be viewed from a theme of competition to one of collaboration with mutually beneficial relation with proper planning to get more return from limited natural resources. Legumes have been the common intercrops in any intercropping system owing to their short duration, N-fixing ability etc. Even though non-leguminous vegetables require longer duration than legumes and are non-N-fixers, they can also be suitable as intercrops because of their high profitability and higher yields.

15.2 PRINCIPLES OF CROPPING SYSTEM

1. Replication of crops having common diseases and pests should be avoided.
2. The deep-rooted crops should be followed by shallow rooted ones.
3. Heavy feeding crops should be followed by low feeding crops
4. Leguminous vegetable crops should be included in the cropping sequence which not only upgrade the protein status of the farm produce but also enhance soil fertility.
5. Green manure crops should be accommodated in the rotation in order to increase the organic matter status of the soil.

15.3 BENEFITS OF CROPPING SYSTEM

1. Maximum land utilization
2. Maximum input utilization
3. Better harvest of solar radiation
4. Yield continuum

5. Regular income
6. Sustainable production
7. Maintain and enhance soil fertility
8. Boost crop growth
9. Less incidence of pest, disease and weed
10. Increase soil cover
11. Reduce risk for crop failure
12. Use resources more efficiently

15.4 INTERCROPPING

The term cropping system is often used interchangeably with multiple cropping, which in fundamental nature represents an idea of maximum production per unit area of land within a year or some other relevant time unit with minimum land degradation (Singh, 1972).Therefore, to maximize land use and production, intercropping is an advanced agronomic technique that allows two or more crops to yield from the same area of land. Better utilization of resources and reduced weed competition minimize the risk of food shortages by enhancing yield stability.

15.5 CRITERIA FOR THE SELECTION OF AN INTERCROP

1. It should not be tall growing and spreading type.
2. It should not be exhaustive.
3. It should not function as alternate host for common pest and diseases.
4. Water requirement schedule should match or phenology of crop should match so that operation could be synchronized.
5. It should match for climatic requirements with main plantings.
6. Separate provision for nutrients should be made for inter-crop to avoid competition.
7. Intercrops should be preferably legumes or shallow rooted vegetables.

15.6 TYPES OF INTER-CROPPING

1. ***Parallel Cropping***: Cultivation of such crops which have different natural habit and zero competition. (Black gram/Green gram + maize)
2. ***Companion Cropping***: Such inter cropping where the production of both intercrops is equal to that of its solid planting. (Mustard/Potato/Onion + Sugarcane)
3. ***Multiple Cropping/Multitier/Multilevel Cropping***: It is one of the cropping systems, wherein two or more crops are grown in succession within a year.
4. ***Synergistic Cropping***: The yields of both crops are higher than of their pure crops on unit area basis. (Sugarcane + Potato)

15.7 EXAMPLES OF INTERCROPPING IN VEGETABLES

Table15. 1 Examples of intercropping in vegetables followed in different parts of India

Main crop	Intercrops	Place of work
Okra	Beet root, knol-khol, Pea	Banglore (Karnataka)
Capsicum	Beet root, Knol-khol,Pea	Banglore (Karnataka)
Cabbage	Radish, Turnip, Methi,	Hisar (Haryana), Akola
Cauliflower	Radish, Palak	Akola (Maharashtra)
Chilli	French bean, Onion	Dharwad (Karnataka)
Tomato	Spinach, Onion, Radish	Hisar (Haryana)
Pigeonpea	Urd, Moong, Cowpea, Okra	New Delhi

Source: Singh (1997)

15.8 SUITABLE INTER-CROPS FOR DIFFERENT FRUIT CROPS

Banana: Green gram, cowpea, cauliflower, cabbage, yam, elephant foot, onion, black gram, turmeric, brinjal, *Colocasia*, *Dioscorea*, chilies, lady's finger.

Ber: Green gram, moth, cluster bean, cowpea, cumin, chilies.

Citrus: Beans, carrot, berseem, onion, potato, chilies, pulses, cucurbits, lady's finger, gram, peas, tomato, cabbage.

Date palm: Citrus medica, guava, sapota.

Grape: Vegetables relevant to area.

Guava: Cauliflower, peach, French bean, cowpea, cluster bean, black gram, lady's finger, onion, turmeric, garlic, cabbage, chillies, papaya.

Litchi: Turmeric, ginger, pointed gourd, sweet potato, tomato, radish, cabbage, turnip, brinjal, cucurbits, green gram, black gram, cowpea.

Mango: Phalsa, papaya, guava, banana, peach, strawberry, pineapple.

Papaya: Cabbage, cauliflower, chilies, radish, tomato.

Pomegranate: Berseem, lucerne, cowpea, green pea, cucurbit, cabbage, cauliflower, bean, peas, tomato, carrot, onion, potato, brinjal.

Sapota: Banana, papaya, pineapple, broad bean, tomato, brinjal, cabbage, cauliflower, spider lily.

Source: *http://ecoursesonline.iasri.res.in*

15.9 ADVANTAGES OF INTERCROPPING

The advantages of intercropping are risk minimization, effective use of available resources, efficient use of labour, increased crop productivity, erosion control and food security (Owuor *et al*, 2002). There is reduction of insect-pest populations due to the diversity of crops grown and reduction of plant diseases because the distance between plants of the same species is increased due to the planting of other crops between them, alteration of more beneficial insects especially when flowering crops are included in the cropping system, increase of total farm production and profitability and reduction of weed population through allelopathy and efficient crop production (Maguguda *et al*, 1979). Further, when the intercrop provides a good soil cover, soil temperature will stay relatively low. This prevents burning of the organic matter in the soil and loss of nutrients. It also provides a microclimate that can be favourable for associated crops. Intercropping has important advantages in regard to efficient land use, increasing crop productivity and monetary returns as compared to sole cropping. Crops and their agronomic characteristics used in intercropping systems are very crucial. Component crops having different harvest time should be preferred to reduce competition for similar resources at the same time. Moreover, different root and above ground parts of plants used in intercropping systems can be advised for effective utilization of moisture and light. The intercrops should have either synergistic or complementary effect relative to the main crop (Kumar *et al.*, 2010). Success of intercropping systems over sole cropping can be achieved by some agronomic manipulations. Supplemental effects in models of resource use should be considered to get better yield and quality in intercropping systems. Many different intercropping patterns are practiced all of over the world that farmers employ the different crops and management practices to supply their requirements for food, fibre, medicine, fuel, building materials, forage, and cash. Annual crops with other annuals, annuals with perennials, or perennials with perennials can be grown in intercropping practices (Liebman and Dyck, 1993).

15.10 VEGETABLE-BASED INTER CROPPING SYSTEM

Vegetables play an important role in achieving the nutritional security as they come across the malnutrition problems in India and also serve as a source of income for the small and marginal farmers. Vegetables are an excellent choice of cash crops as they can be grown easily, produce good yields and generate high price in the market compared to the cereals. Intercropping of different vegetable crops provided important advantages as well as higher profitability than vegetables grown as sole crops (Nursima, 2009).

15.11 INTERCROPPING WITHIN PERENNIAL HORTICULTURAL CROPS

Fruit tree and other perennial horticultural crop-based production system can offer suitable option for profitable utilization of the uplands. In case of fruit orchard, intercrops are generally planted in between rows of the main or base crop, with a view to obtain some extra yield without sacrificing the main crop or base crop yield. Fruit-based cropping system is a very good option for minimization of crop failure risk at a farm due to anomaly of weather. The intercrops not only generate an extra income but the practice also helps to check the soil erosion through ground coverage and improves the physicochemical properties of the soil.

15.11.1 Mango-Based Intercropping System

Fruit crops like mango and guava constitute a major share of area expansion programme under fruit crops in India. Long juvenile period, heavy mortality of the plants during the summer season due to grazing and lack of irrigation are major factors which discourage the farmers to take up mango orcharding. Again, low productivity of mango under the uplands makes mango cultivation, unprofitable. Hence, development of a profitable mango-based production system with income from the first year onwards can help in alluring the farmers to take up mango orcharding. Mango trees provide enough space even if they are fully grown as they do not cover much area. It is possible to grow a mixed fruit orchard, such as mango intercropped with other fruit crops, vegetables and spices during initial years of establishment. Intercropping in mango with suitable crops brings good income and improves the fertility of the soil. During the first few years, intercropping can be practiced with no shortage of irrigation. Economic advantages of mango-based intercropping systems have been reported by number of workers (Ratha and Swain 2006). During the initial five years of orchard establishment, guava has been found to be the most appropriate filler plant for improving the overall productivity (Nath *et al.* 2007).Research was made by Das *et al.* (2017) at ICAR RCER, Research Centre, Ranchi to analyse the plant growth behaviour, productivity of different component crops, profitability, soil fertility status and carbon sequestration potential of 20 different agri-horticultural systems during young bearing stage (6th to 10th year) of mango plants. They have reported enhanced growth of mango and filler plants with paddy as intercrop.

15.11.2 Banana-Based Intercropping System

Bananas are perennial herbs and may be grown on the same piece of land for up to 50 years. Mahant (2011) opined that intercropping in banana was more productive and profitable than their sole cultivation without loss in yield. He reported that intercropping either with onion or garlic in banana at initial growth stage of planting increase the total profit without affecting the yield. Intercropping with garlic and onion with 60% coverage in banana under drip irrigation gave maximum gross and net returns as well as benefit cost ratio. On farm experiment was conducted at multilocation testing sites by Nazrul (2007) for two consecutive years by taking four intercropping combinations, *viz.,* banana (sole), banana + okra, banana + sweet gourd and banana + bitter gourd. The result showed that the highest banana equivalent yield advantage (20%) than sole banana was recorded from the intercropping combination of Banana + Sweet gourd. This intercropping system also showed the highest benefit cost ratio (2.41) and therefore, it might be more economically viable than sole banana. Nedunchezhiyan *et al.*, (2002) conducted an experiment to study the suitability of elephant foot yam as an intercrop in banana and papaya gardens. They revealed that elephant foot yam can substantially increase the net returns from banana (Rs. 21,100.00) and papaya (Rs. 20,200.00) cultivation without any adverse effect on the main crop. Banana is a very important crop in East Africa but due to the fact that small scale farmers also require food security, it should be intercropped with many of the annual crops being grown by farmers to achieve success. When being established crops like beans, coffee, maize and sweet potatoes are intercropped with the young banana plants. Intercrops which can easily be raised in banana plantation at early stages of growth are radish, cauliflower, cabbage, spinach, chili, brinjal, yam and cucurbitaceous crops. Cassava and banana combination is one of the most efficient cropping systems (Bekunda and Woomer, 1999).

15.11.3 Citrus-Based Intercropping System

Intercropping of *Citrus* with groundnut, cotton or soybean in *kharif* season and with wheat or gram in *rabi* season can be successful. Effective utilization of interspaces between the sweet orange cv. mosambi plants in the orchard would be a viable and profitable intercropping system for sustainable profit. It also lowers the need of external inputs and improves the stability and diversity (Chandra *et al*, 2013) if ecological niches are kept in the mind while selecting the intercrops. When mandarin was intercropped with cucumber there was a high yield of cucumber with minimal interference in the growth of citrus seedlings possibly due to the low growing nature of the latter (Nataraja and Nairk,1992). Ghosh and Pal (2010) obtained highest net return from Mosambi + groundnut combination (Rs. 35,820.0/ha) followed by Mosambi + okra (Rs. 22,520.0/ha) and Mosambi + cowpea (Rs. 22,420.0/ha). Singh *et al.* (2012) reported that if mandarin is intercropped with soyabean in *kharif* and coriander in *rabi* after harvesting soyabean during October-November, the main crop mandarin during March-April, there may be income for almost six months as against 1-3 month under either option of sole cropping.

15.11.4 Guava-Based Intercropping System

The result of the two years of investigation carried out by Ghosh (2001) indicated that in terms of total returns, guava + groundnut combination proved the best. However, the net profit per hectare was maximum under guava + ridge gourd followed by guava + groundnut combination. The highest monetary returns per ha was registered from guava + ground nut (Rs. 39,685) whereas, the net profit per ha was recorded maximum under guava + ridge gourd (Rs. 21,368) followed by guava + ground nut (20,685) combination. Available inter space in the perennial crops can be utilized for cultivation of rhizomatous spice crops, vegetable and tuber crops etc. (Singh *et al.*, 2013). According to Singh *et al.* (2014), vegetable crops can be successfully grown under guava tree plantation without showing adverse effect on flowering, fruiting and yield of guava. They obtained maximum fruit yield (46.84 kg/tree) sole crop during the rainy season, followed by intercropping of bunda (46.45 kg/tree). While during the winter season crop, the maximum fruit yield (24.74 kg/tree) is recorded by intercropping of *arvi* followed by *suran* (23.12 kg/tree) as compared to other cropping systems. It is apparent from the results that the fruit yield of rainy season guava is significantly higher than winter season guava crop, which does not show adverse effect of intercropping.

15.11.5 Aonla-Based Cropping System

The tree canopy of aonla (*Emlica officinalis*) allows filtered light and permit intercropping even after it has made full growth. Two models have been developed

(Dhaker *et al.* 2013). 1. (i) M_1 = Aonla – Ber – Brinjal –Mothbean – Fenugreek and (ii) M_2 = Aonla – *Prosopis* – Suaeda –Mothbean –Mustard. The highest gross return of Rs. 25662 per ha was obtained under M_2 model. NRC for Agroforestry initiated a field trial during 1989 with 4 cultivars of aonla *viz.* Chakaiya, Kanchan, Krishna and NA-7 as fruit trees, subabul (*Leucaena leucocephala*) as multipurpose tree and blackgram (*Vigna mungo*) as intercrop in rainfed areas. The subabul was planted on both sides of the fruit trees at 2 m distance. The aonla planted at 10 × 6 m spacing proved to be ideal spacing. At the age of 6 years, on an average the fruit yield from a plant was up to 93 kg. Besides fruit yield of aonla, 256 kg grain (on an average) was obtained from blackgram every year. Introduction of *Leucaena* in the system provides organic matter in form of leaf litter and it also fixes atmospheric nitrogen in the soil.

15.11.6 Ber-Based Intercropping System

Ber is planted with wider spacing (5-6 m apart) which provides ideal conditions for growing of field crops in the inter space. Ber needs pruning every year which is usually done in April-May and it takes about four to five months to develop full canopy which provides unrestricted interspace during this period. Among the intercropping systems the highest net income (Rs.16,330/ha) and benefit cost ratio (3.85) was obtained in ber + Bengal gram followed by ber + safflower (Yaragattikar and Itnal, 2003). In-situ budded ber cultivar Gola as over storey component and groundnut wheat, cluster bean-mustard and Indian aloe as ground storey component were integrated into the system. Among different crop combinations, the highest fruit yield of ber was recorded with Indian aloe while with groundnut-wheat and cluster bean-mustard, the yield level was almost *at par* and minimum yield was recorded under sole plantation. The highest B:C ratio (3.67) was obtained in ber + cluster bean-mustard system (Dhaker *et al.* 2013).

15.11.7 Cashewnut-Based Intercropping

Pattanaik *et al.* (2007) studied the economics of intercrops in cashew plantation with vegetable crops and found that the yield of main crop varied from 60.1 to 8.70 q per ha under various crop combinations. Highest returns of Rs.19,350/- per ha was obtained from colocasia and lowest recorded in cowpea. Net highest return of Rs.44,908/- per ha was obtained when cashew was intercropped with colocasia, followed by brinjal. Minimum net return of Rs.21,350/- per ha was obtained from the sole crop, cashew cv Vengurla-4.

15.11.8 Coconut-Based Intercropping System

Effective utilization of available space, both horizontally and vertically, is the modem concept of cropping system. Growing coconut as monocrop is not the most efficient way of using natural resources. Adoption of coconut-based intercropping system is one of the ways to utilize the natural resources effectively. Adoption of coconut-based multiple cropping system emerges as the viable way for improving the economic status of coconut farmers. Intercropping system under coconut is more profitable than mono-cropping which promises to the farmers a lot besides generating additional employment opportunity (Nath, 2002). Ghosh and Hore (2011) noticed that the cropping system model (coconut + lime + ginger + okra) was found best among all. Further, planting with 25-30 g seed rhizome at 20 cm × 15 cm spacing may be recommended for ginger as intercrop in coconut plantation for maximizing the yield. Black pepper performed well under coconut (Ghosh, 2009). However, Ghosh and Bandopadhyay (2011) studied on the productivity and profitability of coconut-based cropping systems with fruits and black pepper in West Bengal. Six coconut based cropping models (Model I: Coconut + Black pepper + guava, Model II: Coconut + Black pepper + lime, Model III :Coconut+ Black pepper+ lemon, Model IV :Coconut+ Black pepper+ Pineapple, Model V: Coconut+ Black pepper +Banana and Model VI : Coconut+ Black pepper) were evaluated at HRS, Mondouri, BCKV, West Bengal during the year 2003 to 2008 in a 26 years old coconut plantation. Economic assessment of models revealed that out of 6 models, Model-V (consists of coconut, black pepper, pineapple) was more remunerative showing highest net return of Rs. 45600/- per ha followed by Model-IV (Rs. 36050/-) per ha. It was also observed that multiple cropping in coconut plantation under recommended package of practice of both main crop and intercrop, had no adverse effect on production of coconut. Fruit based cropping system with coconut, comprising of coconut, black pepper, pineapple was found best under West Bengal condition.

15.11.9 Arecanut-Based Intercropping System

Acharya and Singh (2005) in a study on arecanut-based cropping system at CPCRI, Kahikuchi, Guwahati, claimed that the average cost of production of vegetables ranged between Rs. 6750/- and Rs. 12600/- per ha in case of radish and potato, respectively, whereas the highest gross return of Rs. 42750/- was obtained by growing cabbage followed by brinjal (Rs. 40600/-) and cauliflower (Rs. 38000/-). Similarly, highest gross return of Rs. 123089 was obtained when gladiolus was grown as intercrop followed by chrysanthemum and marigold. The highest gross return in gladiolus was obtained due to sale of spikes and corms. It has also been studied that the value for return per rupee invested was maximum (1 5.77) for brinjal followed by cabbage and radish. Ghosh *et al.*, (2004) studied colocasia as an intercrop in arecanut cv Mohitnagar plantation and found beneficial effect of colocasia on the growth of arecanut at Nadia, West Bengal. They found increment in plant height (65.73%) and number of leaves per plant (72.82%) due to intercropping. Ray *et al.* (2007) studied different vegetables and flowering crops in pre-bearing arecanut garden. They reported that vegetable crops like radish, cabbage, cauliflower and brinjal performed well and resulted in higher net return per rupee invested during winter season and among flowering crops, marigold and chrysanthemum performed well and resulted in higher B:C ratio (2.05 to 2.43) compared to gladiolus.

15.12 MULTI-STOREY CROPPING SYSTEM

Multi-tier cropping system involves combination of plants with various morpho-phenological features to maximize the natural resource use efficiency and enhanced total factor productivity. Two or more crops of different heights are grown simultaneously on a piece of land in a certain period. Sun light, nutrient, land, water and space are utilized in the most efficient and economical way in this system. It improves soil health, soil physical property and enhances higher production stability. Horticultural crops particularly fruit and plantation crops have self-sustainable system where solar energy can be harvested at different heights, soil resources are used efficiently and can increase cropping intensities. The system consists of three main components, *viz*., main crops, filler crop and intercrops which occupy three different tiers in space of the production system. To ensure sustainable productivity and high returns from underutilized and stressed lands and to improve the soil characteristics multi-storey cropping system have been found successful in tropical rain forest, semi-arid and dry land conditions. Multi-storey cropping system in horticulture is found to be a perspective approach for sustainable productivity in fruit crops (Mango, Ber, Amla, Pomegranate) and Plantation crops (Coconut, Areca nut, Coffee, Cashew) by which natural resources are utilized efficiently to enhance productivity of main crops (15-20%) and high revenue realization per unit area (50-90%) besides maintaining soil health and balancing environment.

Multi-storey cropping system involves the growing of crops of different height, rooting pattern, and duration simultaneously on the same piece of land. This is most common in coconut-based cropping system in Kerala to meet diversified needs of farming community for fodder, food and fuel, besides increasing net return per unit area, e.g., coconut+ pepper + pineapple + grass. Taller trees, which has greater requirement for solar light, are able to trap more solar light when grown as top storey and those having requirement of lower light are raised as ground storey. In this system, annual and perennial crops are grown side by side in different tiers by exploiting soil and vertical space more efficiently. Inter-cropping and mixed cropping with compatible crops in coconut plantations has been found to give increased returns to the farmers, without affecting yield of the main crop.

Table 15.2 Some fruit crop-based system with their intercrops (Horti-Agri System)

S. No.	Crop	Main crop with intercrops	References
1.	Aonla	Aonla + Fenugreek + chick pea + mustard + cumin (Rabi season); Moth bean (Kharip season)	Awasthi *et al.*(2009)
		Aonla + Ber + Karonda + Cluster bean + Brinjal; Aonla + Ber + Karonda + Mothbean + Mustard	Arya *et al.* (2010).
2.	Ber	Ber (Jharber and Bordi) + Mung, Clusterbean, Pearl millet, Mothbean, gram, Vegetable crops	Faroda (1998)
		Ber (*Zizyphusmauritiana*) + Green gram	Gupta *et al.* (2000)
		Ber (Cv. Gola) + Clusterbean+ Mustard + Indian aloe; Wheat + Groundnut	Saroj *et al.* (2003)
		Ber + Wheat; Ber + Chick pea; Ber + Kulthi	Pareek and Nath (1996); Kumar *et al.* (2005)

(Contd.)

S. No.	Crop	Main crop with intercrops	References
3.	Mango	Mango + *Phaseolus acutifolius* Genotype "Frijol Escumite" + *Cajanus cajan* (Pigeon Pea) + *Phaseolus vulgaris* Genotype "Frijol de Vara"	Agreda *et al.* (2006)
		Mango + Guava + Cowpea Guava + Ground nut + Black gram	Nath *et al.* (2007)
4.	Mandarin	Mandrin + Ginger	Upadhyay and Subba (1994)
		Mandarin + Pea	Ghosh et al. (1985)
5.	Lemon	lemon + French bean	Vanlalhruaia *et al.* (2013)
6	Banana	Banana +Sweet potatoes	Bekunda and Woomer (1996)
7.	Fig	Fig+ Strawberry	Mendonca *et al.* (2012)

According to Jakhar *et al.* (2012), among all the cropping systems, papaya + *Gliricidia* based multitier system along with ginger-pigeon pea intercropping (8:2) gave maximum net returns of Rs 2.77 lakhs from one hectare of land with B: C ratio of 2.55 followed by sole ginger with net returns of Rs 2.42 lakhs per ha. Cultivation of pigeon pea along with papaya fruit tree increased net returns by Rs 16,958. Being a pulse crop pigeon pea fetches good market price as well as improves soil fertility due to atmospheric N-fixation through nodulation. Sole ragi cultivation on sloping lands was not economical because it gave poor returns as well as it permitted soil erosion. Hence integration of remunerative multitier components is required to make it a profitable enterprise. Integration of papaya+ Gliricidia in sole ragi and ragi-pigeon pea intercropping (6:2) gave net returns to the tune of Rs 12,533 and Rs 32,181 with B: C ratio of 1.21 and 1.55, respectively.

Coconut + Elephant Foot Yam

Coconut + Cassava

Multi Cropping system in Coconut

Aonla + Guava+ Urd bean

Mango+ Pineapple

Mango + Chilli

Mango + Radish

Cashew + Turmeric

Multistoried cropping system (Khejri + Ber + Grasses)

Intercropping in Banana orchard

REFERENCES

Acharya GC and Singh LS 2005. Arecanut based cropping system: alternative pathway to achieve sustainability in North Eastern India. *Indian J. Arecanut, Spices and Medicinal Plants*, **16** (1): 23-26.

Awasthi OP, Singh IS and More T 2009. *Indian Journal of Agricultural sciences,* 79 (8), 587-91.

Agreda FM, Pohlan J, Marc J and Janssens J 2006. Effects of Legumes Intercropped in Mango Orchards in the Soconusco, Chiapas, Mexico. In: *Conference on International Agricultural Research for Development*, 1-6.

Arya R, Awasthi OP, Singh J and Arya CK 2010. *Indian Journal of Agricultural Scienes,* 80 (5), 423-6.

Bekunda M and Woomer PL 1996. *Agric. Ecosys. Environ*, 59(3), 171-180.

Bekunda M 1999. Farmers responses to soil fertility decline in banana-based cropping systems in Uganda. *Managing Africa's soils.* No. 4 Russell Publishers, Nottingham.

Chandra A, Kandari LS, Negi VS, Maikhuri RK and Rao KS 2013. Role of intercropping on production and land use efficiency in the central Himalaya, India. *Environment International Journal Science Technology,* **8**: 105-113.

Das B, Dhakar MK, Sarkar PK, Kumar S, Nath V, Dey P, Singh AK and Bhatt BP 2017. Performance of mango (*Mangifera indica*) based agri-horticultural systems under rainfed plateau conditions of eastern India. *Indian J. Agril. Sci.*, **87** (4): 521-7.

Dhakar MK, Sharma BB and Prajapat K 2013. Fruit Crop-Based Cropping System: A Key for Sustainable Production. *Popular Kheti*, **1**(2): 39-44

Fininsa C and Yuen J 2000. Temporal progression of bean common bacterial blight in sole and intercropping. *European J. Plant Pathology.***108** (6): 485-495.

Faroda AS 1998. Arid zone research: an overview. (In) Faroda, A. S. and Manjeet Singh Ed. Fifty Years of arid zone research in India. CAZRI, Jodhpur. pp. 1-16.

Gahukar RT 1989. Pest and disease incidence in pearl millet under different plant density and intercropping patterns. *Agriculture. Ecosystems & Erosion,* **26** (1): 69-74.

Ghosh D, Chattopadhyay N, Bandhyopadhyay A and Hore JK 2004. Evaluation of colocasia as an intercrop in arecanut. *Haryana J. Hort. Sci.* **33**(3-4): 269-271.

Ghosh D, Mitra SK and Bose TK 1985. Fruit growing in India. In: *Proceeding of the 3rd National Citrus Seminar*, Calcutta, India, pp. 1-2.

Ghosh DK 2009. Performance of black pepper (*Piper nigrum* L) as an intercrop with coconut in the alluvial plains of West Bengal. *Indian Coc. J.* **51:** 4-7.

Ghosh DK and Bandopadhyay A 2011. Productivity and profitability of coconut-based cropping systems with fruits and black pepper in West Bengal. *Journal of Crop and Weed,* **7** (2):134-137.

Ghosh DK and Hore JK 2011. Economic of coconut based intercropping system as influenced by spacing and seed rhizome size of ginger. *Indian J. Hort.*, **68** (4): 449-452.

Ghosh SN and Pal PP 2010. Effect of intercropping on plant and soil of Mosambi sweet orange orchard under rainfed conditions. *Indian J. Hort.*, **67**(2):185-190.

Gupta JP, Joshi DC and Singh BB 2000. Management of arid agrosystem: In: National resource management for agriculture production in India (Eds.0 (J.S.P. Yadava and G.B. Singh). International conference on managing natural resource for sustainable agriculture production in the 21st Century.

Jakhar P, Barman D, Gowda HC and Madhu MM 2012. Multitier cropping system for profitable resource conservation and sustainable management of sloping lands of Eastern India. *Indian J. Agric. Res.*, **46** (4): 309: 316,

Kumar D, Seema J, Batra PK and Gill AS 2005. *Indian Journal of Agricultural Sciences,* **75**(4): 222-224.

Kumar SR, Kumar H and Kumar SA 2010. *Brassica* based Intercropping Systems-A Review. *Agricultural Reviews*, **31**(4): 253-466.

Liebman M and Dyck E 1993. Crop Rotation and Intercropping Strategies for Weed Management. *Ecological Applications*, **3**(1): 92-122.

Mahanta HD, Patil SJ, Bhalerao PP, Gaikwad SS and Kotadia HR 2012. Economics and land equivalent ratio of different intercrops in banana (*Musa paradisiacal* L.) cv. Grand Naine under drip irrigation, *Asian J. Hort*.,**7**(2):330-332.

Magaguda GT, Haque I, Godfrey W, Fendu I and Masina GT 1979. Intercropping studies in Swaziland: Present status and future projections. Proc. Intl. Workshop on intercropping 10-13 Jan., 1979, Hyderabad, India, Pp. 98-104.

Nath V, Das B, Yadav MS, Kumar S and Sikka AK 2007. *Acta Hort.* 735, 277-295.

Nienow AA 2012. In: *The Phyllochron of Strawberry Intercropped with Fig Trees in a Greenhouse. Proc. XXVIIIth IHC – International Berry Symposium. Acta Hort*. 926, 547-550.

Nath V, Das B, Yadav MS, Kumar S and Sikka AK 2007. Guava- A suitable crop for second floor in multistoried cropping system in upland plateau of eastern India. *Acta Horticulturae*, DOI: 10.17660/735.39.

Owuor C, Tenywa JS, Muwanga S, Woomer P L and Esele P 2002. Performance of a sorghum-legume intercrops in response to row levels of nitrogen. Program and extended abstracts. *FORUM Working Document. No. 5. The Forum on Agriculture Resource Husbandry,* Nairobi, Pp. 542-544.

Pattnaik AK, Lenka PC, Mohapatra KC and Mohaptra RN 2007. Intercropping in the cashew plantation with vegetable crops, *National Seminar on Research, Development and Marketing of Cashew.* 20th – 21st November 2007:61.

Pareek OP and Nath V 1996. Ber. In coordinated Fruit Research in Indian Arid Zone-A two decades profile (1976-1995). National Research Centre for Arid Horticulture, Bikaner, India: 9-30.

Natarajaa M and Nairk DM 1992. Competitive effects of short duration, bush type cowpea when intercropped with cotton in Zimbabwe *Expt. Agric*., **28:** 41-48.

Nedunchezhiyan M, Misra RS and Shivlinga Swamy TM 2002. Elephant foot yam as an intercrop in banana and papaya. *The Orissa J. Hort*., **30** (1): 80-82.

Nursima KA 2009. Profitability of intercropping corn with mungbean and peanut. *USM. R&D,* **17**(1): 65-70.

Ratha S and Swain SC 2006. Performance of intercrops in mango orchard in Eastern Ghat High Land Zone of Orissa. *Indian J. Dry land Agric. Res. Dev.*, **21**(1): 12–5.

Ray AK, Borah AS, Maheswarappa HP and Acharya GC 2007. Economics of intercropping vegetables and flowering crops in pre-bearing arecanut garden under Assam conditions, *J. Plantn. Crops*, **35**(2): 84-87.

Saroj PL, Dhandar DG, Sharma BD, Bhargava R and Purohit CK 2003. *Indian Journal of Agroforestry,* 5 (1&2), 30-35.

Singh J, Arya CK, Bhatnagar P, Jain SK and Pande SBS 2012. Intercropping in Orchard is better option, *Indian Horticulture*, 5-7.

Singh SP 1997. *Cropping systems in vegetable crops, principles of vegetable production.* Agrotech Publishing Academy.

Singh A 1972. Conceptual and experimental basis of cropping patterns. *In: Proceedings of symposium on Cropping Patterns in India*. ICAR, New Delhi.

Singh SR, Banik BC and Hasan MA 2013. Effect of different intercrops on yield, quality and shelf-life in mango cv. Himsagar (*Mangifera indica* L). *Intl. J. Agric. Env. Biotech.*, **6** (1):121-126.

Singh SK, Raghuvanshiz M, Singh PK and Prasad J 2014. Performance of vegetable crops as intercrops with guava plantation. *Res. Environ. Life Sci.*, 7 (4) :259-262.

Vanlalhruaia H, Gopichand B and Lalnunmawia F 2013. *Sci Vis.*, 13(1), 40-44.

Willey RW, Singh RP and Reddy MS 1989. Cropping systems for Vertisols in different rainfall regimes in semiarid tropics. *In*: *Management of Vertisols for improved agricultural production.* ICRISAT, India.

Yaragattikar AT and Itnal CJ 2003. Studies on Ber Based Intercropping Systems in the Northern Dry Zone of Karnataka. *Karnataka J. Agric. Sci.*, **16** (1): 22-25.

Upadhyaya RC and Subba JR 1994. Annual Report 1994. Department of Horticulture, Government of Sikkim, Krishi Bhawan, Tadong, Sikkim.

http://ecoursesonline.iasri.res.in

OUTCOMES ASSESSMENT

PART A

Answer the following questions (True or False).

1. Intercropping of groundnut in *Citrus* orchard in *Kharif* season is not successful. (True/False)
2. Multi-tier cropping system involves combination of plants with various morpho-phenological features. (True/False)
3. Growing coconut as monocrop is the most efficient way of using natural resources. (True/False)
4. Bananas are perennial herbs and may be grown on the same piece of land for up to 50 years. (True/False)
5. Multi-storey cropping system involves the growing of crops of different height. (True/False)

PART B

Answer the following questions.

1. Define cropping system.
2. What is multitier cropping?
3. What do you understand by intercropping?
4. Multi-tier cropping and multi-storey cropping are same/different.
5. Name different types of intercropping.

PART C

Write a brief note on each of the following.

1. Mango-based intercropping system.
2. Benefits of cropping system.
3. Intercropping in guava orchard.
4. Multi-storey cropping.
5. Aonla-based intercropping system.

Chapter 16

Mulching

16.1 INTRODUCTION

Mulch, a technical term, means 'covering of soil'. A mulch is a layer of material applied to the surface of soil. It may be applied to bare soil or around existing plants. Mulches of manure or compost will be incorporated naturally into the soil by the activity of worms and other organisms. Mulching is the process or practice of covering the soil/ground to make more favourable conditions for plant growth, development and efficient crop production. While natural mulches such as leaf, straw, dead leaves and compost have been used for centuries, during the last 60 years the advent of synthetic materials has altered the methods and benefits of mulching. The research as well as field data available on effect of synthetic mulches make a vast volume of useful literature. In this manner it plays a positive role in water conservation. The suppression of evaporation also has a supplementary effect; it prevents the rise of water containing salt, which is important in countries with high salt content water resources.

16.2 MATERIALS USED FOR MULCHING

A variety of **materials** are used as **mulch**:

1. *Organic Residues*: grass clippings, leaves, hay, straw, kitchen scraps comfrey, shredded bark, whole bark nuggets, sawdust, shells, woodchips, shredded newspaper, cardboard, wool, animal manure, etc.
2. *Inorganic Residues*: rock, stone, lava rock, crusher dust, pulverized rubber, landscape fabrics, and other man-made materials.

16.3 AREAS OF APPLICATION

Mulching is mainly employed for:

- Moisture conservation in rainfed areas.
- Reduction of irrigation frequency and water saving in irrigated areas.
- Soil temperature moderation in greenhouse cultivation.
- Soil solarization for control of soil borne diseases.
- Reduce the rain impact, prevent soil erosion and maintain soil structure.
- In places where high value crops only to be cultivated.

16.4 BASIC PROPERTIES OF MULCH

1. Air proof so as not to permit any moisture vapour to escape.
2. Thermal proof for preservation of temperature and prevention of evaporation
3. Durable at least for one crop season.

16.5 SELECTION OF MULCH

The selection of mulches depends upon the ecological situations and primary and secondary aspects of mulching

- Rainy season- Perforated mulch
- Orchard and plantation-Thicker mulch
- Soil solarization-Thin transparent film
- Weed control through solarization-Transparent film
- Weed control in cropped land-Black film
- Sandy soil-Black film
- Saline water use-Black film
- Summer cropped land-White film
- Insect repellent-Silver colour film
- Early germination-Thinner film

16.6 TYPES OF MULCH

16.6.1 Organic Mulch: Wood Chips, Nuggets, or Bark

Both hardwood and softwood bark, chips and nuggets-byproducts of the lumber (wood) and paper industries, typically aged and dried, and sometimes even dyed red or black, fall under this category (Figure 16.1). Hardwood works best around trees, shrubs, and in perennial beds, while softwood (typically made from pine) should be reserved for use around large trees and shrubs. Pine tends to be slightly more acidic and therefore takes longer to decompose than other organic mulches. Sometimes freshly ground tree mulch is available free of cost in municipality. This fresh material is neither dried or aged, so use it only for walkways, as it leaches large amounts of nitrogen from the soil as it decomposes.

Fig.16.1 Organic mulch: wood chips, nuggets or bark

16.6.2 Organic Mulch: Straw

Clean wheat, barley or oat straw is ideal for lightly mulching newly seeded lawns. The straw mulch keeps the grass seed from washing away, deters feeding birds and rodents, and, until it decomposes, conserves the moisture the seeds need for good germination. Straw mulch (Figure 16.2) may be used for vegetable crops. Straw should not be confused with hay. The hay which contains seeds that could sprout up as weeds in the garden should be avoided.

Fig. 16.2 Organic mulch: straw

16.6.3 Organic Mulch: Grass Clippings or Shredded Leaves

Grass clippings or shredded leaves can be used as organic mulch. Leaf mulch is ideal for use in garden beds and around trees and shrubs, while grass clippings (Figure 16.3) may be spread in thin layers across vegetable and perennial beds and then turned into the soil at the end of the growing season. Care should be taken not to apply in thick layers, or else the material will mat. The clippings from lawns that have been treated with herbicides or insecticides should not be saved.

Fig. 16.3 Organic mulch: grass clippings

16.6.4 Organic Mulch: Newspaper or Cardboard

Shredded black-and-white newspaper or undyed natural cardboard can be used as an effective weed suppressant (Figure 16.4). Apply two to three layers at a time, then cover with another heavier organic material, such as leaves or grass clippings, to hold the lightweight mulch in place. Take care not to mix in colour newspaper pages or coated cardboard; these do not decompose readily and may even expose your garden to toxic dyes.

Fig. 16.4 Organic mulch: newspaper or cardboard

16.6.5 Organic Mulch: Cocoa Chips

Popular for their rich colour and pleasant scent, cocoa bean hulls are lightweight, easy to handle, and appropriate for all planting areas (Figure 16.5). Do not apply more than one inch or water excessively, because cocoa chips already decompose quickly—and since they are a costlier option, you will not want to have to do more than an annual application. If you have pets or wildlife, you should avoid cocoa mulch, because chocolate and its byproducts can be fatal to animals if consumed.

Fig. 16.5 Organic mulch: cocoa chips

16.6.6 Organic Mulch: Composted Animal Manure

Nothing beats well-composted, nutrient-rich animal manure when it comes to mulch for vegetables (Figure16.6). Fresh manure burns plant roots, and dog, cat, and pig manure can harbour disease-causing organisms. So, avoid them.

Fig. 16.6 Organic mulch: composted animal manure

16.6.7 Inorganic Mulch: Rock or Crusher Dust

Lava rock, crushed gravel or crusher dust, marble chips, and pea gravel will not break down, making them a popular option for walks and pathways, thanks to their one-time investment of cost and labour (Figurel6.7). Avoid using stones around trees, shrubs, and other plants, however, because they will not effectively retain moisture and can cause heat stress on plants through reflection as well as ground heating, which can burn roots.

Fig.16.7 Inorganic mulch: rock

16.6.8 Inorganic Mulch: Landscape Plastic or Fabric

Plastic polyethylene film is impermeable, which means that water and other nutrients cannot pass through. While this quality makes it ideal as a short-term weed killer, plastic is not suitable for long-term use. If you employ it to warm the soil around fruit and vegetable plants, you will have to install an irrigation system under the plastic or water your plants by hand to make sure they get adequate moisture. Remove the plastic at the end of the growing season to keep it from deteriorating in the sunlight, and then replace it next year.

16.6.9 Inorganic Mulch: Rubber

Rubber mulch—manufactured from recycled, pulverized tires—is inexpensive and highly durable, which makes it perfect for high-traffic areas, such as playgrounds (Figurel6.8). Leave it out of home landscaping projects; rubber mulch does not decompose, and some studies indicate that toxins found in the rubber can actually leach into the soil.

Fig.16.8 Inorganic mulch: rubber

16.7 ADVANTAGES

Mulch has number of advantages as discussed below in brief:

- **Protects from frost**

  Mulch protects plants against frost and extreme temperatures in winter. The plants are very sensitive to cold, so when the frost starts a good mulch can prevent the roots from freezing and dying.

- **Reduces evaporation**

  Padding reduces evaporation of water in summer. It retains the humidity of the soil accumulated during the spring rains that help to hydrate the plant. In this way we will not need to water them so much.

- **Avoids weeds**

  Avoids the proliferation of weeds, since it prevents the passage of light.

- **Serves as fertilizer.**

  If organic materials are used, they nourish the soil little by little while they decompose.

- **Protects from erosion**

  Protects the soil from erosion by rain and wind while maintaining the soil structure.

- **Less probability of rotting of aerial fruits** when not being in contact with soil moisture
- **Less soil compaction**
- **Roots develop more laterally and superficially** on land taking advantage of all the space they have in the soil
- **Roots are more numerous** by the microclimate generated under the padding
- **More effectiveness and safety** in the use of fungicides, fertilizers or other products either natural or chemicals, but that is placed on the ground. This is in case of plastic padding
- **Saving water** Less water evaporation so we do not need to water more regularly
- **Provides organic matter** when the material of the mulch is integrated into the soil, e.g., in case of leaf mulch, straw mulch.
- **Eliminates the risk of cracking or breaking the soil** (which damages the roots) in all types of mulch.
- **Reflective mulches will repel certain insects.**
- **Maintains a warm temperature** even during nighttime which enables seeds to germinate quickly and for young plants to rapidly establish a strong root growth system.
- **Develops a microclimate** underside of the sheet, which is higher in carbon-di-oxide due to the higher level of microbial activity.
- **Synthetic mulches** play a major role in soil solarization process.
- **Moisture is preserved** for several days and increases the period between two irrigations.

16.8 DISADVANTAGES

Mulch has some disadvantages that are discussed below in brief:

- **Cost of mulching** depending on the method one uses can be costly.
- **One needs to know the correct use** of the padding according to the purpose one wants to achieve.
- **Attraction of some types of pests**, according to the type of padding. The one that attracted the most was dry twigs and straw.
- The mulching of dry leaves can facilitate the presence of **spiders and mites.**
- **Disappearance of beneficial plants** such as trebol, especially in the case of plastic mulch.
- **In the case of straw mulch**: It cannot be stacked, nor can it be known in the case of drip irrigation, when some of the irrigation outlets do not work
- In the pumpkins and zucchini, the hot humidity creates under the propitious plastic padding **powdery mildew**
- In some types of mulch, such as that of straw, when it is fresh, it can germinate its seeds and **become weeds**
- The mulch of herbs, leaves, or straw **can attract rodents.**
- In the case of dry leaf mulch, the wind can cause **great loss of the material**, when this is not well settled.
- **Reptile movement and rodent activities** are experienced in some places.
- **More runoff**.
- Environmental **pollution**.
- **Difficult in machinery movement**.
- **Cannot be used for more than one season** using thin mulches.
- **Toxic** to livestock

16.9 REMOVAL OF MULCH

In case the mulch film needs to be used for more than one season (thicker film) the plant is cut at its base near the film and the film is removed and used.

By compounding appropriate additives into the plastics, it is possible to produce a film, which, after exposure to light (solar radiation) will start to breakup at a pre-determined time and eventually disintegrated into very small friable fragments. The time period can be 60, 90, 120 or 150 days and for maize a 60-day photodegradable mulch is used. However, there are still some further problems to resolve. It has been observed that the edges of the mulch, which are buried to secure the mulch to the soil, remain intact and become a litter problem when brought to the surface during the post-harvest ploughing. Currently much development effort is being made to find a satisfactory solution to this problem. In direct contrast, in developing countries which have available agricultural labour a different approach can be made. For example, in the People Republic of China trials have been made using a plastic mulch of 15-micron thickness on a sugarcane crop. After the cuttings have been planted through the mulch, they are left to grow for a period of one month. Then the mulch is removed by hand and wound up so that it can be utilized for a second season. A yield increase of 26% was obtained.

These two examples not only demonstrated the diversity of mechanisms available for resolving the problems of mulch removal, but also illustrate the different techniques, which have been developed in different countries. It also indicates the necessity for each country to adapt and develop mulching technique to meet its own specific requirements.

16.10 COST ECONOMICS OF MULCHING

The cost economics of mulching is an important aspect. In a levelled field if mulching is to be done, the film area required will be almost equal to that of field itself. In fields with ridges and furrows mulching material required will be sizably more than the field area. However, mulching is carried out in strips covering 50-60% of field area. In the present era of decreasing rainfall, conserving moisture with mulching transgresses the plan of economic analysis in the sense that the real cost analysis would be even meaningless in the case of a precious commodity like water.

REFERENCES

Plastic mulching for crop production -www. Agritech.tnau.ac.in

Types of mulching and their uses in dryland condition (October 2017) -by S.G Telkar, Deepak Kumar, Pratap Singh Solanki, Kamal Kant, A.K Singh

Types of mulching, advantages of mulching in farming –www. Agrifarming .in

Mulch- Wikipedia

Pictures- google sites

OUTCOMES ASSESSMENT

PART A

Answer the following questions (True or False).

1. Mulching is the process of uncovering the soil. (True/False)
2. Mulching is employed in rainfed areas. (True/False)
3. Lava rock may be used as organic mulch. (True/False)
4. Mulch helps seeds germinate quicker. (True/False)
5. Soil solarization is not possible with any type of mulch. (True/False)

PART B

Answer the following questions.

1. Define mulch.
2. Why cost economics analysis of mulching is meaningless?
3. What is an organic mulch?
4. Mention the three basic properties of mulches.
5. Cite two examples of inorganic mulches.

PART C

Write a brief note on each of the following.

1. Advantages of mulches.
2. Removal of mulch.
3. Types of mulches.
4. Selection of mulch.
5. Areas of application of mulches.

Chapter 17

Bearing Habit of Fruit Trees

17.1 INTRODUCTION

Bearing habit indicates position of flower bud with respect to vegetative growth of the plant after cessation of juvenility. Different fruit trees have different bearing habits. The knowledge of bearing habits will be of much importance for regulating the bearing of any plant by means of pruning. Terminal bearers are not normally pruned, whereas in case of lateral bearers pruning certainly increase the bearing, as it encourages the sprouting of lateral buds. Fruit trees may bear the fruits either terminally on a long or short growth, laterally on present or past season growth or adventitiously from any point on the trunk or root. *The relative position of a fruit with reference to its potential bud giving rise to flower or inflorescence in the shoot is known as bearing habit.* Knowledge on the bearing habit is a pre-requirement before performing pruning in any fruit crop (particularly in ber, guava, grape etc.). Proper understanding of bearing habits helps a lot in manipulating time and period of flowering and fruiting which consequently influence harvesting and yield of better-quality fruits.

17.2 FLORAL PARTS

A flower is a modified shoot. It contains the reproductive structures of the plant (Figure 17.1). A typical flower consists of four whorls: calyx, corolla, androecium and gynoecium (also called pistil). The androecium and gynoecium are upper essential or reproductive whorls, and the calyx and corolla are lower non-essential or accessory whorls. The calyx and corolla consist of a few sepals and petals respectively. Sometimes, when calyx and corolla are not differentiated, they are called perianth (e.g., banana, onion, garlic, ginger, turmeric, mango ginger, lily, tuberose, etc.) and its members tepals.

The stamens (also called microsporophylls) are individual units of androecium (male organ). They have two parts: (i) filament, and (ii) anther within which pollen grains (also called pollens) develop. They may be free (e.g., mango, guava, custard apple, onion, rose, lotus, cashew nut, etc.) or united (coherent). The gynoecium (female organ) consists of one or more carpels (also called megasporophylls) that bear female spores (also called megaspores or embryo-sacs). It may be simple (made of one carpel), e.g., pea, bean, cowpea, etc., or compound (made of two or more carpels). In a compound pistil, either the carpels may be free (apocarpous), e.g., rose, lotus, *Michelia, Artabotrys*, etc., or all the carpels may be united together (syncarpous), e.g., tomato, onion, etc. The gynoecium has three parts: (i) ovary, (ii) style and (iii) stigma.

The ovary may be superior (e.g., brinjal, China rose, etc.), inferior (e.g., sunflower, guava, gourd, cucumber, apple, pear, etc.) or half-inferior (e.g., rose, plum, peach, *Lagerstroemia*, gold mohur, etc.). Consequently, the flower may be hypogynous, epigynous or perigynous respectively. The stigma is receptive to pollen deposited on it. Within the ovary are found the ovules, which, after embryo formation within, develop into mature seeds. The number of ovules within an ovary may vary from one, as in mango or cashew nut, to many, as in brinjal or guava.

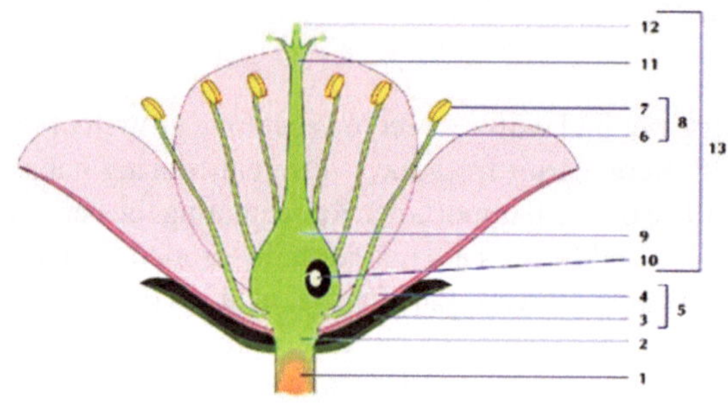

Fig. 17.1 1. Pedicel, 2. Receptacle, 3. Calyx, 4. Corolla, 5. Accessory or non-essential whorls, 6. Filament, 7. Anther, 8. Stamen, 9. Ovary, 10. Ovule, 11. Style, 12. Stigma, 13. Reproductive or essential whorls. (Courtesy: http://courses.botany.wisc.edu)

17.3 TYPES OF FLOWERS

Different types of flowers are found in horticultural crops:

1. **Complete *versus* Incomplete Flower**: If all the four whorls are present in a flower, it is called complete flower, e.g., pea, cowpea, potato, tomato, betel, hollyhock, rose, guava, etc. whereas, incomplete flower (e.g., male flowers of mango and cashew nut, *Colocasia*, *Caladium*, *Anthurium*, lemon grass, citronella, vetiver, etc.) lacks one or more whorls.
2. **Perfect *versus* Imperfect Flower**: If both the essential whorls, androecium and gynoecium, are present in a flower, it is called perfect flower and if either androecium or gynoecium is present, it is known as imperfect flower. In other words, perfect flowers are bisexual (also called hermaphrodite), but they may be complete (e.g., pea, cowpea, potato, tomato, brinjal, chilli, betel, hollyhock, rose, guava, etc.) or, incomplete (e.g., lemon grass, citronella, vetiver, etc.), and imperfect flowers are unisexual and, consequently, incomplete (e.g., male and female flowers of cucurbits, male flowers of mango and cashewnut, etc.).
3. **Unisexual Flower**: The unisexual (imperfect) flower is male (also called staminate) when it bears only androecium (stamens) but no gynoecium or pistil (carpels), and it is female (also called pistillate) when it bears only pistil (gynoecium) but no androecium (stamens).
4. **Monoecy *versus* Dioecy**: When both staminate and pistillate flowers are borne on the same plant either in the same inflorescence or separate inflorescences, it is called monoecy (also called monoecism) and if they are borne on different plants,

it is known as dioecy (also called dioecism). Consequently, the crop plants are called monoecious (e.g., *Casuarina equisetifolia*, pine, *Begonia*, *Petunia*, etc.) and dioecious respectively. The typical dioecious plant may be either male (also called androecious) or female (also called gynoecious), i.e., unisexual plants, e.g., papaya, date palm, palm, pointed gourd (*parwal*), ivy gourd or little gourd (*kundru*), *Juniperus*, *Polygonum cuspidatum*, etc.

17.4 ANTHESIS

The first opening of a flower is called anthesis. It usually happens in the morning, and is influenced by environmental factors such as temperature, humidity, etc. At the time of anthesis the stigma is completely receptive. Pollen grains are shed at this time from the anthers. The relative timing of opening of a flower and pollen shedding from anthers affects the mode of pollination of a crop species.

17.5 POLLINATION

The transference of pollen grains from the anther to the stigma is called pollination. The pollens should be live and the stigma receptive. They germinate and pollen tubes enter the stigma. A pollen tube passes through the style and finally enters the ovule through the micropyle. Eventually, it releases the two sperms into the embryosac.

17.6 FERTILIZATION

Fusion of one male gamete (sperm) with the egg cell is called fertilization (also called syngamy). It gives rise to a zygote (2n). The other sperm unites with the secondary nucleus (2n) yielding a primary endosperm nucleus (3n). It is named as triple fusion. The fertilization and triple fusion are combinedly called double fertilization. The double fertilization must occur in the embryosac to enable the ovule to develop into a seed. The zygote divides mitotically to produce an embryo. The primary endosperm nucleus divides repeatedly to produce endosperm. Eventually, the ovule develops into a seed and the ovary into a fruit. The two integuments develop into two seed coats – the outer testa and the inner tegmen. Embryo is the *only* part in seed which serves to continue the hereditary stream of plant characteristics of species. Endosperm, when present, may show influence traceable to the pollen. The effect of pollen on embryo and endosperm is called xenia. Example of xenia can be seen in maize (corn).

17.7 MODE OF POLLINATION

The manner in which pollination takes place is called mode of pollination. It is of two types–(i) self-pollination and (ii) cross-pollination.

17.7.1 Self-Pollination

The transference of pollen grains from the anther of a flower to the stigma of the same flower, evidently bisexual, is called self-pollination (also called autogamy or natural inbreeding). No external agencies are instrumental in achieving self-pollination.

Mechanisms Encouraging/Enforcing Self-Pollination

There are several mechanisms which encourage/enforce self-pollination in crop plants. They are briefly discussed below:

i. *Bisexuality*: Presence of androecium and gynoecium in the same flower is known as bisexuality. The presence of bisexual flowers is a must for self-pollination. All the self-pollinated plants have bisexual (hermaphrodite) flowers.

ii. *Homogamy*: Maturation of anthers and stigma of a flower at the same time is known as homogamy. As a rule, homogamy is essential for self-pollination, e.g., *Lathyrus odoratus*, lupin, aster, calendula, etc.

iii. *Cleistogamy*: The situation where flowers never open is called cleistogamy and hence, pollen grains may only pollinate the stigma of the same flower. Cleistogamy is found in *Commelina bengalensis*, some species of pansy (*Viola*), balsam (*Impatiens*), wood-sorrel (*Oxalis*), sage (*Salvia*), etc.

iv. *Chasmogamy*: Opening of flowers only after the completion of pollination is known as Chasmogamy. It also promotes self-pollination and found in crops like lettuce, pansy, *Clitorea ternatea*, etc. In such cases, some cross-pollination may occur.

v. *Position of Anthers*: In some species, stigmas are surrounded by anthers in such a way that self-pollination is ensured. Such situation is found in tomato, brinjal, potato, etc.

   In some legumes, the anthers and stigma are enclosed in keel (carina) ensuring self-pollination, e.g., pea, French bean, etc.

vi. *Sticky Stigma*: The stigma becomes sticky as in okra, hollyhock, China rose, etc. It promotes self-pollination.

vii. In some species, stigmas become receptive and elongate through the staminal column. This results in predominant self-pollination.

17.7.2 CROSS-POLLINATION

The transference of pollen grains from one flower to another borne on the same or different plant of the same or allied species, or even allied genera, regardless of whether the flower is bisexual or unisexual, is called cross-pollination (also called allogamy). It is of three types – (a) xenogamy when pollination takes place between flowers borne by two different plants of the same species; (b) geitonogamy when it takes place between two flowers borne by the same plant; and (c) hybridism when it takes place between two flowers borne by two different plants of allied species or even allied genera. In autogamy and geitonogamy only one parent plant is concerned in producing the offspring, while in xenogamy (cross-pollination in strict sense) two parent plants are concerned and, therefore, a mingling of two sets of parental characters takes place resulting in healthier offspring. Both the methods are wide spread in nature.

Cross-pollination is brought about by external agents such as insects (bees, flies, moths, etc.), animals (birds, snails, etc.), wind and water. It is the rule in unisexual flowers, while in bisexual flowers it is of general occurrence. Nature favours cross-pollination and, therefore, adaptations in flowers to achieve it through external agents are many and varied.

A. Mechanisms Encouraging Cross-Pollination

i. *Dichogamy*: In many bisexual flowers the anther and the stigma often mature at different times. This condition is known as dichogamy. Because the anther and the stigma mature at different times, dichogamy often prevents self-pollination. There are two types of dichogamy:

   (a) *Protogyny*: It occurs when the gynoecium matures earlier than the anthers of the same flower. Here the stigma receives the pollen grains brought from another flower. Common examples are fig, custard-apple (*Annona*), some palms, etc.

   (b) *Protandry*: It occurs when the anthers mature (burst and discharge their pollens) earlier than the stigma of the same flower. Here the pollen grains are carried over to the stigma of another flower. Common examples are China rose, lady's finger (okra), sunflower, marigold, coriander, rose, carrots, raspberries, beets, etc.

ii. *Monoecism*: Monoecism, male and female flowers borne separately on the same inflorescence or on the same plant, favours cross-pollination much more than self-pollination. Common examples are mango, banana, coconut, cucurbits, walnut, chestnut, strawberries, rubber, grapes, cassava, etc.

iii. *Herkogamy*: There may be some sort of barrier standing between stamens and pistil of the same flower. The keel in alfalfa (lucerne) and a hood covering the stigma in pansy, iris, etc. are the common forms of such natural barriers. These barriers may be broken down by insects which can take some pollen grains to other plants also, thus causing cross-pollination.

B. Mechanisms Enforcing Cross-Pollination

i. *Male Sterility*: Due to irregular microsporogenesis or inadequate nourishment, spores (pollen grains) are rendered non-functional, or, due to non-dehiscence of anthers, pollen grains are not released in some crop plants. So, the normal stigma needs to be pollinated by foreign pollens only. Male sterility is genetically or cytoplasmically controlled. This system has greatly exploited in crops such as pearl millet, maize, onion, etc.

ii. *Self-Incompatibility*: In a large number of crops, although pollen grains and stigma both are normally functional, pollen grains from the flowers of a plant have no fertilizing effect on the stigmas of the same flower or different flowers borne on the same plant. So, for producing the next generation, cross-pollination is required. This situation is called self-incompatibility (also called self-sterility). It is most frequent in several species of *Brassica* (mustard, rai, cauliflower, etc.) radish, rye and many grasses.

iii. *Dioecism*: In dioecy, male and female flowers are borne on different plants. Hence, *cross-pollination is indispensable* for the production of seeds. Common examples are papaya, date palm, palm, mulberry, spinach, pointed gourd, little gourd, spine gourd, asparagus, etc. In dioecy, there is considerable wastage of gametes in the absence of suitable pollinating agent.

17.8 CLASSIFICATION OF CROPS BASED ON POLLINATION

On the basis of mode of pollination, the horticultural crops are grouped into four categories, *viz.*, (i) naturally self-pollinated crops, (ii) often cross-pollinated crops, (iii) naturally cross-pollinated crops, and (iv) dioecious cross-pollinated crops. They are discussed in brief below:

i. ***Naturally (Normally) Self-Pollinated Crops***: Many crops show more than 95% natural self-pollination and they are called naturally (normally) self-pollinated (also called self-fertilized, selfers, inbreeders, autogamous). In such crops, the amount of natural out-crossing (cross-pollination) is less than 5% and affected by the cultivar, temperature, humidity, location, etc.

ii. ***Often Cross-Pollinated Crops***: When the amount of natural cross-pollination is more than 5% and may be as high as 30%, the crops are called often cross-pollinated (also called often cross-fertilized, intermediates). The genetic make-up of such crops is intermediate between that of self-pollinated and cross-pollinated crops.

iii. ***Naturally (Normally) Cross-Pollinated Crops***: Many crops exhibit natural cross-pollination, and are called naturally (normally) cross-pollinated (also called cross-fertilized, outbreeders, allogamous). In many of such crops, a small amount of natural self-pollination (up to 5-10%) may occur. Such crops are either monoecious or have bisexual flowers and are not self-incompatible. Wherever self-incompatibility occurs, the crops will be *completely cross-pollinated.* Cross-pollinated crops are highly heterozygous, show high inbreeding depression.

iv. ***Dioecious Cross-Pollinated Crops***: As in dioecy male and female flowers are borne on separate plants, cross-pollination is indispensable and 100% in these crops.

17.9 BEARING HABIT

The bearing habit of a species can be described by the location and types of buds which produce flower and fruit (Gardner *et al.,* 1952). In various plant species, different buds located at different locations (on a shoot) varies to attain this condition. All the buds can be considered as potential flower buds. So, flowers, inflorescence and finally fruits will bear wherever buds are borne.

They can be located:

1. Terminally on long or short growths
2. Laterally in the axils of the current or past season leaves
3. Adventitiously from any point on the exposed bark of limbs, trunks or roots (Gardner *et al.,* 1952).

17.10 CLASSIFICATION OF FRUIT PLANTS ACCORDING TO BEARING HABITS

Fruits may be classified under six groups according to their bearing habits. Six distinct bearing habits, the classification being based upon the location of the fruit buds and the type of flower-bearing structure to which they give rise (Gardner *et al.,* 1952) are:

	Fruit buds borne terminally	**Fruit buds laterally**
Flower bud containing flower parts only	**Group-I:** Fruit buds borne terminally. It contains flower parts only and produces inflorescence without leaves. Examples: Mango, Loquat	**Group-IV:** In this group, fruit buds produce laterally. They contain flower parts only and giving rise to inflorescence without leaves or if leaves are present their size is reduced. Examples: Peach, Plum, Apricot, cherry, Almond, Walnut (staminate flower), Pecan (staminate flower), Date palm, Coconut, Citrus fruit,
Flower bud mixed, flowering shoot with terminal inflorescence	**Group-II:** Fruit buds borne terminally, unfolding to produce leafy shoots that terminate in flower clusters. This bearing habit is characteristics of most of the pome fruits. Examples: Apple, Pear, Quince	**Group-V:** Fruit buds borne laterally, unfolding to produce leafy shoots that terminate in flower clusters. E.g.,: Litchi, Grape, Filbert, Cashew nut, Pond apple)
Flower bud mixed, flowering shoot with lateral inflorescence	**Group-III:** Fruit buds borne terminally, after unfolding they produce leafy shoots with flowers or flower clusters in leaf axils. This might be called an incomplete terminal bearing habit for the fruit itself is not borne terminally, but is lateral to the growths upon which it appears. However, the flower buds are terminal. The terminal buds of the flowering shoots may differentiate flower parts for the following year's production or new buds may develop from lateral leaf buds. E.g., Pomegranate, guava, tropical almond, olive, *Jamun.*	**Group-VI:** Fruit buds borne laterally unfolding to produce leafy shoots with flower clusters in the leaf axils. Examples: Ber, Persimmon, Mulberry, Fig, Chestnut, Pistachio nut, Avocado.

Source: Gardner *et al.*, (1952)

17.11 CLASSIFICATION OF BEARING HABIT CONSIDERING ONLY THE POSITION OF FRUITS

Fruits can be classified on the basis of position of fruits on a shoot i.e. whether they are situated terminally, axillary and cauliflorous..

1. **Terminal bearing:** Mango, litchi, pineapple, banana, loquat etc.
2. **Axillary bearing:** Guava, ber, apple, papaya, orange, coconut etc.
3. **Cauliflorous bearing:** Cocoa, Jackfruit, Carambola

17.12 CLASSIFICATION ON THE BASIS OF SHOOT MATURITY ON WHICH FRUIT BUDS ARE GOING TO BE PRODUCED

Another classification of bearing habit is based on the bearing on shoot maturity i.e. old season shoots or current season shoots.

1. **Bearing on old season shoots:** Mango, Litchi, Apple, pear, peach, plum etc.
2. **Bearing on current season shoots:** Guava, orange, papaya etc.

17.13 CLASSIFICATION OF BEARING HABIT ON THE BASIS OF GROWTH HABITS TO POSITION OF FRUIT BUDS

There is very close relation between growth habit and fruiting habits. Plants in which fruit buds are borne either terminally (apple) or laterally (sweet cherry) on short growths or spurs are generally more compact as compared to those (like the peach or grape) in which fruit buds are borne on long shoots.

In a concise way, bearing habit of fruit trees may be considered as:

1. Terminal bearing
   (a) Bearing on new shoots: e.g.. Loquat, Bael
   (b) Bearing on old shoot: e.g., Mango, Litchi
2. Lateral or axillary bearing
   (a) Bearing on new shoot: e.g., Aonla, ber, fig, guava, karonda, phalsa, grape, mulberry
   (b) Bearing on old shoot: e.g., Apple, pear, peach, apricot, plum, walnut, hazelnut etc.

17.14 CLASSIFICATION OF BEARING HABIT OF SOME OTHER FRUIT CROPS ON THE BASIS OF PRUNING BEHAVIOUR (AS DESCRIBED BY THAKRE AND NAYAN DEEPAK, 2015)

1. **Grapes:** Under north Indian conditions, in grapes pruning is practiced from mid-December to mid-January. It bears on current season shoots from old one. So,

particular number of buds are kept to facilitate growth of new shoots. Number of buds varies as per cultivar. The cultivars Perlett, Delight require 3-4 buds per cane; Pusa Urvashi, Pusa Navrang, Pusa Aditi, Pusa Trisar require 4-6 buds per cane and Pusa Seedless, Kishmish Charni, Thomson Seedless require 9-12 buds per cane.

2. **Ber:** It bears on current season shoots. Here, one single node can produce several flowering branches. These branches bear solitary flowers in the leaf axils.
3. **Apple:** In India, maximum spur bearing cultivars are commercially grown. Generally, spur remains productive up to 4-5 years. If a cultivar becomes productive after 4 years, then 25% of the total number of spurs should be pruned. So that, after 4 years grower will get sufficient spurs for production.
4. **Peach:** Peach bears on the current season shoot of last year. If, one shoot of peach is pruned in December, after breaking of dormancy, it will produce vegetative shoot (bear on basal Part As this shoot is of previous season) that will bear subsequently in summer. The tip and basal portion of peach shoot is of vegetative in nature. So, the tip should be pruned to produce shoot for next year bearing and the middle portion of that shoot will produce fruits in the same season.

17.15 BEARING HABIT OF SOME IMPORTANT FRUIT CROPS

17.15.1 Mango

Mango bears on past season shoots and produces panicles terminally. Flowering is the first of several events that set the stage for mango production each year. A better understanding of the nature of flowering induction in mango is necessary not only for yield sustainability but also for yield increase. Flower initiation is very important because it is the first step towards attaining fruit and it is very complex phenomena in mango (Usharani, 2018). The potential of shoot to form flower buds will depend on the floriferous condition of the tree, which in turn will be determined by the amount of fruit load carried by the tree in the previous year (Singh, 1971).Generally, moderate blossoming is one of the principal conditions of annual fruit bearing in fruit trees. Various factors such as cultivar, flush maturity, shoot orientation etc. directly affected in flower bud differentiation of mango which finally affect production. Bearing habit is major problem in this manner. But few cultivars are gaining popularity to meet the problem, *viz.*, Amrapali, Mallika etc. Flush age is another factor that influenced flowering and fruit set of mangoes. A positive relationship between the fruits behaviour of shoots in the preceding season and its vegetative growth in the subsequent season has also been established, i.e., shoot that carried fruits maturity produced less vegetative growth and smaller number of shoots than those which failed to flower or had shed the flowers (Naik, 1949). Shoot orientation is the factor which also affected the sunlight as a result it may perform the subsequent vegetative and reproductive growth of mango (Ghosh, 1998).

17.15.2 Guava

The flowers are borne on the current season growth in the axils of leaves. The flowers are solitary or in cymes of two to three. The shoots bearing the flowers are terminal as well as lateral. Flowers appear on two to three months old shoot. It is hardly possible to forecast whether a shoot would bear flowers or not till the flowers actually emerge. In fact, on the same shoot buds in the axils of some leaves produce flowers, while, others do not. It was further observed that axillary flower buds are not produced all over the shoot. These may appear scattered. The duration of flowering varied from 33-41 days in the spring season while it was 38-45 days in the autumn season. The floral bud passed through eight arbitrary stages from dormant to fully opened stage and took 36-42 days to complete all the stages in rainy and winter season respectively. Seventh stage of floral bud was most suitable for emasculation. All the cultivars bore flower solitary or in cymes of 2-3 flowers on the current season growth in the axils of the leaves. The flower size of all the cultivars was found to be significant in both the season, i.e., April-May and August-September (Sharma *et al.*, 2017).

17.15.3 Citrus

Citrus trees usually have several flushes of growth during the year. The number of flushes and their importance are determined by cultivar characteristics, crop load and climate. The newly formed shoots arise from lateral resting buds and may form either leaves (vegetative shoots), flowers (generative shoots also called leafless inflorescences), or both flowers and leaves (mixed-type inflorescences). Under subtropical climate conditions most citrus cultivars have three main flushes of growth. The main one occurs in late winter or early spring (spring flush). Two additional ones occur at the end of the June drop (summer flush) and late in September (autumn flush). In most cases, only vegetative shoots are formed in the summer and the autumn flushes of growth. Inflorescences are formed in the spring flush of growth as a response to the low winter temperatures. The shoots developed during the flushes of growth arise mainly from the axillary buds present in the vegetative shoots formed in previous flushes of growth. Bud behaviour in spring depends critically both on the time the buds were formed (bud age) and in the position of the bud along the shoot. Citrus species show a relatively long juvenility period (two to five years) before the trees reach the mature stage to produce flowers. The inflorescence developed in citrus may be either leafless or leafy and these may carry a single flower or several of them (Goldschmidt and Huberman, 1974). Citrus exhibits a long juvenility phase and important citrus species including sweet oranges, mandarins, lemons and grapefruits exhibited some degree of apomixis. Furthermore, many of them are parthenocarpic, sterile or self-incompatible and/or develop defective pollen (Baldwin, 1993; Davies and Albrigo, 1994). Citrus species usually produce a large number of flowers over the year. The floral load depends on the cultivar, tree age and environmental conditions (Monselise, 1986). It has been reported, for example, that sweet oranges (*C. sinensis*) may develop 250,000 flowers per tree in a bloom season although only a small amount of these flowers (usually less than 1%) becomes mature fruit (Erickson and Brannaman, 1960; Goldschmidt and Monselise, 1977). Citrus buds on previous summer shoots and buds at apical positions produce more flowers than older or lateral buds (Valiente and Albrigo, 2004). It is also well known that the fruit load has a strong negative effect on spring sprouting of both vegetative and generative buds and therefore constitutes a major inhibitor of flowering. Conversely, the absence of fruit or a scarce fruiting induces huge flowering intensities in the next season. Thus, crop load is likely the main cause of the "alternate bearing" behaviour of many citrus species and cultivars including many seeded mandarins that alternate reduced flowering and fruiting ("off year") with increased flower induction and fruit production ("on year"). In seeded citrus cultivars, fruit development is linked to the presence of seeds and therefore, it depends upon pollination and fertilization. Self-pollination usually takes place in the unopened or opening flower, often allowing pollination before anthesis. Cross-pollination occurs between plants of different genetic background by insect transport of pollen. However, many current citrus cultivars are mostly seedless cultivars showing high parthenocarpy, in many instances due to gametic sterility. Generative sterility can be relative or absolute. Relative gametic sterility may be due to self-incompatibility as in Clementine and to cross incompatibility. On the other hand, absolute gametic sterility is associated with pollen and/or embryo-sac sterility. Some cultivars such as Washington Navel oranges and Satsuma mandarins have both, although even in these two cultivars a few embryo sacs may often reach maturation. there were only two main flowering seasons, the first and the major one occurring in December-February and second one in June-July. In acid lime, majority of the shoots that bore flowers were normal in vigour as measured in terms of length of shoot (Ghawade *et al.*, 2002). The Kagzi lime bore flower mostly on lateral shoots and in the apical region of shoots. In lime, two main flowering seasons were observed the first constituting more than 50 per cent of total number of flowerings produced in the year was during December-February period. The highest number of flowers was produced in the month of January (Ambia bahar). The second main flowering season (Constituting about 25 per cent of total number of flowers) was during June-July (Mrig bahar). Besides these two main flowering seasons, sparse flowering occurred during other months of the year except in the month of April, May and November (Singh and Tomar,1949).

17.15.4 Papaya

In papaya, flowers occur in leaf axils. Older leaves die and fall as the tree grows. Flower type is determined by the presence or absence of functional stamens (male parts), and stigma and ovary (female parts). Within cultivars, flower type is usually identified by flower size and shape. Three types of plants are recognized based on flower type: female, hermaphrodite, and male. Female plants always produce female flowers. If no male or hermaphrodite plants are nearby to provide pollen, female plants usually fail to set fruit. Unpollinated female plants occasionally set parthenocarpic fruits, lacking seeds. Male plants are distinguished by their long flower stalks bearing many flowers. Usually they do not produce fruit, but on rare occasions there is female expression in the flowers, and they may set fruits. Hermaphrodite plants may have male flowers, hermaphrodite flowers, or both, depending on environmental conditions and the time of year. Hot, dry weather may cause suppression of the ovary and the production of female-sterile (i.e., male) flowers. Male flowers on hermaphrodite plants are borne on short peduncles. Hermaphrodite plants tend to produce self-pollinated seeds, which result in relatively uniform progenies.

17.15.5 Jackfruit

Fruit bearing habit in Jackfruit include percent and duration of fruit set, fruit set in different sections of plant. Ullah and Haque (2008) studied on fruiting, bearing habit and fruit growth of jackfruit at orchard of Jackfruit Research Project, Department of Horticulture, Bangladesh Agricultural University (BAU), Mymensingh and reported that the bearing habit of jackfruit was cauliflorous i.e., fruits are borne on trunk and branches. The germplasm under study bore fruits on trunk, primary, secondary, tertiary, fourth, fifth and sixth branches of jackfruit trees. On an average, the maximum fruits were borne on primary branches (33.0%) followed by those on trunk (31.5%), secondary (12.3%), and fourth branch (8.4%), while it was the lowest on sixth branch (2.0%). The growth pattern of fruit of jackfruit is characterized by single sigmoid curve.

17.16 PROBLEM OF UNFRUITFULNESS

'Unfruitfulness' is a major problem in many fruit crops and their cultivars result in a huge loss to growers and make fruit cultivation less profitable. 'Fruitfulness' refers to the state where a plant is not only capable of flowering and bearing fruit but also takes these fruits to maturity. The inability to do so is known as 'unfruitfulness' or 'barrenness'. In spite of adequate flowering, low fruit yields in orchards have been experienced because of low initial fruit set and subsequently higher fruitlet abscission. In an orchard, all the trees do not bear fruits equally or regularly and sometimes fail to flower and bear fruit under similar conditions where another fruit tree bears heavily. This failure to fruit may be attributed to unfruitfulness. Any interference with the development of sex cells and organs leads to unfruitfulness (Wani *et al.*, 2010)

17.17 FACTORS AFFECTING UNFRUITFULNESS

Unfruitfulness can be due to lack of balance between vegetative growth and fruiting, lack of flowering and poor fruit set, which is a result of the unfavourable environment. It can also be due to heavy cropping, leading to inhibition of fruit bud production and poor crop in the following year. Sterility also leads to unfruitfulness due to impotence, incompatibility or the abortion of embryo. The causes of unfruitfulness can be broadly grouped into two categories: (i) Internal factors and (ii) external factors.

17.17.1 Internal Factors

There are a number of internal factors which are associated with unfruitfulness. They have further been divided or grouped into 3 major categories: (a) evolutionary tendencies (b) genetic influence (c) physiological factors.

Evolutionary Tendencies

The evolutionary factors leading to unfruitfulness are:

1. *Imperfect/Defective Flowers*: A perfect flower possesses both male (stamens) and female (pistil) parts, whereas an imperfect flower may be either staminate (functional male having only stamens) or pistillate (functional female having only a pistil or pistils). If staminate and pistillate flowers are borne on the same plant but in different locations, the species is termed as 'monoecious' for example, walnut, pecan nut, cashew nut and chestnut. If staminate and pistillate flowers occur on different plants, the species is termed 'dioecious' for example, pistachio nut, papaya, kiwifruit, fig and mulberry. One can see the ramifications for pollination and orchard design, in that dioecious species such as pistachio or kiwifruit must be planted in an orchard with male plants near females for pollination. Aside pollination, the male plants are useless since they do not possess ovaries that will ripen into fruit. Generally, in monoecious fruit plants, there is no or very little problem of pollination, fruit setting and fruitfulness. A number of sex forms have been reported in papaya. The evolution of unisexual flowers cuts down chances of self-fertilization completely. Of the unisexual plants, the pure staminate flowering ones are essentially barren.
2. *Heterostyly*: The presence of short style with long filaments (Thrum) for example, sapota and pomegranate, or long style and short filaments (Pin type) for example, almond and carambola is dimorphism, a type of heterostyly. The occurrence of flowers with variable length of styles is common in prunus fruit crops (Suranyi, 1976). Three types of flowers, namely thrum, homostylous and pin, were found in Pomegranate cvs Ganesh-1 and Kandhari (Singh *et al.*, 2006).
3. *Structural Peculiarities*: When stigmatic receptivity period does not coincide with pollen viability in monoecious plants, it is known as 'dichogamy'. In dichogamy, self-pollination is prevented in perfect flowered plants, due to maturity of two sex organs at different times. If the stamens are ripened before the stigmas become receptive, the flowers are known as 'protandrous' and if stigmas become receptive before the stamens produce viable pollens, it is known as 'protogynous'. Dichogamy has been reported in hermaphrodite (avocado, mango, ber and annona), monoecious (cashew, pecan nut, walnut, chestnut) and dioecious (persimmon, fig and pistachio) species. In walnut, occurrence of both protandrous and protogynous cultivars prevents self-pollination, which warrants cross-pollination. In mango, the duration of stigma receptivity is only 2 to 4 hours, following anthesis.
4. *Stigmatic Receptivity*: It is the ability of the stigma to support pollen germination and it limits the effective pollination period (it is defined as the number of days during which the pollination is effective in producing a fruit and is determined by the longevity of ovules minus the time lag between pollination and fertilization) in fruit crops. The stigmatic receptivity could be an important factor limiting pear flower receptivity (Sanzol *et al.*, 2003).
5. *Abortive Flowers or Aborted Pistils or Ovules*: This occurs in the developing flowers, pistils and stigmas. Interference either in the development of the

flower or in the full development of sex elements and their function may lead to unfruitfulness. Floral abortion is more common in indeterminate inflorescence as compared to determinate inflorescence. Pistil degeneration leads to unfruitfulness in certain cultivars of plum and ornamental pomegranate, while in strawberry, pistil abortion is so late that unfruitfulness does not take place.

6. *Non-Viable Pollen*: It is due to non-functional pollen or the ovule. Non viability or impotence of pollen results in unfruitfulness. Unfruitfulness in the case of *Muscadine* grape is due to defective pollen. In grapes, sterile pollen results from degeneration processes in the generative nucleus or arrested development prior to mitosis in microspore nucleus. In general, late flowering in apricot genotypes showed lower pollen viability than early flowering genotypes (Ruiz and Egea, 2008).
7. *Sterility*: Some species have genes that prevent development of the pollen or the ovule. Generally, sterility is due to failure to obtain normal development of pollen, embryo sac, embryo and endosperm. Morphological sterility is due to rudimentary pistil or abortion of sex organs ovule degeneration. If one of the sexes is inactivated, then cross-pollination has to take place. Pollen sterility is common in peach cv. J.H. Hale. The proportion of sterile ovules varies between 22% in apricot and 98% in avocado (Verma and Jindal, 1997).

Genetic Influences

1. *Unfruitfulness Due to Sterile Hybrids*: Hybridity is associated with sterility as well as unfruitfulness. The degree of sterility increases with wider crossing. The popular tangelo is a hybrid produced by crossing a grapefruit (*Citrus paradisi*) with a tangerine (*Citrus reticulata*). They are seedless or they produce seeds only with nuclear embryos (Gardner 1952).
2. *Incompatibility*: Incompatibility is defined as failure of viable pollen to grow down the style of flower of the same cultivar (self-incompatibility) or of the different cultivars (cross- incompatibility). Many cross-pollinating species exhibit self-incompatibility, so that fertilization by their own pollen is prevented through physical or biochemical factors. There are different degrees of self-incompatibility, and many self-incompatible species will produce a few fruits even when self-pollinated. Horticulturists have coined the terms "self-fruitful" and "self-unfruitful" to describe cultivars that can set commercial crops or cannot set commercial crops (respectively) when self-pollinated. Self-incompatibility is more common in fruit crops like apple, pear, sweet cherry, almond, avocado, fig, mango, citrus, olive, etc. than cross- incompatibility (apple, pear, sweet cherry, European plums and almond). In fruit production, incompatibility may create a platform to create variation leaving no scope for inbreeding. In mango, self-unfruitfulness is reported in cvs. Dashehari, Chausa and Langra (Ram *et al.*, 1976). Most pear (*Pyrus communis* L.) cultivars are impaired to set fruit under self-pollination because self-fertilization is prevented by gametophytic self-incompatibility system (Sanzol, 2007). In loquat cultivars, improved golden yellow, pale yellow and golden yellow, the pollen tube penetrated the stylar canal up to 1/4th to 1/3rd of its length and did not go further

below, even after 72 hours of pollination. As such, this suggests incompatibility in loquat (Singh and Rajput, 1964).

Physiological Influences

1. *Premature or Delayed Pollination*: Premature or delayed pollination leads to unfruitfulness and is reported that premature pollination followed by germination and tube growth causes fruit drop due to toxicity in pistil. However, in case of oranges, premature pollination did not have any deleterious effect. Low setting due to premature pollination was noticed in persimmon, pear, plum and peach. Similarly, if pollination is delayed, the flower falls without setting.
2. *Nutritive Condition of Plant*: Nutrition of plant controls the percentage of defective pistils. Defective pistils are formed especially on exhausted or weakened plants caused by overbearing, drought and poor nutrition. Nutrition also determines the percentage of flower carried for setting, maturity and also pollen viability.
3. *Fruit Setting of Flowers in Different Positions*: Fruit borne on terminal growth have more competition in many fruit crops and its maturity is set under normal nutritional conditions, but the percentage of the set is small. This positional competition takes place between fruits and branches as well as between different fruits influencing fruitfulness. The apical flower in the bud of apple (*Malus domestica*) develops first, followed by the lateral flower which develops in sequence, beginning at the base of the spur. Some fruits like plum and cherry, bear on both shoots and spurs and it is to be expected that slightly different nutritive conditions are obtained in the different tissues and a distinctly heavier June drop occurs in shoot borne fruits (Gardner, 1952). In apple, old spurs are less likely to initiate flowers than the young ones (Jackson, 2003).

External Factors

Environmental Factors

1. *Temperature*: Among the environmental factors, temperature has a great importance. It affects flowering and fruit set in several ways.
   (a) *High Temperature*: Above 32°C, desiccation of the stigmatic surface and more rapid deterioration of embryo sac occurs (Jindal *et al.*, 1993). As temperature increases, ovule senescence becomes faster in plum. At 15°C, only one ovule per flower remains viable by 8 DAFB for Italian, whereas for 'brooks' cultivar, higher temperature results to a decrease in ovule longevity (Moreno *et al.*, 1992).
   (b) *Low Temperature*: In plum, cherry, apple, pear, etc., the temperature of 4.4°C or lower, completely check the blooming, fertilization and fruit set (Jindal *et al.*, 1993). Fruit set and production of fruit in peach cv. Granda were absent or very low in green house grown trees in orchards. Nava *et al.* (2009) found that a rise of the diurnal air temperature in the interior of the greenhouse promoted a significant reduction in the production of pollen grains.
2. *Humidity*: Low atmospheric humidity causes drying of stigmatic secretions. Wet and humid weather favours anthracnose and poor fruit set in mango. The poor germination of pollen in almonds is attributed to damp weather during fruit set.

3. *Rain*: It directly affects fruit setting by disturbing the process of pollination and germination of pollen grains and staminal fertilization. In almond washing, treatment decreased the number of germinated pollen grains on the stigma mainly when the flowers were immersed before pollination and it seems to affect adhesion in forthcoming pollinations (Ortega *et al*., 2007). Rains at the time of blooming period causes unfruitfulness by washing pollen grains, inhibit pollinators and cause spread of diseases and pests. Every fruit species needs a specific time period without rains at the time of blooming for successful pollinations.
4. *Light*: Light affects fruitfulness indirectly by its effect on photosynthesis. Light is a pre-requisite for photosynthesis and low light intensity or its duration reduces the carbohydrates reserves in the trees. In addition to this, poor light conditions promote fruit abscission.
5. *Wind*: Wind also affects the fruitfulness indirectly by its effect on the pollinating agents, i.e., it controls pollination either by promoting wind pollination or by checking insect pollination. It is desirable in wind pollinated fruit trees like hazelnut and walnut, but if its speed is too high, it is harmful because it results in low fruit set on exposed sides and heavy crop on other sides. Excessively, speedy winds cause ovary abortion (Gardner, 1952) and also make the stigma dry.
6. *Chemicals and Pesticides*: The use of pesticides can kill bees, therefore reducing pollination. Some pesticides can also be toxic to delicate flowers causing abortion and loss of fruit. Pesticides spray effects on receptive stigmatic surface, showed varying degree of injury and range from minor surface wrinkling to degeneration of stigma papillae. Controlled pollination for 1 hour, after pesticides sprays, results in an inhibition of pollen germination and tube growth (Wetzstein, 1990). Commercial fungicides containing captan, dinocarp, sulphur and triforine sprayed on undehisced anthers of several apple cultivars reduced the viability of pollen, impaired pollen release, kill pollen when sprayed onto dehisced anthers (Ruth *et al.*, 1978).Pollen germination and tube growth were drastically suppressed by the spray and almost no fruit set was observed on the treated inflorescence and there was a highly significant difference between fruit set on exposed and protected cluster (Legge and Williams, 1975)

17.18 REMEDY

17.18.1 Pruning and Thinning

The main techniques for controlling the vigour of fruit trees and increasing their relative fruitfulness are the use of dwarfing rootstocks, compact, short internodes scions and of trees training and pruning system which give horizontal or wide-angled branches. As such, growth retardants are also used. The influence of pruning varies with the amount, season and kind of pruning. Judicious and proper pruning is needed to improve the fruit set and fruit retention on the trees and also on the removal of dead results in loss of carbohydrates reserves along with pruned wood. Also, severe pruning promotes too much vegetative growth and hence reduces the productivity.

In mango, the highest number of new flushes per shoot was achieved with severe pruning and spraying of GA_3 at 100 ppm (Sheban, 2009). Singh and Daulta (1986) found that pruning up to 12- bud, improved fruit set significantly and it is suggested that 'Sharbati' cultivar responds well to light pruning.

17.18.2 Use of Pollinizers

Pollen transfer may present an application problem in fruits which are self-incompatible. Many apple cultivars are at least partially self-fertile, especially under warm weather, but in most cases, fruit-sets are improved by the use of pollinator cultivars and bees. Cherries and plums also need cross-pollination and should be inter-planted with pollinizer cultivars. Breeding and selection of self-fertile cultivars or clones of cherry and apple may reduce difficulty or achieve satisfactory pollination, especially in cool marginal areas of fruit production where temperatures at blossoming time are sub-optimal both for bee activity and for pollen tube growth. Until such improved cultivars become available, provision of suitable pollinizer cultivars and bee-hives can be of great help in ensuring satisfactory fruit-set. There is an increasing tendency for using crab-apples, as pollinisers of these occupy very little space in orchards. It is advisable to have a number of pollinizer cultivars with widespread flowering dates to ensure cross-pollination. One of the basic requirements for setting fruit is an adequate requirement of compatible pollen. With most tree fruit crops, the need for cross pollination is recognized. Pollinating insects are necessary for fruit set on all cultivars, and most cultivars will benefit from cross-pollination. Under general conditions, the closer a tree is to a pollinizer, the better fruit set will be. Cross-pollination necessitates the availability of sufficient quantity of compatible pollen, as pollinizer cultivars flower synchronously with the main cultivar and suitable agent for the successful and effective transfer of pollen. Therefore, suitable pollinizer cultivars must be inter-planted at the time of orchard layout.

17.18.3 Introduction of Pollinators

The population of natural pollinators has gone down due to indiscriminate use of pesticides and deterioration of the ecosystem. The managed bee pollination is very limited and the available bee hives, during bloom, hardly meet 2 to 3% of the demand. In red delicious apple, it was found that sequentially increasing the density of colonies increases the amount of cross- pollination and a high proportion of top workers. However, the increased pollination efficiency results in high fruit set and higher yield (50 to 100%) in treated plots (Stern *et al.*, 2001).

17.18.4 Proper Nutrition

Balanced supply of nutrients is always desirable for realizing optimum fruit production. Generally, it is advocated that for application of fertilizers, a few days before emergence of blossoms is generally believed to favour flowering and fruit set. Nitrogen application after terminal bud formation led to the development of flower with enhanced embryo sac longevity (Jackson, 2003). Pollen tube growth in pear was significantly stimulated by increasing concentration (25-200 mg/l of boric acid) and the values are significantly higher at 200 mg/l concentration of boric acid (Lee *et al.*, 2009). In walnut cv. Local selection, the highest fruit set (32.50), fruit retention (41.35) and nut yield (4.02 kg/tree) was recorded under foliar application of H_3BO_3 + Paras (0.1% + 0.6 ml/L) (Tomar and Singh, 2007). However, excess use of manures and fertilizers may produce abnormal flowers.

17.18.5 Application of Plant Growth Regulators

The unfruitful behaviour of several fruit plants can be overcome by the use of plant growth regulators which may be due to decreased fruit set and abscission at various developmental stages.

17.18.6 Use of Suitable Rootstocks

There can be as much as 50% or more difference in the yield of a given cultivar grown on different rootstocks. The reasons for such an effect can be traced to difference in tolerance to adverse soils, in resistance to pests or in uptake of nutrients. Four rootstocks namely: M9, M7, M4 and M1 induced 50% or more bloom in the fifth year in Starking delicious apple and resulted in higher yield efficiency by controlling tree size. (Wani *et al.*, 2010).

REFERENCES

Baldwin EA 1993. Citrus fruit. *In: Seymour GB, Taylor JE, Tucker GA (eds), Biochemistry of Fruit Ripening,* pp.107-149. Chapman & Hall, London.

Erickson LC and Brannaman BL 1960. Abscission of reproductive structures and leaves of orange trees. *Proc. Am. Soc. Hort. Sci.*, **75**:222-229.

Davis FS and Albrigo LG 1994. *Citrus. In: Crop Production Science in Horticulture Series,* pp.134-135. CAB International, Wallington.

Gardner VR, Bradford FC and Hooker HD 1952. The Fundamentals of Fruit Production. McGraw-Hill Book Company, INC., New York, Toronto, London,536-554.

Gardner VR 1952. The Fundamental of Fruit Production. McGraw Hill Book Company, INC pp. 643-685.

Ghawade SM, Panchbhai DM, Jadhao BJ, Katole SR and Dadmal SM 2002. Flowering behaviour of Kagzi lime (*Citrus aurantifolia,*Swingle) under sub-humid tropical climate. *Agric. Sci. Digest.*, **22** (4): 252 – 254.

Ghosh P 1998. Fruit weather of India. In:R. K. Arora and Ramanatha Rao (eds). *Tropical fruits in Asia.* IPGR.3-5 pp.

Goldschmidt EE and Huberman M 1974. The coordination of organ growth in developing Citrus flowers: a possibility for sink type regulation. *J. Exp. Bot.*, **25**: 534- 541.

Goldschmidt EE and Monselise SP 1977. Physiological assumptions toward the development of a citrus fruiting model. *Proc. Int. Soc. Citrus* **2**: 668-672.

Jackson JK 2003. *Biology of apples and pears.* Cambridge University Press, UK pp :268-340.

Lee SH, Kim WS and Han TH 2009. Effect of post-harvest foliar boron and calcium applications on subsequent season's pollen germination and pollen tube growth of pear (*Pyrus pyrifolia*). *Scientia. Hort.*,**122**: 77- 82.

Legge AP and Williams RR 1975. Adverse effect of fungicidal sprays on the pollination of apple flowers. *J. Hort. Sci.*, **50**: 275-277.

Monselise SP 1986. *Citrus. In: Monetise SP (ed), Handbook of Fruit Set and Development,* pp.87-108. CRC Press, Boca Raton.

Naik A 1949. South Indian fruits and their culture. P. Varandachari & co., Madras.94 pp.

Ruth M, Church, Williams RR 1978. Fungicide toxicity to apple pollen in the anther. *J. Hort. Sci.*, **53**: 91-94.

Ortega E, Dicentia F, Egea J 2007. Rain effect on pollen stigma adhesion and fertilization in almond. *Scientia. Hort.*, **122**: 345-348.

Ram S, Bist LD, Lakhanpal SC and Jamwal IS 1976. Search of suitable pollinizers for mongo cultivars. *Acta. Hort.*, **57**: 253-264.

Ruiz D and Egea J 2008. Analysis of the viability and correlation of floral biology factors affecting fruit set in apricot in a Mediterranean climate. *Sci. Hort.*, **115**:154-163.

Sanzol J, Rallo P and Herrero M 2003. Stigmatic receptivity limits the effective pollination period in 'Agua de Aranjuez' pear. *J. Amer. Soc. Hort. Sci.,* **128**(4): 458-462.

Sanzol J 2007. Self-incompatibility and self-fruitfulness in pear cv. Agua de Aranjuez'. *J. Amer. Soc. Hort. Sci.*,**132**(2): 166-171

Sharma S, Sehrawat SK and Sharma KD 2017. Studies on Time and Duration of Flowering, Floral Bud Development and Morphology of Guava (*Psidium guajava* L.) under Semi-Arid Region of India. *Int. J. Curr. Microbiol. App. Sci.* **6**(12): 4176-4186.

Sheban AEA 2009. Effect of summer pruning and GA3 spraying on inducing flowering and fruiting of zebda mango trees. *World. J. Agri. Sci.*, **5**(3): 337-344.

Singh SN and Tomar BS 1949. *Indian Farming,* **10**: 532-38.

Singh JP, and Rajput CBS 1964. Pollination and fruit set studies in loquat (*Eriobotrya japonica*). *Indian J. Hort.*, **21**(2): 143-147.

Singh LB 1971. The mango; Botany, cultivation and utilization. Leonard Hill Ltd., London.

Singh D and Daulta BS 1986. Effect of pruning severity on growth, flowering, fruit set, fruit drop, the extent of sprouting and flowering at various nodal positions in Peach (*Prunus persica* Batsch) cv 'Sharbati'. Haryana. *J. Hort. Sci.*, **15** (3-4): 200-205.

Singh NP, Dhillon WS, Sharma KK and Dhatt AS 2006. Effect of mechanical deblossoming on flowering and fruit set in pomegranate cvs. 'Ganesh-1' and 'Kandhari'. *Indian. J. Hort.* **63**(4): 383-385.

Stern R, Goldway M, Zisovich AH, Shafir S and Dag A 2004. Sequential introduction of honeybee colonies increase cross pollination, fruit set and yield of Spadona Pear (Pyrus communis L.). *J. Hort. Sci. Biot.,* **79** (4): 652-58.

Suranyi D 1976. Differentiation of self-fertility and self-sterility in Prunus by Stamen Number/ Pistil Length Ratio. *Hort.Sci.*,**11** (4): 406- 407.

Thakre M and Nayan DG 2015. Bearing Habit of Fruit Crops. *In: Fundamental of Fruit production*, Division of Fruits and Horticultural Technology, Indian Agricultural Research Institute, New Delhi, edited by Usha, K., Thakre, M. and Goswami, A.K. and Nayan Deepak, G.

Tomar CS and Singh N 2007. Effect of foliar application of nutrients and Bio-fertilizers on growth, fruit set, yield and nut quality of Walnut. *Ind. J. Hort.*, **64**(30): 271-273.

Ullah MA and Haque MA 2008. Studies on fruiting, bearing habit and fruit growth of jackfruit germplasm. *Bangladesh J. Agril. Res.*, **33** (3): 391-397.

Usha Rani K 2018. Advances in Crop Regulation in Mango (*Mangifera indica* L.). *Int. J. Curr. Microbiol. App. Sci.*, 7 (9): 35-42

Valiente JI and Albrigo LG 2004. Flower bud induction of sweet orange trees (*Citrus sinensis* (L.) Osbeck): effect of low temperatures, crop load, and bud age. *J. Am. Soc. Hort. Sci.*, **129**:158-164.

Verma LR and Jindal KK 1997. *Fruit Crops Pollination. Kalyani Publishers*, Ludhiana.

Wani IA, Bhat MY, Lone AA and Mir MY 2010. Unfruitfulness in fruit crops: Causes and remedies. *African J. Agric. Res*., **5**(25): 3581-3589.

Westwood MN and Bjornstad HO 1974. Fruit set as related to Girdling, Early Cluster thinning and pruning of 'Anjou' and 'Comice' pear. *Hort. Sci*., **9**(4): 342-344.

Wetzstein HY 1990. Stigmatic surface degeneration and inhibition of pollen germination with selected pesticidal sprays during receptivity in pecan. *J. Amer. Soc. Hort.* **115** (4): 656-661.

OUTCOMES ASSESSMENT

PART A

Answer the following questions (True or False).

1. Unfruitfulness is a minor problem in many fruit crops. (True/False)
2. Temperature affects flowering and fruit-set in several ways. (True/False)
3. Jackfruit is cauliflorous. (True/False)
4. Axillary bearing is found in mango. (True/False)
5. In protogyny pistil matures earlier than stamens. (True/False)

PART B

Answer the following questions.

1. What is a perfect flower?
2. What is difference between self- and cross-incompatibility?
3. Give examples of two monoecious fruit crops.
4. Define dichogamy.
5. What is incompatibility?

PART C

Write a brief note on each of the following.

1. What is bearing habit? Discuss its classification.
2. Discuss the bearing habit of mango.
3. Discuss in details the factors influencing unfruitfulness.

Chapter 18

Rejuvenation of Old and Senile Orchards

18.1 INTRODUCTION

The decline in productivity of old and senile orchards existing in abundance has become a matter of serious concern for the orchardists, traders as well as scientists. In India 30-35% area under fruit crops is occupied by old, senile, and pest-and-disease-affected orchards. For overcoming the problem of unproductive and uneconomic orchards existing in abundance, large scale uprooting and replacement with new plantations will be a long term and expensive strategy. Thus, rejuvenation of old and senile orchards is a good alternative.

18.2 ESSENTIALITY OF REJUVENATION

Rejuvenation of old and senile orchard is required because of the following reasons:

1. It helps in restoring the production potential of old, unproductive and diseased orchards in shortest possible duration than any other technique.
2. It helps in restoring the production potential, as well as in maintaining the manageable tree height with open architecture.
3. Sustains the life of farmer without affecting his economy to a great extent.
4. Improves in yield and quality of fruits by converting old and unproductive into young and productive orchards.
5. Enhances the photosynthetic surface area.
6. Avail more productive shoots.
. Ensures more penetration of sunlight by pruning of branches as a result of which the fruits on the interior areas of the tree develop proper colour.
8. Decreasing incidence of insect pests and diseases.

18.3 TECHNIQUE FOR REJUVENATING THE SENILE ORCHARDS

The rejuvenation technique involves the heading back of older trees showing marked decline in annual production and quality of produce to the extent of 1.0 to 1.5 meter height above the ground level during May-June or December-February with the objective of facilitating production of new shoots from below the cut point and allowing the development of fresh canopy of healthy shoots. The newly emerging shoots are allowed to grow up to 40-50 cm length and then further pruned for emergence of multiple shoots below the pruning point to modify the tree structure and maintain canopy size. Profusely emerging shoots in the inner canopy are also pruned out to promote branching. The multiple shoots developed as a result of second pruning are capable of producing flower buds (Baba *et al.*,2011).

18.4 REJUVENATION IN MANGO ORCHARD

Since the orchard establishment is a long term process and cannot be done in days but once the yield is reduced to such an extent that orcharding becomes non-economical, rejuvenation is said to be essential as it helps in restoring the production potential of old, unproductive and diseased orchards in shortest possible duration than any other technique. It also helps in restoring the production potential, as well as in maintaining the manageable tree height with open architecture. (Gurjar *et al.*,2015). The process can be started in the month of December -January. Unproductive orchard is selected, branches are marked followed by beheading and 30cm stubs are kept on 3-6 major limbs. A clean cut should be given to the limbs with a sharp saw to avoid bark splitting. Start beheading from base to the top. Branches are pruned during December. Before undertaking pruning work umbrella like frame work of individual trees is planned. About four to five main branches with outward growth are considered for constituting the basic shape of the tree for development of canopy. These four to five selected branches are marked with white chalk by making a ring around the branches at a distance of 75 cm from the base. Other criss-crossed, intermingling, overcrowding, dried and diseased branches are marked with red chalk for complete removal. During December, the red marked branches should be completely removed by pruning from their base. Later, branches marked with white chalk should be pruned at a distance of about 75 cm from their base. Development of canopy is promoted on these branches. Pruning should be initiated firstly from lower surface of the branch. Immediately after pruning, a fungicidal paste (of copper oxy chloride) should be applied on cut surface. Application of cow dung paste or bio dynamic tree paste, prepared by mixing cow dung, orchard soil, sand in equal proportion with cow horn manure has also proved beneficial. Pruning work in alternate rows of tree in the orchard has proved beneficial and economical. Once pruned trees in the alternate row attain canopy and come into fruiting, the remaining trees are subjected to pruning. Three to four months after pruning i.e. during March-April, there is profuse emergence of shoots on pruned branches. Consequently, dense and bushy canopy of un-healthy shoots with poor bearing potential develops on pruned trees. Selective and regular thinning of shoots is essential for facilitating development of open and spreading canopy of healthy shoots. Only outwardly growing 8-10 healthy shoots are retained per branch and the rest are removed. Thinning operation are undertaken during June and August. Pruned trees need intensive care of nutrition and irrigation, weeding, etc. for survival, emergence of new shoots and development of canopy.

18.5 REJUVENATION OF OLD AND SENILE GUAVA ORCHARD

Guava orchard begins to lose vigour and bearing potential after 15-20 years of age and the occurrence of wilt accelerate the process. The commonly occurring senile orchards, is one of the prominent factors relating to the declining trends in production and productivity of guava (Singh, 2005). The declining yield pattern from old guava orchard over the years is the major cause of shifting the interest of farmers towards other crops and cropping system. The majority of orchards became old and senile characterized by intermingling, overcrowding, infestation of insect and disease in branches and trunk, more wood mass and thin shoots in canopy adversely influencing bearing quality of fruits (Singh and Singh, 2003). The fruits of guava are borne on new wood of 9-11 months old and any treatments that encourage new growth influence the fruiting directly. The farmers are advocated for rejuvenation of old senile orchards to allow new shoots on tree, eliminate infected branches and increase light penetration on floor for field crops along with higher fruit yield. The trees are headed back at 1.0 to 1.5-metre-high above ground level during May-June or December-February. Fresh/newly emerging shoots are allowed to develop to a length of about 40 to 50 cm, which could be attained in 4 to 5 months of rejuvenation pruning. Prune out all newly emerging, growing shoots towards inner side of the freshly developing canopy to 6 newly emerging shoots growing outwards be retained at each cut end. These newly emerging shoots of 40 to 50 cm length be further pruned to about 50 per cent of its total length for emergence of multiple shoots below the pruning point. Second pruning is mainly done to modify the tree structure and maintain canopy size. The multiple shoots developed as a result of second pruning are capable of producing flower buds for rainy season crop. The farmers keen to take rainy season crop (ambe bahar crop) can allow the shoots to bear flower buds and fruits. However, as the winter crop (Mrig bahar crop) has more marketing edge and value due to quality and taste besides fruits being free of pest incidence, it is desired to promote fruit load in winter season. As such, to check the onset of rainy crop, shoot pruning (50%) is also done again in May. New shoots emerge after May pruning, which have high fruiting potential for winter crop. The procedure of sequential and periodic pruning is continued every year (May-June) for proper shaping of tree canopy and to ensure enhanced production of quality fruit during winter season. Fruiting starts on third year after rejuvenation.

Baloda *et al.* (2018) reported that guava plants when headed back at 0.5 m level produced maximum fruit yield (30.70 kg/plant). They also report an of yield of (38 kg/ plant) from fresh plantation (Table 18.1).

Table 18.1 Yield of guava as influenced by heading back at different heights

Treatments	Fruit weight (g)		Fruit length (cm)		Fruit breadth (cm)		Yield (kg/plant)		
(Heading back)	Rainy season	Winter season	Rainy season	Winter season	Rainy season	Winter season	Rainy season	Winter season	Total
0.5 m	87.73	95.70	5.90	6.80	5.45	5.53	16.00	12.70	30.70
1.0 m	86.53	93.66	6.23	6.66	5.41	5.43	14.50	9.20	23.70
1.5 m	83.37	93.00	6.26	6.63	5.25	5.25	9.10	6.50	15.60
Control (unpruned trees)	84.76	86.37	6.12	6.73	5.52	5.60	6.50	4.80	11.30
Fresh plantation	98.27	107.50	6.37	6.97	5.60	5.70	11.00	9.00	38.00
CD at 5%	1.24	1.39	0.25	0.20	0.12	0.11	2.30	1.40	3.60

Source: Baloda *et al.*, (2018).

Table 18.2 Effect of rejuvenation on quality and yield of guava

Treatment	Fruit yield after rejuvenation (kg/tree)			Fruit quality		
	1st Year	2nd year	3rd year	Weight (g)	TSS (0 brix)	Total sugar (%)
Rejuvenated	40.0	82.0	138.0	230.0	13.0	11.16
Un-rejuvenated	28.0	41.0	56.0	119.0	9.0	7.97

Source: (Singh, 2007)

18.6 REJUVENATION IN LITCHI ORCHARD

In general, litchi plant starts bearing after 5-6 years of planting and attains commercial bearing stage at least after 10 years and commercially viable life (period) lies between 11 to 40 years. It has been observed in general that orchards after attaining the age of 30-40 years even spaced at 10 m × 10 m, turns dense with compact top canopy covering most of the branches at the bottom and bearing fruits only on the high-tops. These plants pose problems in proper management such as pest control and harvesting etc. On the other hand, with rising cost of management, it may not be economical to maintain these old senile orchards of above 40 years of age. Such orchards need to be rejuvenated for further higher production of quality produce. The harvest of enhanced quality production like young commercial bearing orchards can further be obtained for at least another 12-15 years. In case of rejuvenation, heavy pruning of litchi tree is done at the height of 2 meters to 3 meters depending upon the girth and type of main trunk. About 3 to 5 main branches with outward growth from the base are marked for pruning at required height, with a plan of developing umbrella like or semi-circular frame work of tree canopy. Pruning can be done either with manual saw or power operated saw. Care should be taken to avoid bark splitting or debarking at the cut end due to falling of heavy branches at the time of pruning. To avoid any external infection at the cut portion, it should be pasted with Bordeaux mixture or Copper-oxychloride (Blitox) immediately after pruning. The best time to go for reiterative pruning for rejuvenation is August-September, i.e., mainly after the rainy season. After pruning, the tree should be fed with optimum fertilizer doses, followed by irrigation near the root zone, just like the commercial orchards. The manures and fertilizers should be applied through ring method. The dose per tree as an adult bearing stage, i.e., 75-80 kg well-rotten FYM, 2 kg Neem/Castor cake, 1.00 kg Urea, 1.50 kg Single Super Phosphate and 500 g Muriate of Potash should be applied preferably in two split doses — once during August-September and again in February-March. The micronutrients like Zinc (ZnSO4, @ 2-3 g/l) and Boron (Borax @ 2g/l) may be applied through foliar application from one year after the reiterative pruning. The pruned branches start putting forth vegetative sprouts just after 25-40 days. Shoot removal should not be done just after their emergence, rather, these should be removed sequentially after 6-8 months of rejuvenation. Irrigation is done if dry spell occurs to avoid drying out of the trees. Irrigation is must be done after the manure and fertilizer application. Mulching during the months of April-May and September-October have been found beneficial in conserving soil moisture beneath the tree canopy, reducing the frequency of irrigation and enhancing growth. Flowering plants and annual fruit crops like papaya and banana as intercrops have been found to give good income for 2-3 years. Plant protection measures are equally important for healthy growth. Intensive care to manage the infestation of important pests like stem/shoot borer, bark eating caterpillar, mite, leaf roller, leaf miner and leaf cutting weevil etc. and the diseases like microbial infestations is required. The yield obtained from the old trees (non-rejuvenated) is high but fetching very less price in the market due to inferior quality particularly with respect to size and wastage due to many physiological disorders and attack of pests-diseases. Fruit yield and physiochemical characters of mature fruits were found to be better in fruits obtained from rejuvenated trees. Maturity period is found to be slightly delayed in rejuvenated plants.

18.7 REJUVENATION OF OLD AND SENILE AONLA ORCHARD

Most of the aonla orchards in India are of seedling nature and need to be rejuvenated by top working. Fruits obtained from these seedling tees are generally smaller and inferior in terms of quality. Such established and healthy orchards can be converted into productive ones by top working with improved and commercial cultivars. Singh and Mishra (2007) standardised the rejuvenation technology in aonla at Precision Farming Development Centre, at CISH, Lucknow. It involves heading back (topping) at height of 2.5 to 3.0 m from the ground level depending on the structure of individual trees in the orchard. Before rejuvenation pruning, branches are marked with white chalk by making a ring around the branches. The selected branches should initially be cut from the underside on the lower side by giving at least 10 cm deep cut. Thereafter, the cutting should be done from the upper surface of the branch. The cut portion of the branches is then pasted with cow dung or copper oxychloride to avoid infection of fungal diseases. Immediately after heading back, the pruned wood needs to be removed from the orchard so as to prevent the damage by trunk borers. The new shoots arise on pruned branches of headed back and a few shoots are retained at proper spacing and grown towards periphery of trees. Successive removal of unwanted shoots considering the vigour and growing direction is important. In this technique, only 4 to 6 shoots developing in outer directions on main limbs should be allowed to develop. Proper development of new canopy in horizontal direction should be kept in mind while practicing thinning of shoots. During May-June the selected shoots are further pruned out to about 50 per cent of its total length for emergence of multiple shoots below the pruning points, mainly done to modify the tree structure and maintain canopy size. Fruiting starts on third year after rejuvenation. Yield levels during initial year are slightly low, while the yield from third year onward is better than the unpruned trees.

18.8 REJUVENATION BY TOP WORKING IN AONLA

Top working involves two steps (i) beheading of the tree to be top-worked and (ii) budding with an elite material on the new beheaded tree. The plants are headed back at 2.5 to 3.0 m above the ground level. Four to six shoots from the outer directions on main limbs should be allowed to develop. During June-July, buds of desired cultvar are budded on these shoots. After bud sprouting, the top portion of the shoot is removed. Numerous side shoots, which emerge on the pruned branches after the budding operation should be removed regularly as and when they emerge, so that tree of pure commercial cultivar is obtained. Since aonla is self-incompatible, i.e., the pollens of same tree/cultivar cannot fertilize its own ovary, the production from mono-culture orchards without appropriate polliniser cultivars, suffers adversely as a result of problem of fruit set. Consequently, the polliniser cultivars are budded on developing shoots of pruned trees to strengthen pollination process and enhancement of fruit set and productivity. Budding with mixed cultivars results in better yield. The best combination is NA-6 with NA-7; NA-7 with NA-10 and Kanchan with Krishna. Adequate care should be taken to manage the insect pest problems as these plants are prone to insect and sometimes wind damage (Singh and Mishra, 2007).

Table 18.3 Quality attributes of fruits obtained from rejuvenated aonla trees

Cultivar	Fruit length (cm)	Fruit breadth (cm)	Fruit weight (g)	TSS (° brix)	Vitamin C (mg/100g)
NA-6	3.06	3.24	20.15	7.8	461.3
NA-7	3.48	3.74	26.80	7.5	448.8
NA-10	3.28	3.76	27.80	8.1	453.3
Kanchan	3.07	3.25	21.50	9.5	493.5
Krishna	3.60	3.80	30.14	9.0	470.8
Chakaiya	3.04	3.52	22.50	9.0	400.0
seedling	1.90	2.11	10.80	10.2	440.6
CD at 5%	0.81	0.55	5.23	NS	NS

Source: Mishra *et al.*, (2007).

18.9 TOP WORKING

It is a technique or method of rejuvenation where in the objective is to upgrade seedling plantations of inferior cultivars with superior commercial cultivars or hybrids suitable for domestic or export market or the desired cultivar of the grower. The technique involves grafting with procured scions of desired cultivar on shoots emerged on pruned branches by adopting softwood grafting during monsoon season. The scion shoots and the emerged shoots should be of same thickness.

Advantages:

1. Increase the tree productivity/orchard productivity.
2. Conversion of old and senile orchards into productive orchards.
3. Conversion of seedling or inferior cultivar plantation/orchard into new orchard with desirable cultivar or cultivars through top working.
4. Possibility of grafting several cultivars on the same plant.
5. Increasing the fruit set of orchards by grafting few shoots with pollinizer cultivars.
6. Additional income by selling the pruned wood during non-bearing season or period.

Disadvantages

1. Chances of death of plant if not done properly or on severe pruning.
2. Need good management during post-pruning period.
3. Loss of crop for 2-3 years
4. Chances of pest and disease occurrence (stem borer, anthracnose etc.).
5. Needs skilled labour for thinning of shoots, removal of side shoots etc.
6. Top working technique can be successfully followed in crops like Mango, Sapota, Aonla, Cashew, Guava, Tamarind, Jackfruit, etc.

18.10 FRAME WORKING

Frame working is a technique of converting old fruit trees to different varieties. This is sometimes necessary if the existing variety is of inferior quality or a poor yielder. The main branch system is left intact and the grafting is done on the smaller secondary branches. Frame working keeps the main branches and add on new fruiting laterals, using scions with six to eight buds and joining the two with side grafting. In contrast, top working cuts limbs back further and in this case generally cleft grafting is done. Frame working is probably a slightly better method than top working as the tree comes to bearing earlier on the new grafts.

In case of frame working, grafting is done throughout the main frame of the tree. Here, a large number of scions is required to replace the small laterals and the scaffold branches. Growth coming out of the previous frame of the tree is removed time to time to favour establishment of the scions. The frame worked plants gave good fruiting. But the technique, being cumbersome and expensive, is not practiced commercially in the fruit orchard.

Rejuvenation of Old and Senile Mangoorchard

Unprunned Mango Orchard

Beheaded mango orchard

Rejuvenated mango orchard

Rejuvenation of Old and Senile Guava Orchard

Rejuvenation of Old and Senile Litchi Orchard

Unprunned Litchi orchard

Rejuvenated Litchi Orchard

Old and unproductive aonla orchard

Heading back of anola

Source: Singh, G and Mishra,R. (2007).

REFERENCES

Baba JA, Akbar PF and Kumar V 2011. Rejuvenation of old and senile orchards: A Review. *Annals of Horticulture*, **4** (1): 37-44.

Baloda S, Sharma JR, Kumar M, Singh S and Malik A 2018. Studies on Performance of Rejuvenated Plants and Fresh Plants of Guava c. v. Hisar Safeda. *Int. J. Pure App. Biosci.* **6** (1): 939-941.

Gurjar T, Patel H and Panchal B 2015. Rejuvenation in Senile Mango orchard. *Popular Kheti*, **3** (1): 41-42.

Singh VK and Singh G 2003. Strategic approaches of precision technology for improvement of fruit production. *In: Precision farming in horticulture; Singh, H.P.; Singh, Gorakh; Samuel S.C. and Pathak, R.K. (ed.). NCPAH,* DAC, MOA, PFDC, CISH, Lucknow, p. 75- 91.

Singh G, Mishra R and Singh GP 2005. *Guava rejuvenation.* Central institute of Subtropical horticulture, Lucknow. p 1-20.

Singh VK and Singh G 2007. Photosynthetic efficiency, canopy micro-climate and yield of rejuvenated guava trees. *Acta Horticulturae*, **735**: 249-252.

Singh G and Mishra R 2007. Aonla rejuvenation. *In: Precision Farming Development Centre, N.C.P.A.H., D.A.C.*, Ministry of Agriculture, Govt. of India. Central Institute for Subtropical Horticulture, Rehmankhera, Lucknow- 227 107

Mishra D, Pandey D, Mishra R and Pathak RK 2007. Performance of improved aonla cultivars during top working of senile trees. *Indian J Hort.*, **64** (4): 396-398.

OUTCOMES ASSESSMENT

PART A

Answer the following questions (True or False).

1. Rejuvenation of old and senile orchard is required. (True/False)
2. Rejuvenation cannot restore the production potential of old orchard. (True/False)
3. Top working upgrades seedling plantations of inferior cultivars. (True/False)
4. The rejuvenation technique involves the heading back of older trees (True/False)
5. The cut portion should be pasted with Bordeaux mixture. (True/False)

PART B

Answer the following questions.

1. What is the usual time of rejuvenation?
2. Why is rejuvenation essential?
3. Mention two advantages of top working.
4. Guava orchard begins to lose vigour and bearing potential after —to—years of age.
5. Aonla is self-incompatible/compatible.

PART C

Write a brief note on each of the following.

1. Describe the essentiality of rejuvenation of old and senile orchards.
2. How mango orchard is rejuvenated?
3. Discuss the rejuvenation in aonla by top working.

Chapter 19

Organic Farming

19.1 DEFINITION OF ORGANIC FARMING

1. According to IFOAM (International Federation of Organic Agriculture Movement), Organic Farming definition includes all agricultural system that promote the environmentally, socially and ecologically sound production system of food and fibres. IFOAM stresses and supports the development of self-supporting system on local and regional level.
2. As per the definition of the USDA study team on organic farming "organic farming is a system which avoids or largely excludes the use of synthetic inputs (such as fertilizers, pesticides, hormones, feed additives etc) and to the maximum extent feasible rely upon crop rotations, crop residues, animal manures, off-farm organic waste, mineral grade rock additives and biological system of nutrient mobilization and plant protection".
3. In another definition FAO suggested that "Organic agriculture is a unique production management system which promotes and enhances agro-ecosystem health, including biodiversity, biological cycles and soil biological activity, and this is accomplished by using on-farm agronomic, biological and mechanical methods in exclusion of all synthetic off-farm inputs".

19.2 HISTORY OF ORGANIC FARMING

1905 to 1924: Organic Agriculture Begins in Central Europe and India

Organic agriculture began more or less simultaneously in Central Europe and India. The British botanist Sir Albert Howard, often referred to as the father of modern organic agriculture, worked as an agricultural adviser in Pusa, Bengal, (now in Bihar), where he documented traditional Indian Farming practices, and came to regard them as superior to his conventional agriculture science. In the United States, J. I. Rodale begins to popularize the term and methods of organic growing, particularly to consumers through promotion of organic gardening.

1939: First Use of the Term "Organic Farming" The first use of the term "organic farming" is by Lord Northbourne. The term derives from his concept of "the farm as organism", which he expounds in his book, "Look to the Land" (1940). Influenced by Sir Albert Howard's work, Lady Eve Balfour did first scientific, side-by-side comparison of organic and conventional farming.

19.3 CONCEPT OF ORGANIC FARMING

Organic agriculture considers the concept that soil, plant, animal and human beings are interlinked. So, its goal is to create an integrated, environmentally sound, safe and economically sustainable agricultural production system. The concept of organic farming is based on the following principles:

1. Soil is a living entity.
2. Entire system is based on intimate understanding of nature's ways.
3. It does not degrade soil in any way for today's need.
4. Nature is the best role model for farming.
5. Soil's living population, i.e., micro- and macro-organisms are significant contributors to its fertility on a sustained basis and must be protected and nurtured at all cost.
6. Total environment of the soil is more important starting from soil structure to soil cover.

19.4 BASIC STEPS AND COMPONENTS OF ORGANIC FARMING

Organic farming approach involves steps like: (i) conversion of land from conventional management to organic management, (ii) management of the entire surrounding system to ensure biodiversity and sustainability of the system, (iii) crop production with the use of alternative sources of nutrients such as crop rotation, residue management, organic manures and biological inputs, (iv) management of weeds and pests by better management practices, physical and cultural means and by biological control system, and, (v) maintenance of livestock in tandem with organic concept and make them an integral part of the entire system.

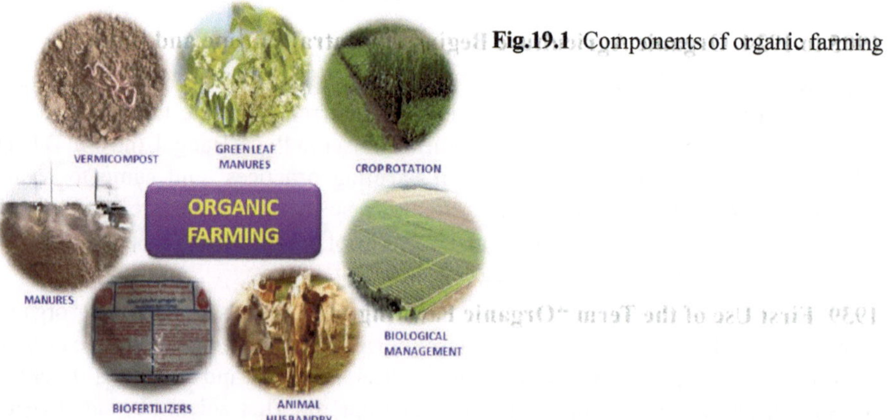

Fig.19.1 Components of organic farming

19.5 PRINCIPLES OF ORGANIC FARMING

The four principles of organic farming are mentioned below (Figure 19.2):

1. **Principle of health:** Organic agriculture should sustain and enhance the health of soil, plant, animals, human being and earth as one and indivisible. The health of people cannot be separated from the health of the ecosystem. Healthy soils produce healthy crops that foster the health of animals and people. Health is the wholeness and integrity of living systems. The role of organic agriculture, whether in farming, processing, distribution, or consumption, is to sustain and enhance the health of ecosystems and organisms from the smallest in the soil to human beings. It intends to produce high quality and nutritious food. So, it avoids the use of chemical fertilizers, pesticides, drugs and food additives that may have adverse health impact.
2. **Principle of ecology:** Organic agriculture should be based on living ecological systems and biological cycles, work with them, emulate them and help sustain them. This principle roots organic agriculture within living ecological systems. It states that production is to be based on ecological processes, and recycling. Nourishment and well-being are achieved through the ecology of the specific production environment. Organic management must be adapted to local conditions, ecology, culture and scale. Inputs should be reduced by reuse, recycling and efficient management of materials and energy in order to maintain and improve environmental quality and conserve resources. It should attain ecological balance through the design of farming systems, establishment of habitats and maintenance of genetic and agricultural diversity.
3. **Principle of fairness:** Fairness is characterized by equity, respect, justice of the shared world both among the people and in their relation to other living being. Organic agriculture should build on relationships that ensure fairness with regard to the common environment and life opportunities. This principle emphasizes that those involved in organic agriculture should conduct human relationships in a manner that ensures fairness at all levels and to all parties —farmers, workers, processors, distributors, traders and consumers. It aims to produce a sufficient supply of good quality food and other products. Natural and environmental resources that are used for production and consumption should be managed in a way that is socially and ecologically just and should be held in trust for future generations. Fairness requires systems of production, distribution and trade that are open and equitable and account for real environmental and social costs.
4. **Principle of care:** Organic Agriculture should be managed in a precautionary and responsible manner to protect the health and well-being of current and future generations and the environment. Organic agriculture is a living and dynamic system that responds to internal and external demands and conditions. This principle states that precaution and responsibility are the key concerns in management, development and technology choices in organic agriculture. Organic agriculture should prevent significance risks by rejecting all predictable ones such as GM crops/GM food materials.

Fig.19.2 Principles of organic farming

19.6 IMPORTANT GOALS OF ORGANIC FARMING

1. A sufficiently high level of productivity.
2. Compatibility of cultivation with the natural cycles of the production system as a whole.
3. Maintaining and increasing the long-term fertility and biological activity of the soil.
4. Maintaining and increasing natural diversity and agro-biodiversity.
5. Maximum possible use of renewable resources.
6. Creation of a harmonic balance between crops and animal husbandry.
7. Creation of conditions in which animals are kept that correspond to their natural behaviour.
8. Protection of, and learning from, indigenous knowledge and traditional management systems.

19.7 ADVANTAGES OF ORGANIC FARMING

1. **Nutritional, poison-free and tasty food:** The nutritional value of food is largely a function of its vitamin and mineral content. In this regard, organically grown food is dramatically superior in mineral content to that grown by modern conventional methods. A major benefit to consumers of organic food is that it is free of contamination with health harming chemicals such as pesticides, fungicides and herbicides. There are reasonably consistent findings for higher nitrate and lower vitamin C contents in conventional vegetables (Woese et al., 1997). Several studies indicate that 10-60 percent more healthy fatty acids (like CLA's) and omega-3 fatty acids occur in organic dairy (Butler et al., 2008). In crops, vitamin C ranges 5-90 percent more and secondary metabolites 10-50 percent more in organic. Also, less residues of pesticides and antibiotics

are present (Huber and van de Vijver, 2009). Heaton, (2002) reported that organic food contains higher minerals and dry matter and 10-50 percent higher phytonutrients. Decreased cell proliferation of cancer cells was observed on extracts of organic strawberries (Olsson et al., 2006). The Parsifal study showed 30 percent less eczema and allergy complaints and less body weight among 14 000 children fed with organic and biodynamic food in five EU countries (Alfven et al., 2006). In animals, organic feed leads to increased fertility (Staiger, 1988) and increased immune parameters (Finamore *et al.,* 2004). Other studies indicate that the most systematic differences between organic and conventional crops are the contents of secondary metabolites (Brandt and Mølgaard, 2001).

Organically grown food tastes better than that conventionally grown. The tastiness of fruit and vegetables is directly related to its sugar content, which in turn is a function of the quality of nutrition that the plant itself has enjoyed. This quality of fruit and vegetable can be empirically measured by subjecting its juice to brix analysis, which is a measure of its specific gravity (density). The brix score is widely used in testing fruit and vegetables for their quality prior to export. Organically grown plants are nourished naturally, rendering the structural and metabolic integrity of their cellular structure superior to those conventionally grown. As a result, organically grown foods can be stored longer and do not show the latter's susceptibility to rapid mould and rotting.

2. **Lower growing cost:** The economics of organic farming is characterized by increasing profits via reduced water use, lower expenditure on fertilizer and energy, and increased retention of topsoil. To add to this the increased demand for organic produce makes organic farming a profitable option for farmers.
3. **Enhances soil nourishment:** Organic farming effectively addresses soil management. Even damaged soil, subject to erosion and salinity, are able to feed on micro-nutrients via crop rotation, inter-cropping techniques and the extensive use of green manure. The absence of chemicals in organic farming does not kill microbes which increase nourishment of the soil. Biodynamic farms had better soil quality: greater in organic matter content and microbial activity, more earthworms, better soil structure, lower bulk density, easier penetrability, and thicker topsoil (Reganold *et al.,* 1993); agricultural productivity doubled with soil fertility techniques: compost application and introduction of leguminous plants into the crop sequence (Dobbs and Smolik, 1996; Drinkwater *et al.,* 1998; Edwards, 2007).
4. **More energy efficiency:** Growing organic rice was four times more energy efficient than the conventional method (Mendoza, 2002). Organic agriculture reduces energy requirements for production systems by 25 to 50 percent compared to conventional chemical-based agriculture (Niggli *et al.,* 2009).
5. **Carbon sequestration:** German organic farms annually sequester 402 kg Carbon/ha, while conventional farms had losses of 202 kg (Clark et al., 1999; Küstermann et al., 2008; Niggli et al., 2009).

6. **Less water pollution:** In conventional farms, 60 percent more nitrate are leached into groundwater over a 5-year period (Drinkwater et al., 1998).
7. **Environment-friendly practices:** The use of green pesticides such as neem, compost tea and spinosad are environment-friendly and non-toxic. These pesticides help in identifying and removing diseased and dying plants in time and subsequently, increasing crop defence systems. Organic farms' biodiversity increases resilience to climate change and weather unpredictability (Niggli *et al.*, 2008). Organic agriculture reduces erosion caused by wind and water as well as by overgrazing at a rate of 10 million hectare annually (Pimentel *et al.,* 1995).
8. **Organic farming is a source for productive labour:** Agriculture is the main employer in rural areas and wage labour provides an important source of income for the poor. Thus, by being labour intensive, organic agriculture creates not only employment but improves returns on labour, including also fair wages and non-exploitive working conditions. New sources of livelihoods, especially once market opportunities are exploited, in turn revitalize rural economies and facilitate their integration into national economies.
9. Organically grown plants are more resistant to disease and insects and hence only a few chemical sprays or other protective treatments are required.
10. Organic farming helps to avoid chain reaction in the environment from chemical sprays and dusts.
11. Organic farming helps to prevent environmental degradation and can be used to regenerate degraded areas.
12. Since the basic aim is diversification of crops, much more secure income can be obtained than when they rely on only one crop or enterprise.

19.8 DISADVANTAGES OF ORGANIC FARMING

1. **Lower productivity:** An organic farm cannot produce as much yield as a conventional or industrialized farm. A 2008 survey and study conducted by the UN Environmental Program concluded that organic methods of farming result in small yields even in developing areas, compared to conventional farming techniques. Though this point is debatable as the productivity and soil quality of an industrialized farm decreases rapidly over the years.
2. **Requires skill:** An organic farmer requires greater understanding of his crop and needs to keep a close watch on his crops as there are no quick fixes involved, like pesticides or chemical fertilizers. Sometimes it can be hard to meet all the strenuous requirements and the experience to carry out organic farming.
3. **Time-consuming:** Significant amounts of time and energy are required to execute the detailed methods and techniques that are required for a farm to be called an organic farm. Failure to comply with any of these requirements could result in loss of certification, which the farmer will not be able to regain in up to three years. And it can be more time-consuming. Organic farming increases soil fertility by way of compost, and organic fertilizers and mulch. Organic fertilizers

tend to be slow-release. As with control by botanicals, horticultural oils, and insecticidal soaps, organic fertilizers may need several applications before the desired results are brought about.

4. **More labour intensive:** It can be more labour-intensive. Because organic farming considers biological, cultural and mechanical responses to production challenges. It focuses on plant and soil health through proper aeration, drainage, fertility, structure and watering. So, there is more above and below ground grunt work involved.
5. Organic farming methods are not yet as established and widespread as conventional production. So organic control by botanicals such as pyrethrin can be more expensive than conventional controls by the longer established, more available, and wider ranging artificial, commercial, synthetic chemical pesticides.
6. Organic farming also requires a lot more inputs and more red-tape than conventional farming because certain practices must be met in order for a farm to retain the organic label. If anything slips, then the farm loses organic certification just like that.

19.9 BARRIERS TO ORGANIC FARMING

Initially there may be some barriers which inhibit the farmers from adopting organic farming. Land resources can move freely from organic farming to conventional farming; they do not move freely in the reverse direction. In changing over to organic farming an initial crop loss generally occurs, particularly if it is rapid. Biological controls may have been weakened or destroyed by chemicals, which may take three or four years to build up. Organic farmers may be afraid to enter the new market without adequate government support.

19.10 FUTURE PROSPECTS

Organic Farming is being practiced in around 150 countries of the world. The health hazards of chemical agriculture have changed the mindset of consumers of mostly the developed world to buy organic with premium/high price. Policy makers, planners are also promoting Organic Farming for restoration of soil health and a clean and safe environment. The organic agriculture movement started in developed world and now it is gradually picking up in developing countries.

India has been traditionally a country of Organic Agriculture. However, growth of modern, scientific and input-intensive chemical agriculture has pushed this concept into back but with increasing awareness about the safety and quality of foods and long-term sustainability of the system organic farming may emerge as an alternative system of farming. However, considering the level of crop yield and productivity and also availability of organic sources of nutrients, India with a large population to feed, needs a careful consideration for adoption of organic farming under all circumstances. Hence, organic farming should be considered for lesser developed region of the country. It should be started with low volume and high value crops like spices, condiments, medicinal plants and also aromatic plants.

19.11 CONCLUSION

An environmentally sustainable system of agriculture like organic farming will be able to maintain a stable resource balance, avoid over exploitation of renewable resource, conserving inherent soil nutritional quality and soil health, and biodiversity. It will lead us to sustainable agriculture and create a sustainable lifestyle for generations to come.

REFERENCES

Anonymous 2002. Package of Practices for Organic cultivation of ginger, turmeric and chillies. Spice board, Ministry of Commerce & Industry, Government of India.

Palaniappan SP and Annadurai K 2010. *Organic Farming: Theory and Practice.* Scientifi Publishers (India), Jodhpur.

www.researchgate.net,content uploaded by Hari Prakash Meena, ICAR-Indian Institute of

Oilseeds Research, Hyderabad (Formerly: Directorate of Oilseeds Research). Crop

Improvement (Plant Breeding)-Organic Farming: Concept and Components, January 2014

www.gjar.org,Vol-1, Issue -2 Pp.101-102-Sowmya. K.S., Research Associate, Organic Farming Research Centre, UAHS, Navile, Shimoga, Karnataka, India

OUTCOMES ASSESSMENT

PART A

Answer the following questions (True or False).

1. An organic farmer requires greater understanding of his crop. (True/False)
2. Organic farming does not address soil management. (True/False)
3. Soil is a non-living entity. (True/False)
4. India has been traditionally a country of organic agriculture. (True/False)
5. Organic farming is more labour-intensive. (True/False)

PART B

Answer the following questions.

1. Define organic farming.
2. Mention one advantage of organic farming.
3. What is the best role model for farming?
4. What are the four principles of organic farming?
5. Who is the father of modern organic agriculture?

PART C

Write a brief note on each of the following.

1. Advantages of organic farming.
2. Principles of organic farming.
3. Goals of organic farming.
4. Disadvantages of organic farming.
5. Concept of organic farming.

Index

F

G

H

I

R

S

T

U

V

W

Y